Aboriginal Sign Languages of the Americas and Australia

VOLUME 1

NORTH AMERICA
CLASSIC COMPARATIVE PERSPECTIVES

ABORIGINAL SIGN LANGUAGES OF THE AMERICAS AND AUSTRALIA

Volume 1 • North America–Classic Comparative Perspectives

Volume 2 • The Americas and Australia

Aboriginal Sign Languages of the Americas and Australia

VOLUME 1

NORTH AMERICA
CLASSIC COMPARATIVE PERSPECTIVES

Edited by

D. JEAN UMIKER-SEBEOK
and
THOMAS A. SEBEOK
Indiana University, Bloomington

PLENUM PRESS • NEW YORK AND LONDON

Library of Congress Cataloging in Publication Data

Main entry under title:

Aboriginal sign languages of the Americas and Australia.

CONTENTS: v. 1. North America: classic comparative perspectives. – v. 2. The Americas and Australia.

Includes indexes.

1. Indians of North America – Sign language – Addresses, essays, lectures . . 2. Australian aborigines – Sign language – Addresses, essays, lectures. 3. Sign language – Addresses, essays, lectures. I. Umiker-Sebeok, Donna Jean. II. Sebeok, Thomas Albert, 1920-

E98.S5A23 001.56 78.1303

ISBN 0-306-31073-2 (v. 1)

ISBN 0-306-31081-3 (v. 2)

A Division of Plenum Publishing Corporation
227 West 17th Street, New York, N.Y. 10011

Printed in the United States of America

ACKNOWLEDGMENTS

REFERENCES TO REPRINTED ARTICLES

Mallery, Garrick. *Introduction to the Study of Sign Language among the North American Indians as Illustrating the Gesture Speech of Mankind* (Introductions Series No. 3). Washington, D.C.: United States Bureau of American Ethnology, 1880. Reprinted by permission of The Smithsonian Institution Press.

Mallery, Garrick. *A Collection of Gesture-Signs and Signals of the North American Indians with Some Comparisons.* Washington, D.C.: Smithsonian Institution, 1880.

Mallery, Garrick. The gesture speech of man. *Proceedings of the American Association for the Advancement of Science,* 30th Meeting (August, 1881), pp. 283–313, 1882.

CONTENTS OF VOLUME 1

CONTENTS OF VOLUME 2

PART II: SOUTH AMERICA

PART III: AUSTRALIA

INTRODUCTION

1. THE SEMIOTIC CHARACTER OF ABORIGINAL SIGN LANGUAGES

In our culture, language, especially in its spoken manifestation, is the much vaunted hallmark of humanity, the diagnostic trait of man that has made possible the creation of a civilization unknown to any other terrestrial organism. Through our inheritance of a *faculté du langage*, culture is in a sense bred into man. And yet, language is viewed as a force which can destroy us through its potential for objectification and classification. According to popular mythology, the naming of the animals of Eden, while giving Adam and Eve a certain power over nature, also destroyed the prelinguistic harmony between them and the rest of the natural world and contributed to their eventual expulsion from paradise. Later, the post-Babel development of diverse language families isolated man from man as well as from nature (Steiner 1975). Language, in other words, as the central force animating human culture, is both our salvation and damnation. Our constant war with words (Shands 1971) is waged on both internal and external battlegrounds.

This culturally determined ambivalence toward language is particularly apparent when we encounter humans or hominoid animals who, for one reason or another, must rely upon gestural forms of communication. Shakespeare provides a beautiful illustration of this theme in *The Tempest*, where the prospective semiotic colonization of the New World—through the transference to it of the English language—is conceived of as in effect the humanization of that world (Hawkes 1973, p. 200). On the negative side, however, the reader is left with no doubt that such a transformation of the formerly humanlike but speechless creatures who inhabit the island on which Shakespeare's characters have been shipwrecked will lead to the irreversible destruction of the harmonious relationship that exists among themselves and between them and their environment. The existing gestural communication of the islanders is viewed as an important part of their present harmony:

> *Gonzalo* For, certes, these are people of the island,—
> Who, though they are of monstrous shape, yet, note,
> Their manners are more gentle-kind than of
> Our human generation you shall find
> Many, nay, almost any.

Prospero Honest lord,
Thou hast said well; for some of you there present
Are worse than devils.
Alonso I cannot too much muse, [*Aside*]
Such shapes, such gesture, and such sound, expressing,—
Although they want the use of tongue,—a kind
Of excellent dumb discourse.

(*The Tempest*, Act III, Scene 3)

Encounters with people practiced in some form of conventional gestural system have tended, in varying degrees, to evoke conflicting emotions. On the one hand, our interlocutor, whether an Australian aborigine, an Indian of the North American Plains, or even a deaf compatriot, does not strike us as quite "human" because he does not communicate to us in a spoken language which we are able to understand. On the other hand, his use of an elaborate set of significative gestures, while not quite conferring full humanness upon him, appeals to us as a graceful and "dignified" form of communication ideally suited to service as a universal *lingua franca*, a bridge between men speaking mutually incomprehensible tongues, or, in recent years, with other, nonhuman primates, and thus in one way at least superior to spoken language.

Aboriginal sign languages,[1] in particular the Plains Sign Language (hereafter PSL), have long captivated the imagination of the nonscientific world by thus holding out the promise of an escape from what Nietzsche called our prison-house of language. "Probably no other phase of Indian culture," writes Harrington (this book, Vol. 2, p. 145), "has proved so interesting to the American public at large." As every American Boy Scout knows, a simplified version of PSL has long been a part of Boy Scout tradition (see, e.g., Beard 1918). Numerous volumes have been written with the general adult population in mind (e.g., Clark 1885, Cody 1952, 1970, Hadley 1890, 1893, Hofsinde 1956, Seton 1918, Tomkins 1929), and reprintings and new editions of these continue to appear. While PSL is presented to the public as a practical system of communication for use in certain outdoor activities where spoken language is undesirable, it is likely that what actually sells copies of such works is the almost mystical view of the sign language projected by the authors, as, for example, in the following passage on the back cover of Cody 1970:

> For many centuries countless warriors, traders and travelers have refined and developed this beautiful, silent language of the hands until amost any common meaning can be expressed. It is a language that is a part of nature itself, the fluttering of aspen leaves under the touch of the wind, a hawk soaring in the sky, the ponderous movement of buffalo herds, the gestures of wise old Indians of the great warrior days who were one with earth and heaven, the rhythm of waterfalls pouring over cliffs or of clouds drifting over the sacred circle of the blue above us.

In recent years PSL has become a significant part of both the revival of tribal identity among Plains Indians and the correlated renewal of curiosity about the Plains Indian culture among non-Indian Americans (Kroeber, this book, Vol. 2, p. 185). Taylor (this book, Vol. 2, p. 230) reports its continued use in storytelling, where signs accompany the spoken narrative, or vice versa, the sign dance, where dancers carry on a dialogue in sign language (see the description of Dodge cited in this book, Vol. 2, p. 230), and for formal oratory within councils or pow-wows (see the descriptions in West 1960). In these oratorical and ceremonial contexts, the "poetry, dramatic thought, and oratorical fire" (Scott quoted in Weil 1931, p. 8) of PSL have become a self-conscious emblem of Plains Indian culture. Further evidence of this is supplied by West (1960), who noted that among some tribal groups more young people appeared to understand PSL than did older tribal members. Furthermore, PSL is now a part of the curriculum for some tribal school programs aimed at increasing ethnic awareness.

If the strong and immediate appeal of aboriginal sign languages has its roots in their function as a form of mediation between the "natural" and "cultural" sides of "the talking animal," their fulfillment of this function can be attributed to the fact that they are representatives of a type of semiotic system uniquely qualified to fill a gap in our Western conception of the order or structure of the bounded universe of human and animal sign systems. As we shall see below, with the possible exception of sign languages of the deaf the semiotic nature of aboriginal sign languages is unlike that of any other sign system by virtue of the combination of the following characteristics:

1. Aboriginal sign languages are a complex of both natural and conventional sign relations with, however, iconic and indexical elements outweighing symbolic ones (the reverse of the hierarchy prevailing in spoken languages).
2. Aboriginal sign languages are semantically open both in the sense that they may be used to formulate a potentially indefinite number of messages and that the lexicon may be enlarged to suit changing demands on the system (as in spoken languages).
3. Aboriginal sign languages are presumed to take advantage of a species-consistent (nonverbal) competence, but each sign system must be learned (a situation analogous to spoken languages).
4. Aboriginal sign languages make use of a visual channel of communication (and thus the roles played by the principles of successivity and simultaneity differ considerably from those found in spoken languages; cf. Sebeok, in Hailman 1977, p. xv–xix).

The first of the two characteristics of aboriginal sign languages noted above which set them apart from spoken language (numbers 1 and 4) is the highly "natural" relationship between their signifiers and signifieds. Kohl (1860, p. 34), for example, describes the signs of PSL as "natural, characteristic, and easy of comprehension," Webb (this book, Vol. 2, p. 92) as "natural" and "universal," the "mother utterance of nature." They are "figurative" (Humfreville, this book, Vol. 2, p. 70) and possess "picturesque clarity" (Hofsinde, this book, Vol. 2, p. 143) and "picturesque novelty" (Scott, this book, vol. 2, p. 66). Another early observer, Wassell, attributed the "metaphorical" character of the sign language to the poetic nature of the "savages" (this book, Vol. 2, p. 40), explaining that Indians are extremely sensitive to meaning in nature and therefore better qualified to develop sign language (*ibid.*, p. 37). Use of the terms *picturesque, figurative,* and *metaphorical* adumbrate the iconic relationship between signifier and signified in PSL. Another favorite word used to convey the iconic character of aboriginal sign language signs is *pantomime* (see, e.g., Taylor this book, Vol. 2, p. 231). Kroeber described PSL as "overwhelmingly pantomimic," with many signs being transparent, especially if one is familiar with the culture of the signer. Other signs seem "reasonable" once their meaning has been learned by context, association, or explanation. Kroeber suggested that a more accurate term than *pantomimic* for such iconicity would be *cheiromimic* (this book, Vol. 2, p. 196),[2] since the imitation in PSL is in general restricted to the hands and arms rather than the entire body.

The signs of the Australian aboriginal sign language (hereafter AASL) also appear to be highly iconic, "imitating the most conspicuous outlines of an object or the most striking features of an action" (Berndt, this book, Vol. 2, p. 397–398), or, as another author writes, "characteristic of the reference" (Meggitt, this book, Vol. 2, p. 414). La Mont West, the only person to have done intensive fieldwork with both AASL and PSL, implies (this book, Vol. 2, p. 428) that 95% of both systems involve "concrete symbolization" of the world through the selection of one or more characteristic features to represent an object or animal. According to the estimate of Miller (this book, Vol. 2, p. 438), 62% of the signs collected from the aborigines residing at the Warburton Mission were "iconic."

While the terms used to describe natural signs in aboriginal sign languages generally stress iconicity, natural signs, as traditionally distinguished from conventional ones, or as they are sometimes called, symbols, are also based on the semiotic relationship of metonymy or *indexicality*, which plays a very important role in the structure of aboriginal sign languages. Harrington (this book, Vol. 2, p. 118–120), for example, notes that one of the two basic components in the building up of PSL is "indication by gesturing at, or painting," the latter term being used by signers to designate what others might call *outlining*. By pointing, signers designate (1) cardinal directions and regions, (2) personal and demonstrative pronouns ("subjective, objective, indirective, and possessive"), (3) parts of one's own body, and (4) colors "of almost universal occurrence in nature, such as black and white." Gesturing at the locality of occurrence of something "replaces indication of the object or abstraction," as, for example, in the sign for *to think, thought*, where the signer gestures at his own heart then brings his hand forward, thus combining "gesturing at locality of occurrence plus action mimicry." In painting or delineating, the signer "outlines the figure of an object by tracing it with the hand or hands in mid-air" (*ibid.*). West, applying Harrington's categories to his own data, states that such deictic signs account for more than half of his basic units of analysis, or kinemes,[3] and nearly all of those of restricted occurrence. West further concludes that these indexical signs "comprise an open end to the kinemic system, where new kinemes of limited recurence [*sic*] may be freely introduced without disruption of the more closed kinemic system participated in by nondiectic [*sic*] signs" (1960, Vol. I, p. 94).

Perhaps because aboriginal sign languages first attracted amateur scientists and later, beginning in the 1950s, professional linguists caught up in an overly narrow type of formalism, aboriginal sign language studies have largely remained outside the tradition of semiotics and nowhere in the literature does one find a consistent application of a classification of signs which incorporates a coherent and systematic distinction between the interdependent iconic and indexical sign relations prevailing in these sign languages. Harrington's analysis of PSL signs as combinations of two or more of either indexical relations or "representation by substitution or by mimicking" perhaps comes closest to a coherent semiotic classification of sign types and amply illustrates for us the complexity of describing such a visual system of communication. For example, the signs for certain colors are listed under his category of pointing or gesturing at, since, for example, the color black is indicated by pointing at one's eyebrows, hair, or some black object nearby (*ibid.*, p. 120). While this sign does involve the use of the index finger and a pointing gesture, the sign represents black by virtue of a similarity between the color of, for example, the eyebrow, and the quality designated, and thus is an iconic sign rather than an indexical one (cf. Sebeok 1976*a*, p. 129, footnote 3). The pointing at an eyebrow would be an indexical sign if it referred to the eyebrow. Similarly, several of his categories of substitution, that is, where "a body part of the sign user . . . and its posture is made to represent . . . the object or abstraction" (*ibid.*), involve mimicking actions, but the action mimicked is not the signified of the sign, but stands for the signified through an indexical relationship. For example, *bread* is used as an illustration of "preparation mimicry," where "the more strikingly mimicked action of preparation replaces the less strikingly mimicked finished product" (*ibid.*, p. 122). It is indicated by imitating the kneading of a piece of dough. It is not the kneading of the dough which is the signified, nor in fact the dough, but the finished baked bread, which stands in existential relationship to the dough. The same may be said of Harrington's categories "effect mimicry," "characteristic outline for a whole," "characteristic part for a whole," and "characteristic action for a whole."

It is the combination of iconicity and indexicality with a substantial amount of conventionalization which is seen by most authors as determining the great potential of aboriginal sign languages as a means of communication which could rival spoken language in its range of expression and semantic flexibility. Writing about PSL, Dodge (this book, Vol. 2, p. 4–8), for example, noted that it is the ever increasing conventionality, or transformation of natural signs into arbitrary ones, which has enabled PSL to become expressively adaptive. Maclean (this book, Vol. 2, p. 43), discussing gestural communication in general, claimed that it is properly called a language, "as among savage races it has various conventional forms, which are in a measure definite and full." Kroeber warned that, by overestimating the pantomimic transparency of PSL we may overlook the important fact that what makes it an "effective system of communication" is that "it did *not* remain on a level of naturalness, spontaneity, and full transparency, but made artificial commitments, arbitrary choices between potential expressions and meanings" (this book, Vol. 2, p. 200). Placing the cupped hand before the mouth might be universally recognized as a sign for drinking, but its extension to denote water in such meanings as *lake* and *island* and *drown* is a specific convention of PSL (*ibid*, p. 199).

As Kroeber mentions, the important role played by iconic and indexical signs in PSL has not uncommonly given casual observers of sign language performances the impression that messages can be decoded without prior training. Voegelin (this book, Vol. 2, p. 206) also remarked that, in response to sign language films, most casual observers cannot resist guessing at the meaning of sign messages, something they would not think of doing if confronted with a recording of a native spoken text. Discussing claims such as these, West (1960, Vol. II, p. 7) reports that the comprehension tests of "naive" subjects, which he conducted as a part of his own research, resulted in extremely low scores for the identification of individual lexical items and an even lower comprehension of entire sign language texts. This corroborates Humfreville's statement (this book, Vol. 2, p. 70) that, while PSL is "highly significant," it is "necessary to follow closely the thread of conversation, for the wrong interpretation of a single sign was sufficient to break the whole chain of thought."

While the conventional nature of PSL has been debated now for over a century, the following summary statement by one of the early contributors to this debate still rings true:

> Some writers, as Captain H. Stansbury, consider the system purely arbitrary; others, Captain Marcy, for instance, hold it to be a natural language similar to the gestures which surd-mutes use spontaneously. Both views are true, but not wholly true . . . The pantomimic vocabulary is neither quite conventional nor the reverse. (Burton 1862, p. 135)

Regarding AASL, Howitt (this book, Vol. 2, p. 306) noted that it constituted a "recognized and well understood system of artificial language." The conventional nature of AASL was also noted, for example, by Basedow, Miller, Roth, Spencer and Gillen, and West (this book, Vol. 2). Miller estimated that 38% of the signs he collected were "arbitrary." Meggitt (this book, Vol. 2, p. 414) remarked that, while some of the Walbiri signs appeared iconic to him, and some to the natives, others were "characteristic" to neither.

The second feature of aboriginal sign languages noted above—that is, semantic openness—is one which they share with spoken languages. Aboriginal sign languages are not simple semiotic systems in the sense that there is a small number of signifiers, each with one fixed conventional signified. Mallery emphasized that PSL "is not a mere semaphoric repetition of motions to be memorized from a limited traditional list, but is a cultivated art" (1972, p. 346), a "general system instead of . . . a uniform code," with "generic unity . . . [not] . . . specific identity" (this book, Vol. 1, p. 19). As Mounin has pointed out (1973, p. 157), Mallery's notion of an *art* is strikingly similar to current notions of an integrated (linguistic) *system*. If aboriginal sign languages were simple semiotic systems, as is sometimes implied by the watered down versions presented in popular handbooks, they could be easily learned from the appropriate use of code books. But, as one author (Scott, this book, Vol. 2, p. 67) has noted in regard to PSL, those who have tried to acquire through written materials the skills necessary for signing have soon abandoned their plans in favor of on-the-spot training with accomplished sign language users. This pedagogical situation parallels that of foreign language learning.

There is ample evidence that the Plains Indians employed sign language in a great variety of contexts, both casual and formal, and for the expression of a seemingly limitless range of abstract as well as concrete ideas. Some authors (for example, Axtell, this book, Vol. 2, p. 29) even maintained that the PSL was more flexible and expressive than some Indian tongues. Scott (quoted in Weil 1931, p. 9) asserted that anything could be expressed in PSL; Maclean (this book, Vol. 2, p. 51) attested the fact that Indians could maintain "intelligent [PSL] conversations for hours." Harrington (this book, Vol. 2, p. 144) further stated that, using a system of several hundred signs which represented all the parts of speech, the Indians of the Plains conversed with one another with a kinetic fluency which equalled the "articulatory dignity" of spoken language.

Certain Australian aboriginal groups appear to possess or have possessed semantically open sign systems. Stirling (this book, Vol. 2, p. 257), for example, noted that the signs he had observed constituted an extensive system of gestures which was a common part of aboriginal life and capable of indicating a very large number of objects as well as ideas concerning them. The natives accompanying Stirling's party were frequently observed to carry on "a more or less continuous and . . . certainly intelligible silent conversation." A year later, Roth (this book, Vol. 2, p. 274) reported that animals, plants, and other objects of the natural world, manufactured objects, individuals, simple and complex states or actions, number, place, interrogation, and abstract notions could all be expressed with AASL. Howitt (this book, Vol. 2, p. 304) also remarked that some groups had a very extensive code of signs, which were used in such a way as to "almost amount to a medium of general communication," and Basedow (this book, Vol. 2, p. 372–373) reported that an almost inexhaustible number of ideas could be communicated in the form of "coherent speech." Meggitt (this book, Vol. 2, p. 410) specified that the Walbiri had a very wide range of signs which could stand as substantives, verbs, adjectives, and adverbs, which could be arranged in "grammatical expressions" paralleling those of spoken Walbiri to form whole AASL conversations. West (this book, Vol. 2, p. 426) estimated the size of some AASL lexicons to be similar to the impressive PSL lexicons discussed in his dissertation (1960).

Another point that needs to be made regarding the conventional nature of the signs of aboriginal sign languages is that, rather than being wholly "natural" (that is, iconic or indexical) or "conventional," they are always a subtle blend of both, and the hierarchical arrangement between the natural and conventional sign relations is constantly shifting, both at the level of individual signing performances and of the sign language as an abstract system shared by a heterogeneous group of signers.

At the level of sign language as shared by a group of signers, the linguist Sayce (1880, p. 93) remarked that "care must be taken to distinguish between two things which are frequently confused together. *Gestures* and *signs* are wholly different, gestures being natural, signs more or less conventional. A *gesticulation* is a gesture which has become a sign, and the nearer signs approach to gesticulations the more readily and instinctively they will be understood."[4] While new signs may be created using metaphoric and/or metonymic sign relations, to the extent that they are in fact a part of the sign system, that is, are used repeatedly by a community of sign language practitioners, they must be conventional at least to a certain degree. This is acknowledged by the majority of authors concerned with aboriginal sign languages. Regarding PSL, for example, both Mallery and West cite the linguist Whitney (1867) concerning the relative degrees of mimicry and the idea that there are no truly "representational" sign language gestures (see Mallery 1880, 1884, 1972, and this book, Vol. 1; West 1960).[5]

The conventional nature of aboriginal sign language signs does not, however, prevent them from being usable as a mode of communication between individuals who come into contact with practitioners of sign languages other than their own. Although, as noted above, it is doubtful whether persons unfamiliar with sign language are capable of comprehending sign language messages without a good deal of background training, the case is less clearcut concerning reports that persons practiced in some form of conventional sign language are capable, after minimal practice, of communicating effectively with one another through gesture. West proposes the following explanation for the possible validity of such reports:

> It is the combination of self-evident pantomime in the shape and motion and relationship kinemes and the general similarity of combination conventions that makes possible the ready communication between any two sign language users, whatever the provenience and degree of similarity of their respective sign systems. Another important factor in such inter-communication is the high redundancy of all sign languages. . . . We are faced with two general possibilities for communication between adepts of different languages. Each party to the communication may select the most highly pantomimic signs in his own or the other's system and use these for the duration of the conversation . . . but Scott and Bell point up the other alternative, which involves abandonment of the signs conventionally accepted in either or both sign languages and the creation of new, spontaneously pantomimic signs acceptable to the other party, by the application of generally shared principles of pantomimic construction. In such a case we have to deal not with a language at all, but pantomimic artistry. . . . In practice, of course, both the alternatives are extensively utilized. Thus the "gesture language of mankind" turns out to be not a language at all, but a combination of generally shared techniques for selecting, improving, redundantly supplementing or abandoning and substituting new pantomimes for the bodies of conventionalized signs available in the respective sign languages of the conversationalists. Since these very techniques are part of the grammatical systems of each of the sign languages investigated by the writer, communication with a representative of an alien sign language is simply an extension and intensive use of the resources of the home sign language. This adequately explains the reported ability of intelligent and experienced sign language adepts to communicate with each other across language boundaries and the usual inability of poor sign talkers or individuals with no sign language experience to do so.
>
> (West 1960, Vol. II, p. 55–56)

One can see at work here the forces leading to the constant shifting of the balance between natural and conventional sign relations. Because, as Mallery states, "meaning does not adhere to the phonetic presentation of thought, while it does to signs" (this book, Vol. 1, p. 9), the opportunity is present in every interaction involving sign language, whether between interlocutors trained in the same or different sign languages, that the pragmatics of the communication context will favor emphasis of one type of sign relation over another, thus shifting the hierarchical arrangement among sign aspects with respect to the particular context of interaction and potentially the sign system as a whole.

2. SIGN LANGUAGES AND SPOKEN LANGUAGES: A SEMIOTIC COMPARISON

Despite his amateur standing as a scientist and his failure to carry out a consistent and systematic semiotic analysis of aboriginal sign languages, Mallery nevertheless approached his subject from a broad, comparative semiotic point of view. His use of the term *semiotics* (as in *native semiotics*, this book, Vol. 1, p. 8) is to our knowledge one of the earliest. In addition, he employed the expressions *semiotic code* (e.g. 1972, p. 320; this book, Vol. 1, p. 43), *semiotic execution* (e.g. 1972, p. 74), *semiotic syntax* (1972, p. 396), and *semiotic expression* (e.g. 1972, p. 88). We share the puzzlement of Mounin (1973, p. 154–156) about the source of Mallery's terminology. As he points out, Mallery gives no indication of having known of Charles Sanders Peirce and Mallery is never named by Peirce.[6] Whatever the sources of his inspiration were, the fact is that Mallery's comparison of aboriginal sign languages, especially PSL, with other auditory and visual systems in terms of a broad semiotic typology has in many respects yet to be surpassed.

First, Mallery's discussion of the hierarchical arrangement between natural and conventional sign relations anticipates Stokoe's (1974*a*, p. 127) conclusion that, while the American sign language of the deaf "can serve human language capacity as fully as can an sSign language, . . . even in as developed a language as ASL," there remains an important element of iconicity and indexicality not found in spoken language (i.e., an sSign language). It would appear that, if one were to position these three types of semiotic systems along a continuum from most natural to most conventional, sign languages of the deaf would fall somewhere between the other two.

Mallery also compares aboriginal and deaf sign languages with regard to the varying degrees to which each is influenced by spoken language. His analysis, while showing the similarities between the two visual systems in this regard, nevertheless reveals some critical ways in which aboriginal sign languages bear the imprint of spoken language where sign languages of the deaf do not. Mallery pointed out (e.g., this book, Vol. 1, p. 62) that the meanings of PSL signs are context sensitive and, furthermore, not simple representations of spoken words. "So far from the signs representing words as logographs, they do not in their presentation of the ideas of actions, objects, and events, under physical forms, even suggest words" (*ibid.*, 61). Stokoe's much later (1974*a*, p. 118) caution that the gSigns of sign languages of the deaf cannot be relegated to simple "code representations, either as speech surrogates (1 gSign = 1 word) or of alphabetical symbols" would appear to apply as well to aboriginal sign languages. Moreover, as Buyssens (1967) notes, unlike even as speech independent a writing system as Chinese, one does not have to know the spoken language of the PSL signer in order to understand signed messages.

On the other hand, Mallery anticipated Stokoe's (1972, p. 17) definition of PSL as a speech surrogate owing to the fact that spoken language is the "primary" semiotic system of aboriginal signers, while for the deaf, it is their sign language which is the primary semiotic system and therefore a true sign "language." Discussing a meeting between Indian signers and deaf-mutes at the Pennsylvania Institution for the Deaf and Dumb in 1873, he remarked that the deaf signers greatly surpassed the Indians in "pantomimic effect." This was due, he concluded, to the fact that

> what is to the Indian a mere adjunct or accomplishment is to the deaf-mute the natural mode of utterance, and that there is still greater freedom from the trammel of translating words into action—instead of acting the ideas themselves—when, sound of words being unknown, they remain still as they originated, but another kind of sign, even after the art of reading is acquired, and do not become entities as with us (this book, Vol. 1, p. 44)

While it is obvious that there is always present the possibility of intersemiotic translation between the sign language and spoken language of hearing sign language users, views on the extent to which signs represent elements of spoken language vary from one extreme, for example that of Harrington ("the signs are everywhere based on spoken language and reflect it at every turn"—this book, Vol. 2, p. 117), to its opposite, for example West's conclusion that there is "no evidence that Amerindian sign language is in any sense derivative of spoken language" (1960, Vol. I, p. 97). Kroeber saw PSL as being heavily indebted to spoken language. Like writing, he thought of PSL as a substitute for speech rather than an independent or original method of communication. The concepts communicated by means of PSL were to him "concepts already developed in speech but translated into a nonspoken medium" (this book, Vol. 2, p. 97). A more moderate view is taken by Taylor who, in the most recent review of discussion about this question, states that many of the features of PSL word order supposed to derive from spoken language could be accounted for more economically by describing PSL as a system operating according to the rule that specification of the general precedes specification of the particular (this book, Vol. 2, p. 241). He also notes, however, that spoken and sign lexicons do not appear to be as independent as word order. New signs are known to have been created in sign language as imitations of words in the spoken vocabulary, and, in turn, through back translation from signs, new spoken words are sometimes introduced as a result of a particular sign (*ibid.*).

Regarding AASL, Miller (this book, Vol. 2, p. 439–440) expressed the view that the sign language system he studied was a surrogate in the sense that some signs appeared to represent spoken words and sign order tended to adhere to word order. It was unlike spoken language, however, in that dialect variation in words and signs did not covary. Roth (this book, Vol. 2, p. 273), on the other hand, referred to AASL signs as *idea-grams*, and Seligmann and Wilkin (this book, Vol. 2, p. 317) as *ideo-grams*, in order to stress the direct expression of ideas through signs.

It would appear that on the whole aboriginal sign languages act more as a substitute for spoken language than do sign languages of the deaf (excluding signed English and finger spelling, of course). Stern refers to the signs of PSL as *lexical ideographs*, or signs which represent a lexical unit of the spoken language without "reference to the phonemic structure of the base language" (1957, p. 127). Such signs would fit Buyssens's definition of a speech surrogate as "un sème dont la signification est constituée par le signifiant d'une autre sémie" (1967, p. 45), instigating a particular process of what Jakobson (1971, p. 261) calls *transmutation*, or intersemiotic (as opposed to intra- or interlingual) translation, which is the "interpretation of verbal signs by means of signs of nonverbal sign systems."

On the other hand, none of the aboriginal sign language dialects have been described as purely substitutive, and each is to a varying degree "independent of but translatable into natural language" (Sebeok and Umiker-Sebeok 1976, p. xiii). In fact, it would appear that intersemiotic translation is to a certain extent a barrier to the attainment of fluency in the use of sign language, as alluded to above with regard to communication between PSL practitioners and deaf signers. In many respects, the sign language is considered a separate additional communications channel which is not subordinated to the vocal-auditory (Taylor, this book, Vol. 2, p. 229). The greater the degree of conventionalization and standardization and the less the individual signer has to rely upon translation from spoken language, the greater the fluency of sign language performances.

Mallery noted that the conventionalization of natural signs introduced by an individual is the result of an expansion of their radius of communication (cf. Stokoe 1974*b*), or diffusion of the signs among an ever larger group of people on the one hand, and, on the other, the increasing frequency with which they are used in different contexts, both types of situations placing demands on sign users to abbreviate the sign. PSL differs from deaf sign language and spoken language precisely because it was used by a restricted number of people—either intratribally by a subgroup within a Plains tribe, or intertribally, during sporadic contacts between representatives of different tribes—and because the variety of communicative contexts in which it was employed was small compared with these other sign systems. Mallery's way of looking at sign systems from the point of view of their meaning and function in relation to the individuals and groups which use them gives his approach a quite modern taste, as well as recalling Charles Sanders Peirce's semiotic pragmaticism (Peirce 1965–1966). With Mallery's framework, we can envision a time when aboriginal sign languages could, given favorable historical circumstances, become as conventional and almost as direct an expression of ideas as the deaf sign languages of today.

Neither AALS nor PSL have developed into uniform systems of communication, however. West's 1956 dialect survey of the Northern Plains sign language revealed that only from 20 to 40 percent of PSL signs were actually held in common by different tribes, and he concluded that the sign language could be considered universal "not in sharing of lexical items, but the application of general pantomimic principles in the formation of whatever signs are selected for conventionalization" (1960, Vol. II, p. 54). Despite the emphasis placed by most authors on the intertribal origin and development of aboriginal sign languages, there is good reason to believe that, in fact, their primary use was for communication among members of a single tribe. Dodge (this book, Vol. 2, p. 8), among others, noted that the Plains Indians used sign language even in their own camps, in everyday conversation with people who spoke the same language. Webb (this book, Vol. 2, p. 95–97) suggested that the primary context in which the sign language developed was hunting and warfare, where spoken language was of limited use for communicating over great distances. West's survey in the 1950s supports the view that PSL was particularly important *within* Plains tribes. He found no correlation to exist between sign language use and the presence of a spoken *lingua franca*, for example Mexican Spanish in the U.S. Southwest. Furthermore, the sign language was at that time described by West as "a going concern," long after it had ceased to serve as a *lingua franca* (1960, Vol. II, p. 63). Its ceremonial and oratorial function continued even after long distance communication had been replaced by modern means of communication, such as radio and telegraph. Among the Crow and Cheyenne, moreover, "more young than older people knew sign language" (*ibid.*) and in some areas, for example, Saskatchewan, the sign language appeared at that time to be spreading. Mallery attributes the divergent patterns of conventionalization of PSL signs to the frequent and varied intratribal application of sign language. The precise intratribal "mode of semiotic expression" as well as the amount of its general use are always fluctuating, depending upon the degree of fluency of those tribal members who act as PSL custodians and teachers. Some signs, invented by one or more of these individuals, may be ultimately forgotten, perhaps to be reinvented at some later date; others may be taken up and "conventionalized among members of the same tribe and its immediate neighbors" but not among them and more distant tribes because the form of abbreviation or conventionalization which seems natural to one set of people may not appear as such to another (Mallery, this book, Vol. 1, p. 18–19). Yet another factor contributing to the high degree of dialect variation of PSL is the increase in ceremonial usage noted above. It is well known that there is a tendency for the language of ritual to change at a slower rate than that used in everyday conversation. As the PSL becomes more closely tied to rituals associated with tribal identity, we would expect it to retain its "picturesque" nature and become progressively more archaic, each tribe freezing its sign dialect at a certain stage of development.

The situation becomes even more complex if one considers that both PSL and AASL were not uncommonly used simultaneously with spoken language, as a kind of "embroidery" (Meggitt, this book, Vol. 2, p. 410); (cf. in this book, Vol. 2, Humfreville, Kohl, Maclean, Miller, Taylor, Webb; also West 1960). The high degree of iconicity and indexicality of PSL and AASL, the former frequently described as "poetry of motion" and "graceful," made them particularly suitable for use with "formal oratory" and "impassioned or emphatic conversation" (Mallery, this book, Vol. 1, p. 42). Unlike the gesticulations accompanying most speech, aboriginal sign language signs, when used with spoken language, could be compared with pictorial illustrations of written texts. Both words and signs have their own meanings in reference to their respective semiotic system, yet each influences the interpretation of the other, the visual image amplifying the verbal and *vice versa.* This type of intersemiotic activity must be viewed as a unique and complex process of mutual interpretation.

Turning to AASL, Stirling (this book, Vol. 2, p. 258) asserted that "a good deal of variation exists in different parts" of Australia, with some "blending" of signs at intertribal meeting places, such as Tempe Downs, where representatives of the Luritcha and Arunta traditionally came together. Later, Meggitt (this book, Vol. 2, p. 419), using Roth's (1908–1910) collection of signs, found that the number of signs held in common between groups varied inversely with an increase in distance between them. West (1960) maintained that the amount of dialect variation was roughly similar for both AASL and PSL.

We also find that intratribal uses of AASL are as prevalent, if not more so, than intertribal ones (for mention of the latter, see, e.g., in this book, Vol. 2: Berndt, Howitt, Spencer and Gillen, Stirling, Warner). There are reports of the employment of AASL for everyday conversation (e.g., in this book, Vol. 2: Basedow, Meggitt, Stirling, Strehlow), long-distance communication, as while stalking game (e.g., in this book, Vol. 2: Basedow, Berndt, Haddon, Howitt, Meggitt, Mountford, Spencer and Gillen), and communication with the deaf (e.g., in this book, Vol. 2: Meggitt, Warner). AASL is also a convenient form of expression either for persons ritually defined as "dead" in the sense of being in a liminal stage between ritual statuses, as, for example, novices during initiation ceremonies or widows in mourning (e.g., in this book, Vol. 2: Howitt, Meggitt, Roth, Spencer and Gillen, Warner), or by persons of normal ritual status when conversing about taboo subjects such as sexual intercourse, dead persons, or ceremonial secrets (e.g., in this book, Vol. 2: Meggitt, Strehlow). We have already noted the function of AASL as an accompaniment to speech.

A third criterion for comparison of PSL, deaf sign language, and spoken language used by Mallery was the type of communication channel utilized by each. Although he stressed the similarity in communicative potential of all three semiotic systems, Mallery warned against a glottocentric analysis of the first two which would overlook the critical differences between the requirements of an auditory versus a visual channel of communication (cf. Hailman 1977). PSL and sign languages of the deaf, as visual systems, may be able to be as expressively complex as spoken language, but they will attain such a level of complexity in quite a different way. West (1960, Vol. 1, p. 89) wrote that the key to sign language grammar is to be found in Mallery's notion of "sign picture" (Mallery 1972, p. 114). Mallery remarked (this book, Vol. 1, p. 61) that in "mimic construction" one must consider "both the order in which the signs succeed one another and the relative positions in which they are made, the latter remaining longer in the memory than the former." This idea was taken up by Scott, who compared a conversation in PSL with a series of moving pictures in which the relations between the object's actions are represented by the relative positions and sequencing and are evident to the eye (this book, Vol. 2, p. 64). Miller (this book, Vol. 2, p. 438–439) also found that the psycholinguistic processing of AASL messages varies considerably from the way we decode spoken messages. With AASL, the internal composition, or sequence of signs, is not recalled by the addressee, while the general meaning, or "picture," of the total sequence is retained in the memory. West believed that syntactic order, or successivity, was for PSL a "redundant, non-obligatory stylistic matter" (1960, Vol. 1, p. 90). To the extent that a PSL or AASL dialect has evolved into a complex, general system of communication, we should expect it to have adopted a wide variety of semiotic strategies which are available to a visual mode but not found among spoken languages. Peng has noted three such principles of organization: (1) *simultaneity*, where two signs are produced together (see also Ljung this book, Vol. 2, p. 215); (2) *reversibility*, where "a sign's movement may be reversed so as to form another sign that has the opposite meaning"; and (3) *directionality*, where "a sign may be positioned in different ways, depending upon the relative space occupied by the signer, the viewer, and the content of their conversation, so as to add a subtle connotation in one shift which would otherwise require several words in any oral language to describe" (Peng 1977, p. 28–29). Since 1960, students of the American sign language of the deaf have developed a mode of description which combines a rigorous systemic analysis derived from linguistics proper with a fresh look at sign language on its own terms, or "from inside" (Stokoe 1974*a*, p. 119). This work has begun to reveal some fascinating techniques available to users of a visual communication system such as sign language which were heretofore ignored precisely because they rely on one or more of the three characteristics noted above, part of Mallery's "spatial syntax" (1972, p. 114), where "the obligatory grammatical relationships are established not by temporal order or syntax, but by spatial relationships, both within the execution of a single sign and between positions of execution of succeeding signs" (West 1960, Vol. 1, p. 90). To date, no such revolution has taken place in aboriginal sign language studies. It would be particularly interesting to know in what way the possession of the facility of speech influences the types of communicative strategies developed in aboriginal sign languages, thus contributing to the divergent lines of development between them and sign languages of the deaf.

The analysis of aboriginal sign languages, like that of visual communication systems in general, has been plagued by language-dominated concepts and methods. One of the earliest of these was the analogy made between the lexicon of spoken language and that of the sign languages, where signs were treated as similar to the linguistic concept of word. As noted above, Mallery warned against basing a description of PSL on the false assumption of the identity of visual and auditory units. In addressing potential contributors to his sign language project (see, for example, 1884 and this book, Vol. 1), he admonished collectors to gather native language interpretations as well as English translations of those interpretations along with their appropriate signs. In addition to this, Mallery showed himself to be sensitive to the influence of the discursive context on the meaning of individual signs when he insisted that signs be collected within examples of connected discourse, either informal conversations or formal speeches and stories. This sort of field technique would be necessary, of course, to guard against treating signs as if they functioned like words of spoken language.

Much later, Kroeber (this book, Vol. 2, p. 201), unaware of Harrington's earlier, sketchy attempt to analyze PSL on its own, visual terms, called for a "*systematic analysis of the sign language in terms of itself.*" The study which Kroeber envisioned would distinguish the distinctive patterns of motion or position from "accidentals," providing a verbal description of such patterns together with outline linear sketches. The description would be made in terms of semantic categories plus executional forms, in dictionary fashion. As an illustration of what such an analysis would look like, Kroeber grouped together signs made with two hands according to whether neither hand moves, one hand moves, or both hands are in motion. Signs produced with two hands moving are grouped according to whether the hands are interacting (or crossing) or whether there is a bilaterally symmetrical simultaneous motion involved. Signs in the latter category are divided further still according to whether the motion is centrifugal or centripetal. Many of the descriptive categories mentioned by Kroeber, such as direction of motion, repetition, relation of motion to body parts, and so forth, are mentioned by Mallery (1884, p. 207ff. and this book, Vol. 1, p. 68–69) in relation to his instructions to investigators concerning the notation and description of PSL. Mounin (1973, p. 158) has suggested that Mallery's emphasis on the importance of finding "the radical or essential part" of each sign "by which it can be distinguished from any other . . . sign" (Mallery 1972, p. 84–85) bears a remarkable resemblance to the crucial linguistic notion of distinctive feature. Mallery, however, was unable to take advantage of the advances toward systemic analysis which were fashioned by linguists largely after he had completed his work; (Jost Winteler's *Die Kerenzer Mundart des Kantons Glarus in ihrer Grundzügen dargestellt* was completed in 1875, but his dissertation remained out of Mallery's ken).

While Kroeber saw the importance of an application of linguistic technique on the level of sign morphemes, he considered it unwise to extend its use beyond that level, due to the highly iconic and indexical nature of sign language signs in contrast with spoken language. His student, C. F. Voegelin, took him to task for his refusal to grant the possibility that PSL possessed a phonemic level of structure and duality of patterning such as found in spoken language. Contrary to what he viewed as the fundamental weakness of earlier descriptions of PSL, Voegelin emphasized the importance of subordinating the PSL dictionary to its grammar, as he put it, making the former an appendix of the latter rather than the reverse.

Voegelin's student, La Mont West, spent the second half of the 1950s engaged in the most detailed study of PSL to date, one which resulted in his dissertation in 1960, the first volume being devoted to a formal linguistic analysis of the sign language, the second to issues such as dialect survey, sociological analysis of sign language use, and the like. West's linguistic analysis was specifically designed to set forth a kinemic as well as a morphemic level of structure and thus reveal the dual patterning of this communication system. He claimed that PSL could be described fully in terms of a total of eighty kinemes, which fell into five basic kineme classes, which he compared with units of spoken language (indicated in parentheses) in terms of number and combinatorial privileges: *hand-shapes* (consonants), *directions* (vowels), *dynamics* (stress, tone, length), *motion-patterns* (semi-vowels), and *referents* (kinemes which "specify the body parts, parts of hands or external reference in relation to which the *hand-shapes* or active hand parts are positioned or moved"—Vol. 1, p. 13). The point by point comparison with spoken language was a constant source of frustration to West, who makes clear that he would have greatly preferred to base his analysis of PSL on a sufficiently detailed and rigorous analysis of another gesture system, had such a study been available. He repeatedly warns the reader that the use of a language-based model of analysis could lead to the incorrect conclusion that the structure of sign language mirrors exactly that of spoken language. Since West's dissertation was the first and most comprehensive linguistic treatment of PSL and is difficult to obtain, it is worth including here at length one such admission by the author:

> [I]t must be conceded that many decisions required in the setting up of kinemes have been made upon adherence to single criteria, often rather arbitrarily selected. Had other criteria been rated higher, a somewhat different kinemic inventory could have been set up to account for the same set of distinctions between components and subcomponents. . . . However, a second factor is also at play in giving the kinemic level less stability than the morphemic in sign language. The system is partially open-ended on the kinemic level, as well as on the morphemic. . . . On the phonemic level for spoken language the system is at least as firmly closed as the minor morphemes, except for a marginal category of onomatopoetic sounds and exclamatory sounds, which may be simply excluded from the system if they fail to pattern with it. In this sign language description, as in those for most spoken language grammars, the morphology is chiefly restricted to a treatment of major morpheme classes qua classes and of minor morphemes in paradigms and individually. Since this much of sign language morphology also forms a nearly closed system, it gives the same, rather misleading, impression of self-contained structural integration at the morpheme level that is characteristic of spoken language morphological statements. However, the kinemic level of sign language is far less parallel with the phonemic level of spoken language, than is the case for the respective morphemic levels. In sign language the onomatopoetic (pantomimic) cannot be read out of the system, since it comprises 98% of the system at the morphemic level and a less high, but still very imposing percentage at the kinemic level, due to the high incidence of mono-kinemic morphemes and an even higher incidence of "fl"-like sub-morphemic form-meaning relationships. This high onomatopoetic content of sign language makes possible and encourages the constant development of new signs. In most cases such signs do utilize kinemic material already well established in the system. In fact, the *motion-pattern*, *direction* and *dynamic* classes of kinemes participate in a nearly closed system in the idiolect of sign language upon which this analysis is based. However, the *hand-shape* class of kinemes is slightly open-ended and the *referent* class is markedly so. In most cases the new or rare hand-shapes and referents can be included as allokines of kinemes already well established, since the very variety of their occurrence insures complementarity with one or another of the more frequent kinemes. In most cases, even, some common element of shape can be discovered between the new candidate kine and one or another of the kinemes with which it is in complementary distribution. All too often, however, the complementarity in such cases is dependent upon rather special and unique environments, and such uniqueness robs the statement of allokine distribution in sign language of the authority characteristic of statements possible for allomorphs.
>
> (1960, Vol. I, p. 29–30)

The claims by Voegelin and those he influenced (West 1960 and, in this book, Vol. 2: Kakumasu, Ljung, and West) concerning the duality of aboriginal sign languages are, as a recent reviewer has noted, "not persuasive" (Mounin 1973, p. 161). Mounin points out that in overlooking the visual nature of PSL, where the simultaneity of certain minimal units make them appear to resemble the distinctive features of phonemes, there arose a confusion between phonetic transcription, based on minimal descriptive units, and a phonological transcription, based on distinctive features (*ibid.*, p. 161–162).

Despite the pressures exerted on West to remain within the then popular methodological frame for phonemic analysis and ignore questions of semantic structure, he admitted, after a summary of earlier semantically oriented descriptions of sign language, or what he called "the conceptual approach to sign language," that it "does throw considerable light upon the ease of learning and communication characteristic of the sign language" (1960, Vol. I, p. 88–89). West would have done well to heed the advice of his chief benefactor, Kroeber, who, as noted above, stressed the importance of combining a formal grammatical analysis with a semantic one. Ljung, who applied a linguistic technique derived from Hjelmslev's and Uldall's glossematics, also known in some quarters as stratificational grammar, to PSL, also found that meaning kept intruding upon the purely formal analysis which he was attempting. His "gestemes" were "more tainted with meaning"[7] than are phonemes, as, to cite his own example, the common circle element in *sun, star, coin,* etc., or the growth image recurring in *grass, grow, tree*" (this book, Vol. 2, p. 221). However, pointing to the fact that the circle element also could be found in signs for *want* and *drink,* Ljung reaffirmed Mallery's idea that, had PSL been allowed to develop freely, it would "no doubt have become more arbitrary as time went by" (*ibid.*). The very limited success of the analysis of PSL using meaning-free descriptive techniques borrowed from linguistics points up the need for a new description of PSL or AASL, one which would take into account recent developments in the semiotics of nonverbal iconicity and indexicality as well as meaning-sensitive formal theories of conventional sign systems. Anyone attempting a new look at aboriginal sign languages should ponder the experience of Stokoe, who produced his first linguistic treatment of the American sign language of the deaf during the same period as West's dissertation, and under a comparable commitment to the structuralist framework of the day, but who soon recognized the severe limits of a structuralist analysis. In 1972, he noted, for example, that, with respect to ASL *cheremes,* "the ways in which they differ from phonemes in operation are as important as the similarities" (p. 20). Only a semiotic analysis can provide an adequate account of the whole system, with its high degree of iconic and indexical signs (1974*a,* p. 127). Such a statement would be even more appropriate in a discussion of aboriginal sign languages.

Latest available reports (Taylor and Miller, this book, Vol. 2) suggest that there are enough PSL and AASL practitioners active today to make fresh analyses of aboriginal sign languages feasible. In our distillation of the voluminous materials pertaining to aboriginal sign languages, our aim has been to focus attention on what makes them unique as systems of representation and communication and therefore of paramount importance for the study of other visual codes, spoken language, and, ultimately, semiotic theory. It would be regrettable were this rich and uncommon species of semiotic organization to become extinct without the benefit of another long, hard look by well-trained and appreciative eyes.

There are a number of topics which are related to the principal, semiotic concern of this introduction but which could not be included in the present discussion (but will be taken up elsewhere):

1. While we have outlined some of the points of similarity and difference between aboriginal sign languages, sign languages of the deaf, and spoken language, there are many interesting discussions in the literature (especially Mallery 1884, 1972, this book, Vol. 1; West, this book, Vol. 2) concerning the relationships between aboriginal sign languages and other gestural systems, both formal (e.g. Hindu dance gestures) and informal (e.g. monastic, Neopolitan, and Armenian signs), writing systems (particularly Chinese), and pictographs. While much of this material is slanted toward questions of the origin of aboriginal sign languages, a universal "gesture speech of mankind" (e.g. Mallery, this book, Vol. 1), or the gestural origin of spoken language, a careful study, which would make use as well of the modern analyses of these and other such visual systems accomplished since West's dissertation, could make a substantial contribution to the field of nonverbal communication.
2. The same may be said for the equally significant problem of the notation of aboriginal sign languages. Many authors (especially West 1960) have tried to overcome the difficulties involved in the recording and transcription of signs, but with little success. The search for such tools is of great semiotic interest in itself and deserves separate and detailed consideration.
3. Finally, modern poetic, discourse, and narrative analyses of aboriginal sign language texts need to be made, taking into account the framing of visual moving sign "pictures" both when used by themselves and in interaction with facial expression, speech, and other auditory and visual messages, such as music, especially when integrated in multimedia performances (spectacles).

3. ORGANIZATION OF THE BOOK

Aboriginal sign languages have captivated a broad spectrum of observers during the last two hundred years, including early explorers, military personnel and government administrators, missionaries, amateur and professional ethnographers and linguists. We have tried above to account in part for this enduring fascination in terms of the uniqueness of aboriginal sign languages as independent, conventional gestural codes, which, while sharply distinct from spoken language in basic semiotic structure, nevertheless approach it in the range and flexibility of the content and style of their messages. The articles culled for presentation in these two volumes amply exemplify and variously elaborate upon this semiotic theme.

The papers were selected according to the additional criterion that they provide insight into the historical development and variety of approaches to many other aspects of aboriginal sign languages, from their geographical distribution to the scope of their expression in the everyday and ceremonial life of the societies in which they are practiced.

Volume 1 contains three important yet hardly accessible works from the early 1880s by Colonel Garrick Mallery. The first and third are representative of Mallery's sensitive—and in many respects still unsurpassed—comparison of the

North American Plains sign language with several other verbal and nonverbal codes. The second, a collection of Plains signs, stands as the largest published list available. It offers the user a broad data base against which general descriptive and theoretical statements made by Mallery and his successors may be measured.

Volume 2 gives a wide panorama of both American and Australian sign languages, featuring contributions by many and diverse authors ranging from the late 19th century to contemporaneous unpublished papers. All major article- or chapter-length scientific ethnographic and linguistic treatments of American and Australian systems have been included, as well as several of the more significant semipopular papers which deal with the North American Plains sign language.

Bloomington, Indiana
June, 1977

D. Jean Umiker-Sebeok
Thomas A. Sebeok

REFERENCES

Beard, Daniel Carter. 1918. *The American boy's book of signs, signals, and symbols.* Philadelphia, J. B. Lippincott Co.

Bossu, Jean Bernard, 1771. *Travels in the interior of North America 1751–1762.* London, T. Davies. [Reprinted 1962. Norman, University of Oklahoma Press.]

Burton, Richard F. 1862. *The City of Saints and across the Rocky Mountains to California*, pp. 135–144. New York, Harper.

Buyssens, Eric. 1967. *La communication et l'articulation linguistique.* (=Travaux de la faculté de philosophie et lettres 31). Brussels, Presses Universitaires de Bruxelles.

Clark, William Philo. 1885. *Indian sign language.* Philadelphia, L. R. Hamersly and Co.

Cody, Iron Eyes. 1952. *How: Sign talk in pictures.* Hollywood, H. H. Boelter Lithography.

Cody, Iron Eyes. 1970. *Indian talk. Hand signals of the American Indians.* Healdsburg, California, Naturegraph Publishers.

Dunbar, William. 1809. On the language of signs among certain North American Indians. *Transactions of the American Philosophical Society* 6 (pt. 1, no. 1):1–8.

Eco, Umberto. 1976. *A theory of semiotics.* Bloomington, Indiana University Press.

Fisch, Max. 1977. Peirce's place in the history of ideas. *Ars Semeiotica* 1:21–37.

Hadley, Louis F. 1890. *A lesson in sign talk.* Forth Smith, Arkansas, Hadley Pub.

Hadley, Louis F. 1893. *Indian sign talk.* Chicago, Baker and Co.

Hailman, Jack P. 1977. *Optical signals. Animal communication and light.* Bloomington, Indiana University Press.

Hawkes, Terence. 1973. *Shakespeare's talking animals. Language and drama in society.* London, Edward Arnold.

Hofsinde, Robert (Gray-Wolf). 1956. *Indian sign language.* New York, William Morrow & Co.

Jakobson, Roman. 1970*a*. *Main trends in the science of language.* New York, Harper.

Jakobson, Roman. 1970*b*. Language in relation to other communication systems. *Linguaggi nella società e nella tecnica*, pp. 3–17. Milan, Edizione de comunità.

Jakobson, Roman. 1971. On linguistic aspects of translation. *Selected writings 2, (Word and language)*, pp. 260–266. The Hague, Mouton.

Kendon, Adam. 1972. Time relationships between body motion and speech. *Studies in dyadic communication*, ed. by A. W. Siegman and B. Pope, chapter 9. New York, Pergamon.

Kohl, Johann G. 1860. *Kitchi-Gami. Wanderings round Lake Superior*, pp. 137–143. London, Chapman and Hall.

Long, Stephen Harriman. 1823. Account of an expedition from Pittsburgh to the Rocky Mountains 1:378–394. Ann Arbor, University Microfilms.

Mallery, Garrick. 1879. The sign language of the North American Indians. *Proceedings of the American Association for the Advancement of Science* 28: 493–519.

Mallery, Garrick. 1880. The sign language of the Indians of the Upper Missouri, in 1832. *American Antiquarian* 2 (no. 3):218–228.
Mallery, Garrick. 1884. Sign language. *Internationale Zeitschrift für Allgemeine Sprachwissenschaft* 1:193–210.
Mallery, Garrick. 1972. *Sign language among North American Indians compared with that among other peoples and deaf-mutes.* The Hague, Mouton. [Original 1881. Washington, Bureau of American Ethnology.]
Marcy, R. B. 1866. *Thirty years of Army life on the border,* pp. 32–34. New York, Harper and Brothers.
Mounin, Georges. 1973. Une analyse du langage par gestes des Indiens (1881). *Semiotica* 6:154–162.
Peirce, Charles Sanders. 1965–1966. *Collected papers of Charles Sanders Peirce,* ed. by C. Hartshorne, P. Weiss, A. W. Burks. Cambridge, Mass., Harvard University Press.
Peng, Fred C. C. 1977. Sign language and culture. *Sociolinguistics Newsletter* 8 (no. 1):28–29.
Ray, Verne F. 1933. *The Sanpoil and Nespelem: Salishan peoples of N.E. Washington,* pp. 122–123. (=University of Washington Publications 5). Seattle, University of Washington Press.
Roth, Walter E. 1908–1910. Signals on the road. Gesture language. North Queensland ethnography, pp. 82–93. (=Australian Museum Records 7).
Sayce, A. H. 1880. Sign language among the American Indians. *Nature* 22 (no. 553):93–94.
Scott, Hugh Lennox. 1934. Film dictionary of the North American Indian sign language. Washington, D.C., National Archives.
Sebeok, Thomas A. 1976*a*. *Contributions to the doctrine of signs.* Lisse, Peter de Ridder Press.
Sebeok, Thomas A. 1976*b*. Iconicity. *Modern Language Notes* 91:1427–1456.
Sebeok, Thomas A. 1977. Zoosemiotic components of human communication. *How animals communicate,* ed. by T. A. Sebeok, pp. 1055–1077. Bloomington, Indiana University Press.
Sebeok, Thomas A., and D. Jean Umiker-Sebeok, eds. 1976. *Speech surrogates: Drum and whistle systems.* The Hague, Mouton.
Seton, Ernest Thompson. 1918. *Sign talk: A universal signal code, without apparatus, for use in the Army, Navy, camping, hunting, and daily life. The gesture language of the Cheyenne Indians.* Garden City, New York, Doubleday, Page, and Co.
Shands, Harley C. 1971. *The war with words. Structure and transcendence.* The Hague, Mouton.
Steiner, George. 1975. *After Babel. Aspects of language and translation.* New York, Oxford University Press.
Stern, Theodore. 1957. Drum and whistle 'languages': An analysis of speech surrogates. *American Anthropologist* 59:487–506.
Stokoe, William C., Jr. 1972. *Semiotics and human sign languages.* The Hague, Mouton.
Stokoe, William C., Jr. 1974*a*. Motor signs as the first form of language. *Semiotica* 10:117–130.
Stokes, William C., Jr. 1974*b*. Classification and description of sign languages. *Current trends in linguistics 12 (Linguistics and adjacent arts and sciences),* ed. by T. A. Sebeok, pp. 345–372. The Hague, Mouton.
Taylor, Allan R. 1975. Nonverbal communications systems in native North America. *Semiotica 13*: 329–374.
Tomkins, William. 1929. *Universal Indian sign language of the Plains Indians of North America.* San Diego, William Tomkins. [Reprinted 1969. New York, Dover Press.]
Umiker-Sebeok, D. Jean. 1974. Speech surrogates: Drum and whistle systems. *Current trends in linguistics 12 (Linguistics and adjacent arts and sciences),* ed. by T. A. Sebeok, pp. 497–536. The Hague, Mouton.
Weil, Elsie, 1931. Preserving the Indian sign language. *New York Times Magazine* 8 (section 8–July 5).
West, La Mont, Jr. 1960. The sign language, an analysis. Unpublished Ph.D. dissertation in Anthropology, Indiana University, Bloomington.
Whitney, William Dwight. 1867. *Language and the study of language.* New York, Charles Scribner's Sons.
Wied-Neuweid, Prince Maximilian von. 1839–1841. *Reise in das Nordamerika in den Jahren 1832–1834* 2, pp. 645–653. [English version 1906. *Travels in the interior of North America* 3, pp. 300–312. Cleveland, Arthur H. Clark Co.]
Zorn, E. R. 1928. Zeichensprache und Signalwesen bei den Prärieindianvervölkern. *Erdball* 2: 52–56.

NOTES

[1]The most heavily studied conventional and generalized gestural codes employed by native peoples are those found among North American Indians [in the Plains–Plateau–Great Basin–Southwest area usually

simply referred to as the "Plains" sign language (West 1960, Vol. I, p. 60–61; Taylor this book, Vol. 2, p. 227)] and Australian aborigines [scattered throughout the continent but especially prevalent among Central and Northeastern groups (Howitt and Spencer and Gillen this book, Vol. 2)]. The untimely demise of West's extensive Australian survey (this book, Vol. 2) left us without a clear picture of the extent of sign language use in that continent. The Urubú sign language of Brazil, our only example of a South American system, was included in this book (Vol. 2), because although it is primarily a code designed for communication with the deaf, Kakumasu's report showed that it was also practiced by hearing tribal members and that, in regard to grammatical structure, it resembles PSL (this book, Vol. 2, p. 247). The vast majority of articles concerning aboriginal sign languages in Siberia are, of course, in Russian. A Hungarian ethnosemiotician, Mihály Hoppál, has in preparation a reader dealing with this other major group of aboriginal sign languages.

[2]See Sebeok's note (1976*b*, p. 1451) concerning the related pairs of terms *chiral* and *chirality*, *chirologia* and *chironomia*, and their relevance to a discussion of the concept of iconicity [cf. also Fisch's mention of Peirce's "Art Chirography" (1977, p. 28)]. Stokoe (e.g., 1972) has introduced into studies of sign languages of the deaf the terms *cheremes* (patterned after spoken language *phonemes*), *cherology* (cf. *phonology*), and *cheremics* (cf. *phonemics*).

[3]West's adoption of Birdwhistell's terminology brought to aboriginal sign language studies a set of terms parallel to those noted in footnote 2 (cf. *kinesics/cheremics*, *kineme/chereme*). A third set of terms, *gestemics* and *gestemes*, was proposed by Ljung (this book, Vol. 2), following the linguist Lamb's stratificational analysis. His lead was followed by Kakumasu (this book, Vol. 2), who used a modified tagmemic model. Meggitt (this book, Vol. 2) suggested *finguistics* and *mimetics* as a sign language parallel to *linguistics* and *phonetics* (see the discussion in West this book, Vol. 2).

[4]Stokoe (1974*a*, p. 118) suggests the use of the term *gSign*, where *g* is "any gestural manifestation," and *Sign* "a sign vehicle in a semiotic system." Following Kendon (1972), Stokoe would restrict the term *gesticulation* to the flow of corporeal movement accompanying speech. See also Sebeok's discussions of these as well as other terms frequently used to describe nonverbal communication systems (1976*a*, p. 158–162, 1977, p. 1065–1067).

[5]Cf. the similar positions regarding the question of the existence of pure icons taken by Eco (1976) and Sebeok (1976*b*).

[6]However, in a recent personal communication (July 2, 77), Max H. Fisch writes: "Peirce and Mallery must have known of each other and may have been personally acquainted. They were fellow members of The Philosophical Society of Washington. Peirce's active participation (presenting formal papers) was in the early years, 1871–1874. (Peirce's father was also a member during his superintendency of the Coast Survey.) So far as my notes show, Mallery was active at least from 1877. He was president around 1887–1888 and sketched the history of the Society at a meeting in January 1888. He seems to have been present at a meeting on October 23, 1880, at which resolutions memorializing Benjamin Peirce were presented, and several members recited their recollections of him."

[7]Cf. Kakumasu's tentative conclusion that Urubú sign language units carry "a certain functional load, but . . . have no consistent semantic association" (this book, Vol. 2, p. 253).

INTRODUCTION

TO THE

STUDY OF SIGN LANGUAGE

AMONG THE

NORTH AMERICAN INDIANS

AS

ILLUSTRATING THE GESTURE SPEECH OF MANKIND

By GARRICK MALLERY

BREVET LIEUT. COL., U. S. ARMY

SMITHSONIAN INSTITUTION,
BUREAU OF ETHNOLOGY,
Washington, D. C., February 12, 1880.

Eleven years ago ethnographic research among North American Indians was commenced by myself and my assistants while making explorations on the Colorado River and its tributaries. From that time to the present such investigations have been in progress.

During this time the Secretary of the Smithsonian Institution placed in my hands a large amount of material collected by its collaborators relating to Indian languages and other matters, to be used, in conjunction with the materials collected under my direction, in the preparation of a series of publications on North American Ethnology. In pursuing this work two volumes have already been published, a third is in press, and a number of others are in course of preparation.

The work originally begun as an incident to a geographical and geological survey has steadily grown in proportions until a large number of assistants and collaborators are engaged in the collection of materials and the preparation of memoirs on a variety of subjects relating to the North American Indians. The subject under investigation is of great magnitude. More than five hundred languages, belonging to about seventy distinct stocks or families, are spoken by these Indians; and in all other branches of this ethnic research a like variety of subject-matter exists. It will thus be seen that the materials for a systematic and comprehensive treatment of this subject can only be obtained by the combined labor of many men. My experience has demonstrated that a deep interest in Anthropology is widely spread among the educated people of the country, as from every hand assistance is tendered, and thus valuable material is steadily accumulating; but experience has also demonstrated that much effort is lost for want of a proper comprehension of the subjects and methods of investigation appertaining to this branch of scientific research. For this reason a series of pamphlet publications, designed to give assistance and direction in these investigations, has been commenced.

The first of the series was prepared by myself and issued under the title of "Introduction to the Study of Indian Languages;" the second is the present, upon Sign-Language; and a third, by Dr. H. C. Yarrow, United States Army, designed to incite inquiry into mortuary observances and beliefs concerning the dead prevailing among the Indian tribes, will shortly be issued. Other publications of a like character will be prepared from time to time. These publications are intended to serve a somewhat temporary purpose until a manual for the use of students of American Anthropology is completed.

J. W. POWELL.

INQUIRIES AND SUGGESTIONS

UPON

SIGN-LANGUAGE AMONG THE NORTH AMERICAN INDIANS.

BY GARRICK MALLERY.

INTRODUCTORY.

The Bureau of Ethnology of the Smithsonian Institution has in preparation a work upon Sign-Language among the North American Indians, and, further, intended to be an exposition of the gesture-speech of mankind thorough enough to be of suggestive use to students of philology and of anthropology in general. The present paper is intended to indicate the scope of that future publication, to excite interest and invite correspondence on the subject, to submit suggestions as to desirable points and modes of observation, and to give notice of some facilities provided for description and illustration.

The material now collected and collated is sufficient to show that the importance of the subject deserves exhaustive research and presentation by scientific methods instead of being confined to the fragmentary, indefinite, and incidental publications thus far made, which have never yet been united for comparison, and are most of them difficult of access. Many of the descriptions given in the lists of earlier date than those contributed during the past year in response to special request are too curt and incomplete to assure the perfect reproduction of the sign intended, while in others the very idea or object of the sign is loosely expressed, so that for thorough and satisfactory exposition they require to be both corrected and supplemented, and therefore the cooperation of competent observers, to whom this pamphlet is addressed, and to whom it will be mailed, is urgently requested.

The publication will mainly consist of a collation, in the form of a vocabulary, of all authentic signs, including signals made at a distance, with their description, as also that of any specially associated facial expression, set forth in language intended to be so clear, illustrations being added when necessary, that they can be reproduced by the reader. The descriptions contributed, as also the explanation or conception occurring to or ascertained by the contributors, will be given in their own words, with their own illustrations when furnished or when they can be designed from written descriptions, and always with individual credit as well as responsibility. The signs arranged in the vocabulary will be compared in their order with those of deaf-mutes, with those of foreign tribes of men, whether ancient or modern, and with the suggested radicals of languages, for assistance in which comparisons travelers and scholars are solicited to contribute in the same manner and with the same credit above mentioned. The deductions and generalizations of the editor of the work will be separate from this vocabulary, though based upon it, and some of those expressed in this preliminary paper may be modified on full information, as there is no conscious desire to maintain any preconceived theories. Intelligent criticisms will be gratefully received, considered, and given honorable place.

PRACTICAL VALUE OF SIGN-LANGUAGE.

The most obvious application of Indian sign-language will for its practical utility depend, to a large extent, upon the correctness of the view submitted by the present writer, in opposition to an opinion generally entertained, that it is not a mere semaphoric repetition of traditional signals, whether or not purely arbitrary in their origin, but is a cultivated art, founded upon principles which can be readily applied by travelers and officials so as to give them much independence of professional interpreters—a class dangerously deceitful and tricky. Possessing this art, as distinguished from a limited list of memorized motions, they would accomplish for themselves the desire of the Prince of Pontus, who begged of Nero an accomplished pantomimist from the Roman theater, to interpret among his many-tongued subjects. This advantage is not merely theoretical, but has been demonstrated to be practical by a professor in a deaf-mute college who, lately visiting several of the wild tribes of the plains, made himself understood among all of them without knowing a word of any of their languages; nor would it only obtain in connection with American tribes, being applicable to intercourse with savages in Africa and Asia, though it is not pretended to fulfill by this agency the schoolmen's dream of an œcumenical mode of communication between all peoples in spite of their dialectic divisions.

Sign-language, being the mother utterance of nature, poetically styled by LAMARTINE the visible attitudes of the soul, is superior to all others in that it permits every one to find in nature an image to express his thoughts on the most needful matters intelligently to any other person, though it must ever henceforth be inferior in the power of formulating thoughts now attained by words, notwithstanding the boast of Roscius that he could convey more varieties of sentiment by gesture alone than Cicero could in oratory.

It is true that gestures excel in graphic and dramatic effect applied to narrative and to rhetorical exhibition; but speech, when highly cultivated, is better adapted to generalization and abstraction; therefore to logic and metaphysics. Some of the enthusiasts in signs have, however, contended that this unfavorable distinction is not from any inherent incapability, but because their employment has not been continued unto perfection, and that if they had been elaborated by the secular labor devoted to spoken language they might in resources and distinctness have exceeded many forms of the latter. GALLAUDET, PEET, and others may be right in asserting that man could by his arms, hands, and fingers, with facial and bodily accentuation, express any idea that could be conveyed by words. The process regarding abstract ideas is only a variant from that of oral speech, in which the words for the most abstract ideas, such as law, virtue, infinitude, and immortality, are shown by MAX MÜLLER to have been derived and deduced, that is, abstracted from sensuous impressions. In the use of signs the countenance and manner as well as the tenor decide whether objects themselves are intended, or the forms, positions, qualities, and motions of other objects which are suggested, and signs for moral and intellectual ideas, founded on analogies, are common all over the world as well as among deaf-mutes. Concepts of the intangible and invisible are only learned through percepts of tangible and visible objects, whether finally expressed to the eye or to the ear, in terms of sight or of sound.

It will be admitted that the elements of the sign-language are truly natural and universal, by recurring to which the less natural signs adopted dialectically or for expedition can, with perhaps some circumlocution, be explained. This power of interpreting itself is a peculiar advantage, for spoken languages, unless explained by gestures or indications, can only be interpreted by means of some other spoken language. There is another characteristic of the gesture-speech that, though it cannot be resorted to in the dark, nor where the attention of the person addressed has not been otherwise attracted, it has the countervailing benefit of use when the voice could not be employed. When highly cultivated its rapidity on familiar subjects exceeds that of speech and approaches to that of thought itself. This statement may be startling to those who only notice that a selected spoken word may convey in an instant a meaning for which the motions of even an expert in signs may require a much longer time, but it must be considered that oral speech is now wholly conventional, and that with the similar development of sign-language conventional expressions with hands and body could be made more quickly than with the vocal organs, because more organs could be worked at once. Without such supposed development the habitual communication between deaf-mutes and among Indians using signs is perhaps as rapid as between the ignorant class of speakers upon the same subjects, and in many instances the signs would win at a trial of speed.

Apart from their practical value for use with living members of the tribes, our native semiotics will surely help the archæologist in his study of native picture-writing, the sole form of aboriginal records, for it was but one more step to fasten upon bark, skins, or rocks the evanescent air-pictures that still in pigments or carvings preserve their skeleton outline, and in their ideography approach the rudiments of a phonetic alphabet. Gesture-language is, in fact, not only a picture-language, but is actual writing, though dissolving and sympathetic, and neither alphabetic nor phonetic.

Though written characters are in our minds associated with speech, they are shown, by successful employment in hieroglyphs and by educated deaf-mutes, to be representative of ideas without the intervention of sounds, and so also are the outlines of signs. This will be more apparent if the motions expressing the most prominent feature, attribute, or function of an object are made, or supposed to be made, so as to leave a luminous track impressible to the eye, separate from the members producing it. The actual result is an immateriate graphic representation of visible objects and qualities which, invested with substance, has become familiar to us as the *rebus*, and also appears in the form of heraldic blazonry styled punning or "canting." The reproduction of gesture-lines in the pictographs made by our Indians seems to have been most frequent in the attempt to convey those subjective ideas which were beyond the range of an artistic skill limited to the direct representation of objects, so that the part of the pictographs, which is still the most difficult of interpretation, is precisely the one which the study of sign-language is likely to eludicate. In this connection it may be mentioned that a most interesting result has been obtained in the tentative comparison so far made between the gesture-signs of our Indians and some of the characters in the Chinese, Assyrian, Mexican, and Runic alphabets or syllabaries, and also with Egyptian hieroglyphs.

While the gesture-utterance presents no other part of grammar to the philologist besides syntax, or the grouping and sequence of its ideographic pictures, the arrangement of signs when in connected succession affords an interesting comparison with the early syntax of vocal language, and the analysis of their original conceptions, studied together with the holophrastic roots in the speech of the gesturers, may aid to ascertain some relation between concrete ideas and words. Meaning does not adhere to the phonetic presentation of thought, while it does to signs. The latter are doubtless more flexible and in that sense more mutable than words, but the ideas attached to them are persistent, and therefore there is not much greater metamorphosis in the signs than in the cognitions. The further a language has been developed from its primordial roots, which have been twisted into forms no longer suggesting any reason for their original selection, and the more the primitive significance of its words has disappeared, the fewer points of contact can it retain with signs. The higher languages are more precise because the consciousness of the derivation of most of their words is lost, so that they have become counters, good for any sense agreed upon; but in our native dialects, which have not advanced in that direction to the degree exhibited by those of civilized man, the connection between the idea and the word is only less obvious than that still unbroken between the idea and the sign, and they remain strongly affected by the concepts of outline, form, place, position, and feature on which gesture is founded, while they are similar in their fertile combination of radicals. For these reasons the forms of sign-language adopted by our Indians will be of special value to the student of American linguistics.

A comparison sometimes drawn between sign-language and that of our Indians, founded on the statement of their common poverty in abstract expressions, is not just to either. Allusion has before been made to the capacities of the gesture-speech in that regard, and a deeper study into Indian tongues has shown that they are by no means so confined to the concrete as was once believed.

Indian language consists of a series of words that are but slightly differentiated parts of speech following each other in the order suggested in the mind of the speaker without absolute laws of arrangement, as its sentences are not completely integrated. The sentence necessitates parts of speech, and parts of speech are possible only when a language has reached that stage where sentences are logically constructed. The words of an Indian tongue being synthetic or undifferentiated parts of speech, are in this respect strictly analogous to the gesture elements which enter into a sign-language. The study of the latter is therefore valuable for comparison with the words of the speech. The one language throws much light upon the other, and neither can be studied to the best advantage without a knowledge of the other.

ORIGIN AND EXTENT OF GESTURE-SPEECH.

It is an accepted maxim that nothing is thoroughly understood unless its beginning is known. While this can never be absolutely accomplished for sign-language, it may be traced to, and claims general interest from, its illustration of the ancient intercommunication of mankind by gesture. Many arguments have been adduced and more may be presented to prove that the latter preceded articulate speech. The corporeal movements of the lower animals to express, at least, emotion have been correlated with those of man, and classified by DARWIN as explicable on the principles of serviceable associated habits, of antithesis, and of the constitution of the nervous system. A child employs intelligent gestures long in advance of speech, although very early and persistent attempts are made to give it instruction in the latter but none in the former; it learns language only through the medium of signs; and long after familiarity with speech, consults the gestures and facial expressions of its parents and nurses as if to translate or explain their words; which facts are important in reference to the biologic law that the order of development of the individual is the same as that of the species. Persons of limited vocabulary, whether foreigners to the tongue employed, or native, but not accomplished in its use, even in the midst of a civilization where gestures are deprecated, when at fault for words resort instinctively to physical motions that are not wild nor meaningless, but picturesque and significant, though perhaps made by the gesturer for the first time; and the same is true of the most fluent talkers on occasions when the exact vocal formula desired does not at once suggest itself, or is not satisfactory without assistance from the physical machinery not embraced in the oral apparatus. Further evidence of the unconscious survival of gesture-language is afforded by the ready and involuntary response made in signs to signs when a man with the speech and habits of civilization is brought into close contact with Indians or deaf-mutes. Without having ever before seen or made one of their signs he will soon not only catch the meaning of theirs, but produce his own, which they will likewise comprehend, the power seemingly remaining latent in him until called forth by necessity. The signs used by uninstructed congenital deaf-mutes and the facial expressions and gestures of the congenitally blind also present considerations under the heads of "heredity" and "atavism," of some weight when the subjects are descended from and dwell among people who had disused gestures for generations, but of less consequence in cases such as that mentioned by Cardinal WISEMAN of an Italian blind man who, curiously enough, used the precise signs made by his neighbors. It is further asserted that semi-idiotic children who cannot be taught more than the merest rudiments of speech can receive a considerable amount of knowledge through signs and express themselves by them, and that sufferers from aphasia continue to use appropriate gestures after their words are uncontrollable. In cases where men have been long in solitary confinement, been abandoned, or otherwise have become isolated from their fellows, they have lost speech entirely, in which they required to be reinstructed through gestures in the same manner that missionaries, explorers, and shipwrecked mariners became acquainted with tongues before unknown to civilization. These facts are to be considered in connection with the general law of evolution, that in cases of degeneration the last and highest acquirements are lost first.

The fact that the deaf-mute thinks without phonetic expression is a stumbling-block to MAX MÜLLER's ingenious theory of primitive speech, to the effect that man had a creative faculty giving to each conception, as it thrilled through his brain for the first time, a special phonetic expression, which faculty became extinct when its necessity ceased.

In conjecturing the first attempts of man or his hypothetical ancestor at the expression either of percepts or concepts, it is difficult to connect vocal sounds with any large number of objects, but readily conceivable that there should have been resort, next to actual touch (of which all the senses may be modifications) to suggest the characteristics of their forms and movements to the eye—fully exercised before the tongue—so soon as the arms and fingers became free for the requisite simulation or portrayal. There is no distinction between pantomime and sign-language except that the former is the parent of the latter, which is more abbreviated and less obvious. Pantomime acts movements, reproduces forms and positions, presents pictures, and manifests emotions with greater realization than any other mode of utterance. It may readily be supposed that a trogdolyte man would desire to communicate the finding of a cave in the vicinity of a pure pool, circled with soft grass, and shaded by trees bearing edible fruit. No natural sound is connected with any of those objects, but the position and size of the cave, its distance and direction, the water, its quality, and amount, the verdant circling carpet, and the kind and height of the trees could have been made known by pantomime in the days of the mammoth, if articulate speech had not then been established, precisely as Indians or deaf-mutes would now communicate the news by the same agency or by signs possessing a natural analogy.

Independent of most of the above considerations, but from their own failures and discordancies, linguistic scholars have recently decided that both the "bow-wow" and the "ding-dong" theories are unsatisfactory; that the search for imitative, onomatopoetic, and directly expressive sounds to explain the origin of human speech has been too exclusive, and that many primordial roots of language have been founded in the involuntary sounds accompanying certain actions. As, however, the action was the essential, and the consequent or concomitant sound the accident, it would be expected that a representation or feigned reproduction of the action would have been used to express the idea before the sound associated with that action could have been separated from it. The visual onomatopœia of gestures, which even yet have been subjected to but slight artificial corruption, would therefore serve as a key to the audible. It is also contended that in the pristine days, when the sounds of the only words yet formed had close connection with objects and the ideas directly derived from them, signs were as much more copious for communication than speech as the sight embraces more and more distinct characteristics of objects than does the sense of hearing.

The preponderance of authority is that man, when in the possession of all his faculties, did not choose between voice and gesture, both being originally instinctive, as they both are now, and never, with those faculties, was in a state where the one was used to the absolute exclusion of the other. With the voice he at first imitated the few sounds of nature, while with gesture he exhibited actions, motions, positions, forms, dimensions, directions, and distances, and their derivatives. It would appear from this unequal division of capacity that oral speech remained rudimentary long after gesture had become an art. With the concession of all purely imitative sounds and of the spontaneous action of the vocal organs under excitement, it is still true that the connection between ideas and words generally depended upon a compact between the speaker and hearer which presupposes the existence of a prior mode of communication.

For the present purpose there is, however, no need to determine upon the priority between communication of ideas by bodily motion and by vocal articulation. It is enough to admit that the connection between them was so early and intimate that the gestures, in the wide sense indicated of presenting ideas under physical forms, had a direct formative effect upon many words; that they exhibit the earliest condition of the human mind; are traced from the farthest antiquity among all peoples possessing records; are universally prevalent in the savage stage of social evolution; survive agreeably in the scenic pantomime, and still adhere to the ordinary speech of civilized man by motions of the face, hands, head, and body, often involuntary, often purposely in illustration or emphasis.

MODERN USE OF GESTURES AND SIGNS.

The power of the visible gesture relative to and its influence upon the words of modern oral speech are perhaps, with the qualification hereafter indicated, in inverse proportion to the general culture, but do not bear that or any constant proportion to the development of the several languages with which gesture is still more or less associated They are affected more by the sociological conditions of the speakers than by the degree of excellence of their tongue. The statement is frequently made that gesture is yet to some highly-advanced languages a necessary modifying factor, and that only when a language has become so artificial as to be completely expressible in written signs—indeed, has been remodeled through their long familiar use—can the bodily signs be wholly dispensed with. The story has been told by travelers in many parts of the world that various languages cannot be clearly understood in the dark by their possessors, using their mother tongue between themselves. The evidence for this anywhere is suspicious, and when it is, as it often has been, asserted about some of the tribes of North American Indians, it is absolutely false, and must be attributed to the error of travelers who, ignorant of the dialect, never see the natives except when trying to make themselves intelligible to their visitors by a practice which they have found by experience to have been successful with strangers to their tongue, or perhaps when they are guarding against being overheard by others. In fact, individuals of those American tribes specially instanced in these reports as unable to converse without gesture, often, in their domestic *abandon*, wrap themselves up in robes or blankets with only breathing holes before the nose, so that no part of the body is seen, and chatter away for hours, telling long stories. If in daylight they thus voluntarily deprive themselves of the possibility of making signs, it is clear that their preference for talks around the fire at night is explicable by very natural reasons without the one attributed. The inference, once carelessly made from the free use of gesture by some of the Numa stock, that their tongue was too meager for use without signs, is refuted by the now ascertained fact that their vocabulary is remarkably copious and their parts of speech better differentiated than those of many people on whom no such stigma has been affixed. All theories, indeed, based upon the supposed poverty of American languages must be abandoned.

The true distinction is that where people speaking precisely the same dialect are not numerous, and are thrown into constant contact on equal terms with others of differing dialects and languages, gesture is necessarily resorted to for converse with the latter, and remains as a habit or accomplishment among themselves, while large bodies enjoying common speech, and either isolated from foreigners, or, when in contact with them, so dominant as to compel the learning and adoption of their own tongue, become impassive in its delivery. The undemonstrative English, long insular, and now rulers when spread over continents, may be compared with the profusely gesticulating Italians dwelling in a maze of dialects and subject for centuries either to foreign rule or to the influx of strangers on whom they depended. King Ferdinand returning to Naples after the revolt of 1821, and finding that the boisterous multitude would not allow his voice to be heard, resorted successfully to a royal address in signs, giving reproaches, threats, admonitions, pardon, and dismissal, to the entire satisfaction of the assembled lazzaroni, which rivalry of Punch would, in London, have occasioned measureless ridicule and disgust. The difference in what is vaguely styled temperament does not wholly explain this contrast, for the performance was creditable both to the readiness of the King in an emergency and to the aptness of his people, the main distinction being that in Italy there was a recognized and cultivated language of signs long disused in Great Britain. As the number of dialects in any district decreases so will the gestures, though doubtless there is also influence from the fact not merely that a language has been reduced to and modified by writing, but that people who are accustomed generally to read and write, as are the English and Germans, will after a time think and talk as they write, and without the accompaniments still persistent among Hindus, Arabs, and the less literate Europeans.

Many instances are shown of the discontinuance of gesture-speech with no development in the native language of the gesturers, but from the invention for intercommunication of one used in common. The Kalapuyas of Southern Oregon until recently used a sign-language, but have gradually adopted for foreign intercourse the composite tongue, commonly called the Tsinuk or Chinook jargon, which probably arose for trade purposes on the Columbia River before the advent of Europeans, founded on the Tsinuk, Tsihali, Nutka, &c., but now enriched by English and French terms, and have nearly forgotten their old signs. The prevalence of this mongrel speech, originating in the same causes that produced the pigeon-English or *lingua-franca* of the Orient, explains the marked scantness of sign-language among the tribes of the Northwest coast. No explanation is needed for the disuse of that mode of communication when the one of surrounding civilization is recognized as necessary or important to be acquired, and gradually becomes known as the best common medium, even before it is actually spoken by many individuals of the several tribes.

IS INDIAN SIGN-LANGUAGE UNIVERSAL AND IDENTICAL?

The assertion has been made by many writers, and is currently repeated by Indian traders and some Army officers, that all the tribes of North America have had and still use a *common* and *identical* sign-language of ancient origin, in which they can communicate freely without oral assistance. The fact that this remarkable statement is at variance with some of the principles of the formation and use of signs set forth by Dr. TYLOR, whose inimitable chapters on gesture-speech in his "Researches into the Early History of Mankind" have in a great degree prompted the present inquiries, does not appear to have attracted the attention of that eminent authority. He receives the report without question, and formulates it, that "the same signs serve as a medium of converse from Hudson Bay to the Gulf of Mexico." Its truth can only be established by careful comparison of lists or vocabularies of signs taken under test conditions at widely different times and places. For this purpose lists have been collated by the writer, taken in different parts of the country at several dates, from the last century to the last month, comprising together more than eight hundred signs, many of them, however, being mere variants or synonyms for the same object or quality, and some being of small value from uncertainty in description or authority, or both.

The result of the collation and analysis thus far made is that the alleged existence of *one* universal and absolute sign-language is, in its terms of general assertion, one of the many popular errors prevailing about our aborigines. In numerous instances there is an entire discrepancy between the signs made by different bodies of Indians to express the same idea; and if any of these are regarded as determinate, or even widely conventional, and used without further devices, they will fail in conveying the desired impression to any one unskilled in gesture as an art, who had not formed the same precise conception or been instructed in the arbitrary motion. Probably none of the gestures that are found in current use are, in their origin, conventional, but are only portions, more or less elaborate, of obvious natural pantomime, and those proving efficient to convey most successfully at any time the several ideas became the most widely adopted, liable, however, to be superseded by yet more appropriate conceptions and delineations. The skill of any tribe and the copiousness of its signs are proportioned to the accidental ability of the few individuals in it who act as custodians and teachers, so that the several tribes at different times vary in their degree of proficiency, and therefore both the precise mode of semiotic expression and the amount of its general use are always fluctuating. All the signs, even those classed as innate, were at some time invented by some one person, though by others simultaneously and independently, and many of them became forgotten and were reinvented. Their prevalence and permanence were determined by the experience of their utility, and it would be highly interesting to ascertain how long a time was required for a distinctly new conception or execution to gain currency, become "the fashion," so to speak, over a large part of the continent, and to be supplanted by a new "mode."

The process is precisely the same as among the deaf-mutes. One of those, living among his speaking relatives, may invent signs which the latter are taught to understand, though strangers sometimes will not, because they may be by no means the fittest expressions. Should a dozen or more deaf-mutes, possessed only of such crude signs, come together, they will be able at first to communicate only on a few common subjects, but the number of those and the general scope of expression will be continually enlarged. They will also resort to the invention of new signs for new ideas as they arise, which will be made intelligible, if necessary, through the illustration and definition given by signs formally adopted, so that the fittest signs will be evolved, after mutual trial, and will survive. A multiplication of the numbers confined together, either of deaf-mutes or of Indians whose speech is diverse, will not decrease the resulting uniformity, though it will increase both the copiousness and the precision of the vocabulary. The only one of the correspondents of the present writer who remains demonstratively unconvinced of the diversities in Indian sign-language, perhaps became prejudiced when in charge of a reservation where Arapahos, Cheyennes, and Sioux had for a considerable time been kept secluded, so far as could be done by governmental power, from the outer world, and where naturally their signs were modified so as to become common property.

Sometimes signs, doubtless once air-pictures of the most striking outline of an object, or of the most characteristic features of an action, have in time become abbreviated and, to some extent, conventionalized among members of the same tribe and its immediate neighbors, and have not become common to them with other tribes simply because the form of abbreviation has been peculiar. In other cases, with the same conception and attempted characterization, another yet equally appropriate delineation has been selected, and when both of the differing delineations have been abbreviated the diversity is vastly increased. The original conception, being independent, has necessarily also varied, because all objects have several characteristics, and what struck one set of people as the most distinctive of these would not always so impress another. From these reasons we cannot expect, without trouble, to understand the etymology of all the signs, being less rich in ancillary material than were even the old philologists, who guessed at Latin and Greek derivations before they were assisted by Sanscrit and other Aryan roots.

It is not difficult to conjecture some of the causes of the report under consideration. Explorers and officials are naturally brought into contact more closely with those persons of the tribes visited who are experts in the sign-language than with their other members, and those experts are selected, on account of their skill as interpreters, as guides to accompany the visitors. The latter also seek occasion to be present when the signs are used, whether with or without words, in intertribal councils, and then the same class of experts are the orators, for this long exercise in gesture-speech has made the Indian politicians, with no special effort, masters of the art only acquired by our public speakers after laborious apprenticeship before their mirrors. The whole theory and practice of sign-language being that all who understand its principles can make themselves mutually intelligible, the fact of the ready comprehension and response among all the skilled gesturers gives the impression of a common code. Furthermore, if the explorer learns to use any of the signs used by any of the tribes, he will probably be understood in any other by the same class of persons who will surround him in the latter, thereby confirming him in the "universal" theory. Those of the tribe who are less skilled, but who are not noticed, might be unable to catch the meaning of signs which have not been actually taught to them, just as ignorant persons among us cannot derive any sense from newly-coined words or those strange to their habitual vocabulary, which linguistic scholars would instantly understand, though never before heard, and might afterward adopt.

In order to sustain the position taken as to the existence of a general system instead of a uniform code, admitting the generic unity while denying the specific identity, and to show that this is not a distinction without a difference, a number of specimens are extracted from the present collection of signs, which are also in some cases compared with those of deaf-mutes and with gestures made by other peoples.

AUTHORITIES FOR THE SIGNS CITED.

The signs, descriptions of which are submitted in the present paper, are taken from some one or more of the following authorities, viz:

1. A list prepared by WILLIAM DUNBAR, dated Natchez, June 30, 1800, collected from tribes then west of the Mississippi, but probably not from those very far west of that river, published in the Transactions of the American Philosophical Society, vol. vi, as read January 16, 1801, and communicated by Thomas Jefferson, president of the society.

2. The one published in 1823 in "An Account of an Expedition from Pittsburgh to the Rocky Mountains, performed in the years 1819–1820. By order of the Hon. J. C. Calhoun, Secretary of War, under the command of Maj. S. H. LONG, of the United States Topographical Engineers." (Commonly called James' Long's Expedition.) This appears to have been collected chiefly by Mr. T. Say, from the Pani, and the Kansas, Otoes, Missouris, Iowas, Omahas, and other southern branches of the great Dakota family.

3. The one collected by Prince MAXIMILIAN von WIED-NEUWIED in 1832–34, from the Cheyenne, Shoshoni, Arikara, Satsika, and the Absaroki, the Mandans, Hidatsa, and other Northern Dakotas. This list is not published in the English edition, but appears in the German, Coblenz, 1839, and in the French, Paris, 1840. Bibliographic reference is often made to this distinguished explorer as "Prince Maximilian," as if there were not many possessors of that christian name among princely families. For brevity the reference in this paper will be "*Wied.*"

4. The small collection of J. G. KOHL, made about the middle of the present century, among the Ojibwas and their neighbors around Lake Superior. Published in his "Kitchigami. Wanderings around Lake Superior," London, 1860.

5. That of the distinguished explorer, Capt R. F. BURTON, collected in 1860–61, from the tribes met or learned of on the overland stage route,

including Southern Dakotas, Utes, Shoshoni, Arapahos, Crows, Pani, and Apaches. This is contained in "The City of the Saints," New York, 1862.

6. A manuscript list in the possession of the Bureau of Ethnology, contributed by Brevet Col. JAMES S. BRISBIN, Major Second Cavalry, United States Army, probably prepared in 1878–79, and chiefly taken from the Crows, Shoshoni, and Sioux.

7. A list prepared in July, 1879, by Mr. FRANK H. CUSHING, of the Smithsonian Institution, from continued interviews with Titchkemátski, an intelligent Cheyenne, now employed at that Institution, whose gestures were analyzed, their description as made dictated to a phonographer, and the more generic signs also photographed as made before the camera. The name of the Indian in reference to this list is used instead of that of the collector, as Mr. Cushing has made other contributions, to be separately noted with his name for distinctiveness.

8. A valuable and illustrated contribution from Dr. WASHINGTON MATTHEWS, Assistant Surgeon United States Army, author of "Ethnography and Philology of the Hidatsa Indians," &c., lately prepared from his notes and recollections of signs observed during his long service among the Indians of the Upper Missouri and the plains.

9. A report of Dr. W. J. HOFFMAN, from observations among the Teton Dakotas while Acting Assistant Surgeon, United States Army, and stationed at Grand River Agency, Dakota, during 1872–73.

10. A special contribution from Lieut. H. R. LEMLY, Third United States Artillery, compiled from notes and observations taken by him in 1877 among the Northern Arapahos.

11. Some preliminary notes lately received from Rev. TAYLOR F. EALY, missionary among the Zuñi, upon the signs of that body of Indians.

12. Similar notes from Rev. A. J. HOLT, Denison, Tex., respecting the Comanche signs.

13. Similar notes from Very Rev. EDWARD JACKER, Pointe St. Ignace, Mich., respecting the Ojibwa.

14. A special list from Rev. J OWEN DORSEY, missionary at Omaha Agency, Nebraska, from observations lately made among the Ponkas and Omahas.

15. A letter from J. W. Powell, esq., Indian superintendent, British Columbia, relating to his observations among the Kutine and others.

16. A special list from Dr. Charles E. McChesney, Acting Assistant Surgeon United States Army, of signs collected among the Dakotas (Sioux) near Fort Bennett, Dakota, during the present winter.

17. A communication from Rev. James A. Gilfillan, White Earth, Minn., relating to signs observed among the Ojibwas during his long period of missionary duty, still continuing.

18. A communication from Brevet Col. Richard I. Dodge, Lieutenant-Colonel Twenty-third Infantry, United States Army, author of "The Plains of the Great West and their Inhabitants," &c., relating to his large experience with the Indians of the prairies.

19. A list contributed by Rev. G. L. Deffenbaugh, of Lapwai, Idaho, giving signs obtained at Kamiah, and used by the Sahaptin or Nez Percés.

20. Information obtained by Dr. W. J. Hoffmann, in assisting the present writer, from Nátshes, a Pah-Ute chief, who was one of a delegation of that tribe to Washington, in January, 1880.

21. Information from Major J. M. Haworth, special agent of the Indian Bureau, relating to the Comanches.

The adjunction to the descriptions of the name of the particular author, contributor, or person from whom they are severally taken (a plan which will be pursued in the final publication) not only furnishes evidence of authenticity, but indicates the locality and time of observation.

INSTANCES OF DIVERSE CONCEPTIONS AND EXECUTIONS.

Some examples have been selected of diverse conceptions and executions for the same object or thought.

Chief. Seven distinct signs.

1. Forefinger of right hand extended, passed perpendicularly downward, then turned upward in a right line as high as the head. (*Long.*) "Rising above others."

2. With forefinger of right hand, of which the other fingers are closed, pointing up, back to forehead, describe the flight of an arrow shot up and turning down again, allowing the hand to drop, the finger pointing down until about the middle of the body. (*Brisbin.*) Same idea of superior

height expressed conversely. Almost the same sign, the hand, however, being moved downward rapidly and the gesture preceded by touching the lower lip with the index, the French deaf-mutes use for "command," "order."

3. The extended forefinger of the right hand, of which the other fingers are closed, is raised to the right side of the head, and above it as far as the arm can be extended, and then the hand is brought down in front of the body, with wrist bent, the back of hand in front, extended forefinger pointing downward and the others closed. "Raised above others." (*McChesney.*)

4. Begin with sign for "man;" then the forefinger of right hand points forward and downward, followed by a curved motion forward, outward, and downward. (*Titchkemátski.*) "He who sits still and commands others."

5. Raise the index of right hand, which is held upright; turn the index in a circle and lower it a little to the earth. (*Wied.*) "He who is the center of surrounding inferiors." The air-picture reminds of the royal scepter with its sphere.

6. Bring the closed right hand, forefinger pointing up, on a level with the face; then bring the palm of the left hand with force against the right forefinger; next send up the right hand above the head, leaving the left as it is. (*Dorsey.*)

7. The Pah-Utes distinguish the head chief of the tribe from the chief of a band. For the former they grasp the forelock with the right hand, palm backward, pass the hand upward about six inches, and hold the hair in that position a moment; and for the latter they make the same motion, but instead of holding the hair above the head they lay it down over the right temple, holding it there a moment. (*Nátshes.*)

Day. Seven signs.

1. Pass the index-finger pointing along the vault of heaven from east to west. (*Kohl.*) Our deaf-mutes use the same sign.

2. Same motion with whole right hand. (*Brisbin.*)

3. Same motion with forefinger of right hand crooked, followed by both hands slightly spread out and elevated to a point in front of and considerably above the head, then brought down in a semicircle to a level below the shoulder, ending with outspread palms upward. (*Titchkemátski.*) This, probably, is the opening out of the day from above, after the risen sun.

4. Simply make a circle with the forefingers of both hands. (*Burton.*) The round disk.

5. Place both hands at some distance in front of the breast, apart, and backs downward (*Wied.*)

6. Bring both hands simultaneously from a position in front of the body, fingers extended and joined, palms down one above the other, forearms horizontal, in a circularly separating manner to their respective sides, palms up and forearms horizontal; *i. e.*, "Everything is open." (*Lemly.*)

7. Both hands raised in front of and a little higher than the head, fingers of both hands horizontal, extended, and meeting at the tips, palms of hands downward, and arms bowed; open up the hands with fingers perpendicular, and at once carry the arms out to their full extent to the sides of the body, bringing the palms up. "The opening of the day from above. The dispersion of darkness." (*McChesney.*)

The French deaf-mutes fold the hands upon each other and the breast, then raise them, palms inward, to beyond each side of the head.

To-day, this day, has four widely discrepant signs in, at least, appearance. In one, the nose is touched with the index tip, followed by a motion of the fist toward the ground (*Burton*), perhaps including the idea of "now," "here." In another, both hands are extended, palms outward, and swept slowly forward and to each side. (*Titchkemátski.*) This may combine the idea of *now* with *openness*, the first part of it resembling the general deaf-mute sign for "here" or "now."

A third observer gives as used for the idea of the present day the sign also used for "hour," viz: join the tips of the thumb and forefinger of the same hand, the interior outline approximating a circle, and let the hand pause at the proper altitude east or west of the assumed meridian. (*Lemly.*)

A fourth reports a compound sign: First make the following sign, which is that for "now." Forefinger of right hand (of which the other fingers are closed) extended, raise the arm perpendicularly a little above the right side of the head, so that the extended finger will point to the center of the heavens and then brought down on a level with the right breast, forefinger still pointing up, and immediately carry it to the position required in making the sign for day as above (*McChesney*), which is used to complete the sign for *to-day*. (*McChesney*.)

Death, dead. Seven signs.

1. Right hand, fingers front at height of stomach, then, with a sort of flop, throw the hand over with the palm up, finger pointing a little to the right and front, hand held horizontal. (*Brisbin*.) "Upset, keeled over."

2. Left hand flattened and held, back upward, thumb inward, in front of and a few inches from the breast; right hand slightly clasped, forefinger more extended than the others, and passed suddenly under the left hand, the latter being at the same time gently moved toward the breast. (*Titchkemátski*.) "Gone under."

3. Hold the left hand flat against the face, back outward; then pass the right hand, held in the same manner, under the left, striking and touching it lightly. (*Wied*.) The same idea of "under" or "burial," quite differently executed. Dr. McChesney, however, conjectures this sign to be that of wonder or surprise at hearing of a death, but not a distinct sign for the latter.

4. Throw the forefinger from the perpendicular into a horizontal position toward the earth with the back downward. (*Long*.)

5. Place the left forefinger and thumb against the heart, act as if taking a hair from the thumb and forefinger of the left hand with the forefinger and thumb of the right and slowly cast it from you, only letting the left hand remain at the heart, and let the index-finger of the right hand point outward toward the distant horizon. (*Holt*.)

6. Palm of hand upward, then a wave-like motion toward the ground. (*Ealy*.)

7. Place the palm of the hand at a short distance from the side of the head, then withdraw it gently in an oblique downward direction, inclining the head and upper part of the body in the same direction. (*Jacker*.)

The last authority notes that there is an apparent connection between this conception and execution and the etymology of the corresponding terms in Ojibwa: "he dies," is *nibo;* "he sleeps," is *niba*. The common idea expressed by the gesture is a sinking to rest. The original significance of the root *nib* seems to be "leaning;" *anibeia*, "it is leaning;" *anibekweni*, "he inclines the head sidewards." The word *niba* or *nibe* (only in compounds)

conveys the idea of "night," perhaps as the falling over, the going to rest, or the death, of the day. The term for "leaf" (of a tree or plant), which is *anibish*, may spring from the same root, leaves being the leaning or down-hanging parts of the plant. With this may be compared the Chahta term for "leaves," literally translated "tree hair".

The French deaf-mute conception is that of gently falling or sinking, the right index falling from the height of the right shoulder upon the left forefinger toward which the head is inclined.

Kill. In one sign the hands are held with the edges upward, and the right strikes the left transversely, as in the act of chopping. This seems to convey particularly the notion of a stroke with a tomahawk or war-club. (*Long.*) It is more definitely expressed as follows: The left hand, thumb up, back forwards, not very rigidly extended, is held before the chest and struck in the palm with the outer edge of the right hand. (*Matthews.*) Another sign: Smite the sinister palm earthward with the dexter fist sharply, in suggestion of going down. (*Burton.*) Another: Strike out with the dexter fist toward the ground, meaning to shut down. (*Burton; McChesney.*) This same sign is made by the Utes, with the statement that it means "to kill" or "stab" with a knife, having reference to the time when that was the most common weapon. A fourth: Pass the right under the left forefinger (*Burton*), "make go under." The threat, "I will kill you," appears in one case as directing the right hand toward the offender and springing the finger from the thumb as in the act of sprinkling water (*Long*), the idea being perhaps causing blood to flow, or perhaps sputtering away the life, though this part of the sign is nearly the same as that sometimes used for the discharge of a gun or arrow.

Fear, coward.

1. Both hands, with fingers turned inward opposite the lower ribs, then brought upward with a tremulous motion, as if to represent the common idea of the heart rising up to the throat. (*Dunbar.*)

2. Head stooped down, and arm thrown up quickly as if to protect it. (*Long.*)

3. Fingers and thumb of right hand, which droops downward, closed to a point to represent a heart, violently and repeatedly beaten against the

left breast just over the heart to imitate palpitation. (*Titchkemátski.*) The Sioux use the same sign without closing the fingers to represent a heart. (*McChesney.*)

The French deaf-mutes, besides beating the heart, add a nervous backward shrinking with both hands. Our deaf-mutes omit the beating of the heart, except for excessive terror.

4. Point forward several times with the index, followed by the remaining fingers, each time drawing the index back (*Wied*), as if impossible to keep the man to the front.

5. May be signified by making the sign for a squaw, if the one in fear be a man or boy. (*Lemly.*)

6. Cross the arms over the breast, fists closed, bow the head over the crossed arms, but turn it a little to the left. (*Dorsey.*)

Woman has four signs; one expressing the mammæ, one indicating shortness as compared with man, and the two most common severally indicating the longer hair or more flowing dress. The hair is sometimes indicated by a motion with the right hand as though drawing a comb through the entire length of the hair on that side of the head (*McChesney*); and sometimes by turning the right hand about the ear, as if putting the hair behind it. (*Dodge.*) The deaf-mutes generally mark the line of the bonnet-string down the cheek.

Quantity, many, much. Six wholly distinct executions and several conceptions.

1. The flat of the right hand patting the back of the left several times, proportioned in number to the quantity. (*Dunbar.*) Simple repetition.

2. Clutching at the air several times with both hands. (*Kohl.*) Same idea of repetition, more objective. This sign may easily be confounded with the mode of counting or enumeration by presenting the ten digits.

3. Hands and arms passed curvilinearly outward and downward as if forming a large globe, then hands closed and elevated as if something were grasped in each, and held up as high as the face. (*Long.*)

4. Hands held scoop-fashion, palms toward each other, about two feet apart, at the height of the lower ribs, finger-ends downward; then with a diving motion, as if scooping up small articles from a sack or barrel,

bring the hands nearly together, fingers closed, as if holding a number of the small objects in each hand, and up again to the height of the breast. (*Brisbin.*) The Sioux make substantially the same sign, with the difference that they begin about a foot and a half from the ground and bring the hands up to the height of the breast. (*McChesney.*)

5. Both hands closed, brought up in a curved motion toward each other to the level of the neck. (*Titchkemátski.*) Idea of fullness.

6. Move the two open hands toward each other and slightly upward (*Wied*); the action of forming or delineating a *heap*.

I, myself, first personal pronoun.

Represented in some tribes by motions of the right hand upon the breast, the hand sometimes clinched and struck repeatedly on the breast—or the fingers or the index alone placed upon it. Others touch the nose-tip with the index, or lay it upon the ridge of the nose, the end resting between the eyes.

Some deaf-mutes push the forefinger against the pit of the stomach, others against the breast, and others point it to the neck for this personality.

Yes, affirmative, "it is so."

One of the signs is somewhat like "truth," but the forefinger proceeds straight forward from the breast instead of the mouth, and when at the end of its course it seems gently to strike something, as if the subject were at an end (*Long*); no further discussion, "'nuff said," as is the vulgar phrase of agreement. Another: Quick motion of the right hand forward from the mouth, first position about six inches from the mouth and final as far again away. In the first position the index is extended, the others closed, in the final the index is loosely closed, thrown in that position as the hand is moved forward, as though hooking something with it. Palm of hand out. (*Deffenbaugh.*)

Others wave both hands straight forward from the face (*Burton*), which may be compared with the forward nod common over most of the world for assent, but that gesture is not universal, as the New Zealanders elevate the head and chin, and the Turks shake it like our negative.

With others, again, the right hand is elevated to the level and in front of the shoulder, the first two fingers somewhat extended, thumb resting against the middle finger, and then a sudden motion in a curve forward and downward. (*Titchkemátski.*) As this corresponds nearly with the sign made for " sit " by the same tribes, its conception may be that of resting upon or settling a question.

Still another variant is where the right hand, with the forefinger (only) extended, and pointing forward, is held before and near the chest. It is then moved forward one or two feet, usually with a slight curve downward. (*Matthews.*)

Good. Six diverse signs.

1. The hand held horizontally, back upward, describes with the arm a horizontal curve outward. (*Long.*)

2. Simple horizontal movement of the right hand from the breast. (*Wied.*) These signs may convey the suggestion of level—no difficulty—and are nearly identical with one of those for "content," "glad." The first of them is like our motion of benediction, but may more suggestively be compared with several of the above signs for "yes," and in opposition to several of those below for "bad" and "no," showing the idea of acceptance or selection of objects presented, instead of their rejection.

3. With the right hand, palm down, fingers to the left, thumb touching the breast, move the hand straight to the front and slightly upward. (*Brisbin.*) The Sioux make the same sign without the final upward motion. (*McChesney.*)

4. Wave the right hand from the mouth, extending the thumb from the index and closing the other three fingers. (*Burton.*)

5. The right hand, fingers pointing to the left, on a level with mouth, thumb inward, suddenly moved with curve outward, so as to present the palm to the person addressed. (*Titchkemátski.*)

These last signs appear to be connected with a pleasant taste in the mouth, as is the sign of the French and our deaf-mutes, waving thence the hand, back upward, with fingers straight and joined, in a forward and downward curve. The same gesture with hand sidewise is theirs and ours for general assent; " very well !"

6. Move the right hand, palm down, over the blanket, right and left several times. (*Dorsey.*)

Bad. The signs most common consist mainly in smartly throwing out the dexter fingers as if sprinkling water, or snapping all the fingers from the thumb. This may be compared with the deaf-mute sign of flipping an imaginary object between the thumb-nail and the forefinger, denoting something small or contemptible The motion of snapping a finger either on or from the thumb in disdain is not only of large modern prevalence in civilization, but is at least as ancient as the contemporary statue of Sardanapalus at Anchiale. Another sign is, hands open, palms turned in, move one hand toward and the other from the body, then *vice versâ*. Another less forcible but equally suggestive gesture for *bad* is closing the hand and then opening while lowering it, as if dropping out the contents (*Wied; McChesney*); "not worth keeping." It becomes again more forcible in another variant, viz: the hand closed, back toward and near the breast, then as the forearm is suddenly extended the hand is opened and the fingers separated from each other. (*Matthews.*) This is the casting away of a supposed object, and the same authority connects it with *contempt* by reporting that the sign for the latter is the same, only still more forcibly made. Another sign for *contempt*, and which is the highest degree of insult, is as follows: The right hand is shut or clinched and held drawn in toward the chest and on a level with it, with the back of the hand down, and the shut fingers and thumb up, and the expression of contempt is given by extending out the hand and arm directly in front of the body, at the same time opening the thumb and fingers wide and apart, so that at the termination of the motion the arm is nearly extended, and the thumb and fingers all radiating out as it were from the center of the hand, and the palm of the hand still pointing upward. (*Gilfillan.*) The Neapolitans, to express contempt, blow towards the person or thing referred to. The deaf-mutes preserve the connection of "bad" and "taste" by brushing from the side of the mouth.

Understand, know, is very variously expressed by manipulations in which the nose, ear, chin, mouth, and breast are selected as objective points, all the motions being appropriate. *Think* or *guess* is also diversely indicated. Sometimes the forefinger is simply drawn sharply across the breast from left to right. (*Burton.*) Some hit the chest with closed fist, thumb over the fist. Again, the right fist is held with the thumb between the eyes and propelled front and downward. We, for show of thought, rest the forefinger on the forehead. There is also a less intelligible sign, in which the right hand, fingers and thumb loosely closed, index crooked and slightly extended, is dipped over toward and suddenly forward from the left shoulder. (*Titchkemátski.*) All the gestures of deaf-mutes relating to intelligence are connected with the forehead.

Animals are expressed pantomimically by some characteristic of their motion or form, and the Indian mimographers generally seem to have hit upon similar signs for the several animals; but to this rule there are marked exceptions, especially in the signs for the *deer* and the *dog*. For the *deer* six signs are noted:

1. Right hand extended upward by the right ear, with a quick puff from the mouth (*Dunbar*), perhaps in allusion to the fleet escape on hearing noise.

2. Make several passes with the hand before the face. (*Wied.*)

3. With the right hand in front of body on a level with the shoulder, and about eighteen inches from it, palm down, make the quick up-and-down motion with all the fingers held loosely together, as of the motion of the deer's tail when running. The wrist is fixed in making this sign. It is very expressive to any one who has ever seen the surprised deer in motion. (*McChesney.*)

4. Forefinger of right hand extended vertically, back toward breast, then turned from side to side, to imitate the motion of the animal when walking at leisure. (*Long.*)

5. Both hands, fingers irregularly outspread at the sides of the head, to imitate the outspread horns. (*Titchemátski.*) This sign is made by our deaf-mutes.

6. Same position, confined to the thumb and two first fingers of each hand. (*Burton.*)

The above signs all appear to be used for the animal generically, but the following are separately reported for two of the species:

Black-tailed deer [*Cariacus macrotis* (Say), Gray].

1. Make several passes with the hand before the face, then indicate a tail. (*Wied.*)

2. Hold the left hand pendant a short distance in front of the chest, thumb inward, finger ends approximated to each other as much as possible (*i. e.*, with the first and fourth drawn together under the second and third). Then close the right hand around the left (palm to back, and covering the

bases of the left-hand fingers) and draw them downward, still closed, until it is entirely drawn away. This sign seems to represent the act of smoothing down the fusiform tuft at the end of the animal's tail. (*Matthews.*)

White-tailed deer [*Cariacus virginianus macrurus* (Raf.), Coues].

Hold the right hand upright before the chest, all fingers but the index being bent, the palm being turned as much to the front as possible. Then wag the hand from side to side a few times rather slowly. The arm is moved scarcely, or not at all. This sign represents the motion of the deer's tail. (*Matthews.*)

For *dog*, one of the signs gives the two forefingers slightly opened, drawn horizontally across the breast from right to left. (*Burton.*) This would not be intelligible without knowledge of the fact that before the introduction of the horse, and even yet, the dog has been used to draw the tent-poles in moving camp, and the sign represents the trail. Indians less nomadic, who built more substantial lodges, and to whom the material for poles was less precious than on the plains, would not perhaps have comprehended this sign, and the more general one is the palm lowered as if to stroke gently in a line conforming to the animal's head and neck. It is abbreviated by simply lowering the hand to the usual height of the wolfish aboriginal breed (*Wied; Titchkemátski*), and suggests *the* animal *par excellence* domesticated by the Indians and made a companion. The French and American deaf-mutes more specifically express the dog by snapping the fingers and then patting the thigh, or by patting the knee and imitating barking with the lips.

INSTANCES OF PREVALENT SIGNS.

Among the signs that are found generally current and nearly identical may be noted that for *horse*, made by the fore and middle finger of the right hand placed by some astraddle of the left forefinger and by others of the edge of the left hand, the animal being considered at first as only serviceable for riding and not for draft. Colonel Dodge mentions, however, that these signs are used only by Indians to white men, their ordinary sign for *horse* being made by drawing the right hand from left to right across the body about the heart, all the fingers being closed excepting the index. It is to be observed that this sign has a strong resemblance to the one given above by Captain BURTON for *dog*, and may have reference to the girth. It is still more easily confused with Captain BURTON'S "think, guess". The French deaf-mutes add to the straddling of the index the motion of a trot. The Utes have a special sign for *horse*—the first and little fingers of the right hand, palm down, extended forward, the balls of the remaining fingers falling down and resting upon the end of the thumb, presenting a suggestion of the animal's head and ears. Our deaf-mutes indicate the ears, followed by straddling the left hand by the fore and middle fingers of the right.

Same, similar, is made not only among our tribes generally, but by those all over the world, and by deaf-mutes, by extending the two forefingers together side by side, backs upward, sometimes moved together slightly forward. When held at rest in this position, *companion* and the tie of fellowship, what in days of chivalry was styled "brothers in arms," can be indicated, and, as a derivative also, *husband*. The French and American deaf-mutes use this sign, preceded by one showing the sex, for "brother" or "sister."

The most remarkable variant from the sign as above described which is reported to be used by our Indians, is as follows: Extend the fore and middle finger of the right hand, pointing upward, thumb crossed over the other fingers, which are closed. Move the hand downward and forward. (*Dorsey*)

An opposition to the more common sign above mentioned is given, though not generally reported, for *he*, or *another person*, by placing one straight forefinger over the other, nearly touching, and then separated with a moderately rapid motion. (*Dunbar*.) The deaf-mutes for "he" point the thumb over the right shoulder.

The principal motion for *surprise, wonder*, consists in placing the right hand before the mouth, which is open, or supposed so to be—a gesture seemingly involuntary with us, and which also appears in the Egyptian hieroglyphs.

The general sign for *sun*, when it is given as distinguished from *day*—made by forming a circle with the thumb and forefinger raised to the east or along the track of the orb—is often abbreviated by simply crooking the elevated forefinger into an arc of a circle, which would more naturally be interpreted as the crescent moon. It appears that some tribes that retain the full descriptive circle for the sun do form a distinguishing crescent for the moon, but with the thumb and forefinger, and for greater discrimination precede it with the sign for *night.* An interesting variant of the sign for *sun* is, however, reported as follows: The partly bent forefinger and thumb of the right hand are brought together at their tips so as to represent a circle; and with these digits next to the face, the hand is held up toward the sky from one to two feet from the eye and in such a manner that the glance may be directed through the opening. (*Matthews.*) The same authority gives the sign for "moon" as that for "sun," except that the tips of the finger and thumb, instead of being opposed, are approximated so as to represent a crescent. This is not preceded by the sign for *night,* which, with some occasional additions, is the crossing of both horizontally outspread palms, right above left, in front of the body, the conception being covering, shade, and consequent obscurity. With a slight differentiation, *darkness* is represented, and with another, *forget, forgotten,* that is, darkness in the memory.

Inquiry, question. What? Which? When?

This is generally denoted by the right hand held upward, palm upward, and directed toward the person interrogated, and rotated two or three times edgewise. When this motion is made, as among some tribes, with the thumb near the face, it might be mistaken for the derisive, vulgar gesture called "taking a sight," "donner un pied de nez," descending to our small boys from antiquity. The separate motion of the fingers in the vulgar gesture as used in our eastern cities is, however, more nearly correlated with the Indian sign for *fool.* It may be noted that the Latin "sagax," from which is derived "sagacity," was chiefly used to denote the keen scent of dogs, so there is a relation established between the nasal organ and wisdom or its absence, and that "suspendere naso" was a classic phrase for hoaxing. The Italian expressions "restare con un palmo di naso," "con tanto di naso," &c., mentioned by the Canon De Jorio, refer to the same vulgar gesture in which the face is supposed to be thrust forward sillily. The same rotation upon the wrist, with the index and middle finger diverged over the heart, among our Indians means specifically *uncertainty, indecision,* "more than one heart for a purpose," and a variant of it appears in one of the signs for "*I don't know.*" The special inquiry "Do you know?" is reported as follows: Shake the right hand in front of the face, a little to the right, the whole arm elevated so as to throw the hand even with the face and the forearm standing almost perpendicular; principal motion with hand, slight motion of forearm, palm outward. (*Deffenbaugh.*)

The Indian sign for "inquiry" is far superior to that of the French deaf-mutes, which is the part of the French shrug with the hunched shoulders omitted.

A sign for a special form of inquiry as to the tribe to which the person addressed belongs is to pass the right hand from left to right across the face, which is answered by the appropriate tribal sign. (*Powell.*)

Instead of a direct question the Utes in sign-conversation use a negative form, *e. g.*, to ask "Where is your mother?" would be rendered "Mother—your—I—see—not."

Fool, foolish. The prevailing gesture is a finger pointed to the forehead and rotated circularly—"rattle-brained." The only reported variance is where the sign for "man" is followed by shaking the fingers held downward, without reference to the head—the idea of looseness simply. French deaf-mutes shake the hands above the head after touching it with the index.

No, negative. The right hand—though in the beginning of the sign held in various positions—is generally either waved before the face (which is the sign of our deaf-mutes for emphatic negative), as if refusing to accept the idea or statement presented, or pushed sidewise to the right from either the breast or face, as if dismissing it or setting it aside One of the signs given for the Pah-Utes by NÁTSHES of oscillating the index before the face from right to left is substantially the same as one reported from Naples by DE JORIO. This may be compared with our shaking of the head in denial; but that gesture is not so universal in the Old World as is popularly supposed, for the ancient Greeks, followed by the modern Turks and rustic Italians, threw the head back, instead of shaking it, for *No.* A sign differing from all the above is by making a quick motion of the open hand from the mouth forward, palm toward mouth. (*Deffenbaugh.*) The Egyptian negative linear hieroglyph is clearly the gesture of both hands, palms down, waved apart horizontally and apparently at the level of the elbow, between which and the Maya negative particle "*ma*" given by LANDA there is a strong coincidence.

Lie, falsehood, is almost universally expressed by some figurative variation on the generic theme of a forked or double tongue—"two different stories"—in which the first two fingers on the right hand separate from the mouth. One reported sign precedes the latter motion by the right hand touching the breast over the heart. (*Hoffman.*) Another instance given, however, is when the index is extended from the two corners of the mouth successively. (*Ealy.*) Still another is by passing the hand from right to left close by and across the mouth, with the first two fingers of the hand opened, thumb and other fingers closed. (*Dodge; Nátshes.*) A further variant employed by the Utes is made by closing the right hand and placing the tips of the first two fingers upon the ball of the extended thumb, and snapping them forward straight and separated while passing the hand from the mouth forward and to the left. In the same tribe the index is more commonly moved, held straight upward and forward, alternately toward the left and right front. "Talk two ways." *Truth, true,* is naturally contradistinguished by the use of a single finger, the index, pointing straight from the mouth forward and sometimes upward—"One tongue; speech straight to the front; no talk behind a man." Sometimes, however, the breast is the initial point, as in the French deaf-mute sign for "sincere." The deaf-mutes also gesture "truth" by moving one finger straight from the lips—"straight-forward speaking"—but distinguish "lie" by moving the finger to one side—"sideways speaking."

Offspring or descendant, child in filial relation—not simply as young humanity—is generally denoted by a slightly varied dumb show of issuance from the loins, the line traced sometimes showing a close diagnosis of parturition. This is particularly noticeable in the following description: Place the left hand in front of the body, a little to the right, the palm downward and slightly arched; pass the extended right hand downward, forward, and upward, forming a short curve underneath the left. (*Hoffman.*) The sign, with additions, means "father," "mother," "grandparent," but its expurgated form among the French deaf-mutes means "parentage" generically, for which term there is a special sign reported from our Indians by only one authority, viz: Place the hand bowl-shaped over the right breast, as if grasping a pap. (*Dodge.*) It is not understood how this can be distinguished from one of the signs above mentioned for "woman."

Possession, mine, my property. The essential of this common sign is clinching the right hand held at the level of the head and moving it gently forward, clearly the grasping and display of property. None of the deaf-mute signs to express "possession, ownership," known to the writer, resemble this or are as graphic. Our deaf-mutes press an imaginary object to the breast with the right hand.

Steal. The prevalent delineation is by holding the left arm horizontally across the body and seizing from under the left fist an imaginary object with the right hand (*Burton*), implying concealment and the transportation that forms part of the legal definition of larceny. This sign is also made by our deaf-mutes. Sometimes the fingers of the right hand are hooked, as if grabbing or tearing. (*Titchkemátski.*) Another sign is reported, in which the left arm is partly extended and held horizontally so that the left hand will be palm downward, a foot or so in front of the chest. Then, with the right hand in front, a motion is made as if something were grasped deftly in the fingers and carried rapidly along under the left arm to the axilla. (*Matthews.*) The specialty of horse-theft is indicated by the pantomime of cutting a lariat. (*Burton.*)

Trade, barter, exchange, is very commonly denoted by a sign the root of which is the movement of the two flat hands or the two forefingers past each other, so that one takes the place before held by the other, the exact conceit of exchange. One description is as follows: The hands, backs forward, are held as index-hands pointing upward, the elbows being fully bent. Each hand is then simultaneously with the other, moved to the opposite shoulder, so that the forearms cross one another almost at right angles. (*Matthews.*) Another: Pass the hands in front of the body at the height of the waist, all fingers closed except the index-fingers. (*Deffenbaugh.*) This is also made by the Comanches (*Haworth*), Bannocks, and Umatillas. (*Nátshes.*) Another instance is reported where the first two fingers of the right hand cross those of the left, both being slightly spread. (*Hoffman.*) Our deaf-mutes use the same gesture as first above mentioned with the hands closed. An invitation to a general or systematic barter or trade, as distinct from one transaction, is expressed by repeated taps or the use of more fingers. The rough resemblance of this sign to that for "cutting" has occasioned mistakes as to its origin. It is reported by Captain BURTON as the conception of one smart trader cutting into the profits of another—"diamond cut diamond." The trade sign is, on the plains, often used to express the *white man*—vocally named Shwop—a legacy from the traders, who were the first Caucasians met. Generally, however, the gesture for *white man* is by designating the hat or head-covering of civilization. This the French deaf-mutes apply to all *men*, as distinct from women.

INSTANCES OF SIGNS HAVING SPECIAL INTEREST.

A few signs have been selected which are not remarkable either for general or limited acceptance, but are of interest from special conception or peculiar figuration.

The relation of *brothers, sisters*, and of *brother and sister*, children of the same mother, is signified by putting the two first finger tips in the mouth, denoting the nourishment taken from the same breast. (*Burton; Dorsey.*) One of the signs for *child* or infant is to place the thumb and fingers of the right hand against the lips, then drawing them away and bringing the right hand against the left fore-arm, as if holding an infant (*Dunbar.*) The Cistercian monks, vowed to silence, and the Egyptian hieroglyphers, notably in the designation of Horus, their dawn-god, used the finger in or on the lips for "child." It has been conjectured in the last instance that the gesture implied, not the mode of taking nourishment, but inability to speak—*in-fans.* This conjecture, however, was only made to explain the blunder of the Greeks, who saw in the hand placed connected with the mouth in the hieroglyph of Horus (the) son, "Hor-(p)-chrot," the gesture familiar to themselves of a finger on the lips to express "silence," and so mistaking both the name and the characterization, invented the God of Silence, Harpokrates. A careful examination of all the linear hieroglyphs given by CHAMPOLLION (Dictionnaire Égyptien), shows that the finger or the hand to the mouth of an adult (whose posture is always distinct from that of a child) is always in connection with the positive ideas of voice, mouth, speech, writing, eating, drinking, &c., and never with the negative idea of silence. The special character for "child" always has the above-mentioned part of the sign with reference to nourishment from the breast. An uninstructed deaf-mute, as related by Mr. Denison of the Columbia Institution, invented, to express "sister," first the sign for "female," made by the half-closed hands with the ends of fingers touching the breasts, followed by the index in the mouth.

Destroyed, all gone, no more.

The hands held horizontal and the palms rubbed together two or three times circularly; the right hand is then carried off from the other in a short horizontal curve. (*Long.*) "Rubbed out." This resembles the Edinburgh and our deaf-mute sign for "forgive" or "clemency," the rubbing out of offense. Several shades of meaning under this head are designated by varying gestures "If something of little importance has been destroyed by accident or design, the fact is communicated by indicating the thing spoken of, and then slightly striking the palms and open fingers of the hands together, as if brushing dust off of them. If something has been destroyed by force the sign is as if breaking a stick in the two hands, throwing the pieces away, and then dusting the hands as before. The amount of force used and the completeness of the destruction are shown by greater or less vigor of action and facial expression." (*Dodge.*)

Done, finished. The hands placed edges up and down, parallel to each other, right hand outward, which is drawn back as if cutting something. (*Dunbar.*) An *end* left after cutting is suggested; perhaps our colloquial "cut short." The French and our deaf-mutes give a cutting motion downward, with the right hand at a right angle to the left.

Glad, pleased, content. Wave the open hand outward from the breast (*Burton*), to express heart at ease—"bosom's lord sits lightly on its throne." Another gesture, perhaps noting a higher degree of happiness, is to raise the right hand from the breast in serpentine curves to above the head. (*Wied.*) "Heart beats high." Another: Extend both hands outward, palms turned downward, and make a sign exactly similar to the way women smooth a bed in making it. (*Holt.*) "Smooth and easy."

Dissatisfaction, discontent, is naturally contrasted by holding the index transversely before the heart and rotating the wrist several times, indicating disturbance of the organ, which our aborigines, like modern Europeans, poetically regard as the seat of the affections and emotions, not selecting the liver or stomach as other peoples have done with greater physiological reason.

To *hide, conceal,* is graphically portrayed by placing the right hand inside the clothing of the left breast, or covering the right hand, fingers hooked, by the left, which is flat, palm downward, and held near the body. The same gestures mean "*secret.*"

Peace, or *friendship,* is sometimes shown by placing the tips of the two first fingers of the right hand against the mouth and elevated upward and outward to mimic the expulsion of smoke—"we two smoke together." (*Titchkemátski.*) It is also often rendered by the joined right and left hands, the fingers being sometimes interlocked, but others simply hook the two forefingers together. Our deaf-mutes interlock the forefingers for "friendship," clasp the hands, right uppermost, for "marriage," and make the last sign, repeated with the left hand uppermost, for "peace." The idea of union or linking is obvious. It is, however, noticeable that while this ceremonial gesture is common and ancient, the practice of shaking hands on meeting, now the annoying etiquette of the Indians in their intercourse with whites, was never used by them between each other, and is clearly a foreign importation. Their fancy for affectionate greeting was in giving a pleasant bodily sensation by rubbing each other's breasts, arms, and stomachs. The senseless and inconvenient custom of shaking hands is, indeed, by no means general throughout the world, and in the extent to which it prevails in the United States is a matter of national opprobrium.

The profession of *peace, coupled with invitation,* is often made from a distance by the acted spreading of a real or imaginary robe or blanket—"come and sit down."

The sign for *stone* has an archæological significance—the right fist being struck repeatedly upon the left palm, as would be instinctive when a stone was the only hammer.

Prisoner is a graphic picture. The forefinger and thumb of the left hand are held in the form of a semicircle opening toward and near the breast, and the right forefinger, representing the prisoner, is placed upright within the curve and passed from one side to another, in order to show that it is not permitted to pass out. (*Long.*)

Soft is ingeniously expressed by first striking the open left hand several times with the back of the right, and then striking with the right the back of the left, restoring the supposed yielding substance to its former shape.

Without further multiplying examples, the conclusion is presented that the gesture-signs among our Indians show no uniformity in detail, the variety in expression among them and in their comparison with those of deaf-mutes and transatlantic mimes being in itself of psychological interest. The generalization of TYLOR that "gesture-language is substantially the same among savage tribes all over the world" must be understood, indeed would be so understood from his remarks in another connection, as referring to their common use of signs and of signs formed on the same principles, but not of the same signs to express the same ideas, even "substantially," however indefinitely that dubious adverb may be used.

GESTURE-SPEECH UNIVERSAL AS AN ART.

The *attempt to convey meaning by signs is*, however, *universal* among the Indians of the plains, and those still comparatively unchanged by civilization, as is its successful execution as an art, which, however it may have commenced as an instinctive mental process, has been cultivated, and consists in actually pointing out objects in sight not only for designation, but for application and predication, and in suggesting others to the mind by action and the airy forms produced by action.

In no other part of the thoroughly explored world has there been spread over so vast a space so small a number of individuals divided by so many linguistic and dialectic boundaries as in North America. Many wholly distinct tongues have for a long indefinite time been confined to a few scores of speakers, verbally incomprehensible to all others on the face of the earth who did not, from some rarely operating motive, laboriously acquire their language. Even when the American race, so styled, flourished in the greatest population of which we have any evidence (at least according to the published views of the present writer, which seem to have been favorably received), the immense number of languages and dialects still preserved, or known by early recorded fragments to have once existed, so subdivided it that but the dwellers in a very few villages could talk together with ease, and all were interdistributed among unresponsive vernaculars, each to the other being bar-bar-ous in every meaning of the term. It is, however, noticeable that the three great families of Iroquois, Algonkin, and Muskoki, when met by their first visitors, do not appear to have often impressed the latter with their reliance upon gesture-language to the same extent as has always been reported of the aborigines now and formerly found farther inland. If this absence of report arose from the absence of the practice and not from imperfection of observation, an explanation may be suggested from the fact that among those families there were more people dwelling near together in sociological communities, of the same speech, though with dialectic peculiarities, than became known later in the later West, and not being nomadic, their intercourse with strange tribes was less individual and conversational.

The use of gesture-signs, continued, if not originating, in necessity for communication with the outer world, became entribally convenient from the habits of hunters, the main occupation of all savages, depending largely upon stealthy approach to game, and from the sole form of their military tactics—to surprise an enemy. In the still expanse of virgin forests, and especially in the boundless solitudes of the great plains, a slight sound can be heard over a vast area, that of the human voice being from its rarity the most startling, so that it is now, as it probably has been for centuries, a common precaution for members of a hunting or war party not to speak together when on such expeditions, communicating exclusively by signs. The acquired habit also exhibits itself not only in formal oratory, but in impassioned or emphatic conversation.

This domestic as well as foreign exercise for generations in the gesture-language has naturally produced great skill both in expression and reception, so as to be measurably independent of any prior mutual understanding, or what in a system of signals is called preconcert. Two accomplished army signalists can, after sufficient trial, communicate without either of them learning the code in which the other was educated and which he had before practiced, one being mutually devised for the occasion, and those specially designed for secrecy are often deciphered. So, if any one of the more approximately conventional signs is not quickly comprehended, an Indian skilled in the principle of signs resorts to another expression of his flexible art, perhaps reproducing the gesture unabbreviated and made more graphic, perhaps presenting either the same or another conception or quality of the same object or idea by an original portraiture. The same tribe has, indeed, in some instances, as appears by the collected lists, a choice already furnished by tradition or importation, or recent invention or all together, of several signs for the same thought-object. Thus there are produced synonyms as well as dialects in sign-language.

The general result is that two intelligent mimes seldom fail of mutual understanding, their attention being exclusively directed to the expression of thoughts by the means of comprehension and reply equally possessed by both, without the mental confusion of conventional sounds only intelligible to one. The Indians who have been shown over the civilized East have also often succeeded in holding intercourse, by means of their invention and application of principles, in what may be called the voiceless mother utterance, with white deaf-mutes, who surely have no semiotic code more nearly connected with that attributed to the plain-roamers than is derived from their common humanity. When they met together they were found to pursue the same course as that noticed at the meeting together of deaf-mutes who were either not instructed in any methodical dialect or who had received such instruction by different methods. They seldom agreed in the signs at first presented, but soon understood them, and finished by adopting some in mutual compromise, which proved to be those most strikingly appropriate, graceful, and convenient, but there still remained in some cases a plurality of fitting signs for the same idea or object. On one of the most interesting of these occasions, at the Pennsylvania Institution for the Deaf and Dumb, in 1873, it was remarked that the signs of the deaf-mutes were much more readily understood by the Indians, who were Absaroki or Crows, Arapahos, and Cheyennes, than were theirs by the deaf-mutes, and that the latter greatly excelled in pantomimic effect. This need not be surprising when it is considered that what is to the Indian a mere adjunct or accomplishment is to the deaf-mute the natural mode of utterance, and that there is still greater freedom from the trammel of translating words into action—instead of acting the ideas themselves—when, the sound of words being unknown, they remain still as they originated, but another kind of sign, even after the art of reading is acquired, and do not become entities as with us.

It is to be remarked that Indians when brought to the East have shown the greatest pleasure in meeting deaf-mutes, precisely as travelers in a foreign country are rejoiced to meet persons speaking their language, with whom they can hold direct communication without the tiresome and often suspected medium of an interpreter. A Sandwich Islander, a Chinese, and the Africans from the slaver Amistad have, in published instances, visited our deaf-mute institutions with the same result of free and pleasurable intercourse, and an English deaf-mute had no difficulty in conversing with Laplanders. It appears, also, on the authority of SIBSCOTA, whose treatise was published in 1670, that Cornelius Haga, ambassador of the United Provinces to the Sublime Porte, found the Sultan's mutes to have established a language among themselves in which they could discourse with a speaking interpreter, a degree of ingenuity interfering with the object of their selection as slaves unable to repeat conversation.

SUGGESTIONS TO OBSERVERS.

The most important suggestion to persons interested in the collection of signs is that they shall not too readily abandon the attempt to discover recollections of them even among tribes long exposed to Caucasian influence and officially segregated from others.

During the last week a missionary wrote that he was concluding a considerable vocabulary of signs finally procured from the Ponkas, although after residing among them for years, with thorough familiarity with their language, and after special and intelligent exertion to obtain some of their disused gesture-language, he had two months ago reported it to be entirely forgotten. A similar report was made by two missionaries among the Ojibwas, though other trustworthy authorities have furnished a list of signs obtained from that tribe. Further discouragement came from an Indian agent giving the decided statement, after four years of intercourse with the Pah-Utes, that no such thing as a communication by signs was known or even remembered by them, which, however, was less difficult to bear because on the day of the receipt of that well-intentioned missive some officers of the Bureau of Ethnology were actually talking in signs with a delegation of that very tribe of Indians then in Washington, from one of whom the Story hereinafter appearing was received. The difficulty in collecting signs may arise because Indians are often provokingly reticent about their old habits and traditions; because they do not distinctly comprehend what is sought to be obtained, and because sometimes the art, abandoned in general, only remains in the memories of a few persons influenced by special circumstances or individual fancy.

In this latter regard a comparison may be made with the old science of heraldry, once of practical use and a necessary part of a liberal education, of which hardly a score of persons in the United States have any but the vague knowledge that it once existed; yet the united memories of those persons could, in the absence of records, reproduce all essential points on the subject.

Even when the specific practice of the sign-language has been generally discontinued for more than one generation, either from the adoption of a jargon or from the common use of the tongue of the conquering English, French, or Spanish, some of the gestures formerly employed as substitutes for words may survive as a customary accompaniment to oratory or impassioned conversation, and, when ascertained, should be carefully noted. An example, among many, may be found in the fact that the now civilized Muskoki or Creeks, as mentioned by Rev. H. F. BUCKNER, when speaking of the height of children or women, illustrate their words by holding their hands at the proper elevation, palm up; but when describing the height of "soulless" animals or inanimate objects, they hold the palm downward. This, when correlated with the distinctive signs of other Indians, is an interesting case of the survival of a practice which, so far as yet reported, the oldest men of the tribe now living only remember to have once existed. It is probable that a collection of such distinctive gestures among even the most civilized Indians would reproduce enough of their ancient system to be valuable, even if the persistent enquirer did not in his search discover some of its surviving custodians even among Chahta or Cheroki, Iroquois or Abenaki, Klamath or Nutka.

Another recommendation is prompted by the fact that in the collection and description of Indian signs there is danger lest the civilized understanding of the original conception may be mistaken or forced. The liability to error is much increased when the collections are not taken directly from the Indians themselves, but are given as obtained at second-hand from white traders, trappers, and interpreters, who, through misconception in the beginning and their own introduction or modification of gestures, have produced a jargon in the sign as well as in the oral intercourse. If an Indian finds that his interlocutor insists upon understanding and using a certain sign in a particular manner, it is within the very nature, tentative and elastic, of the gesture art—both performers being on an equality—that he should adopt the one that seems to be recognized or that is pressed upon him, as with much greater difficulty he has learned and adopted many foreign terms used with whites before attempting to acquire their language, but never with his own race. Thus there is now, and perhaps always has been, what may be called a *lingua-franca* in the sign vocabulary. It may be ascertained that all the tribes of the plains having learned by experience that white visitors expect to receive certain signs really originating with the latter, use them in their intercourse, just as they sometimes do the words "squaw" and "papoose," corruptions of the Algonkin, and once as meaningless in the present West as the English terms "woman" and "child," but which the first pioneers, having learned them on the Atlantic coast, insisted upon as generally intelligible. This process of adaptation may be one of the explanations of the reported universal code.

It is also highly probable that signs will be invented by individual Indians who may be pressed by collectors for them to express certain ideas, which signs of course form no part of the current language; but while that fact should, if possible, be ascertained and reported, the signs so invented are not valueless merely because they are original and not traditional, if they are made in good faith and in accordance with the principles of sign-formation. The process resembles the coining of new words to which the higher languages owe their copiousness. It is noticed in the signs invented by Indians for each new product of civilization brought to their notice. Less error will arise in this direction than from the misinterpretation of the idea intended to be conveyed by spontaneous signs.

The absurdity to which over-zeal may be exposed is illustrated by an anecdote found in several versions and in several languages, but repeated as a veritable Scotch legend by Duncan Anderson, esq., principal of the Glasgow Institution for the Deaf and Dumb, when he visited Washington in 1853.

King James I of England desiring to play a trick upon the Spanish ambassador, a man of great erudition, but who had a crotchet in his head upon sign-language, informed him that there was a distinguished professor of that science in the University at Aberdeen. The ambassador set out for that place, preceded by a letter from the King with instructions to make the best of him. There was in the town one Geordy, a butcher, blind of one eye, a fellow of much wit and drollery. Geordy is told to play the part of a professor, with the warning not to speak a word, is gowned, wigged, and placed in a chair of state, when the ambassador is shown in and they are left alone together. Presently the nobleman came out greatly pleased with the experiment, claiming that his theory was demonstrated. He said, "When I entered the room I raised one finger, to signify there is one God. He replied by raising two fingers to signify that this Being rules over two worlds, the material and the spiritual. Then I raised three fingers, to say there are three persons in the Godhead. He then closed his fingers, evidently to say these three are *one*." After this explanation on the part of the nobleman, the professors sent for the butcher and asked him what took place in the recitation-room. He appeared very angry and said, "When the crazy man entered the room where I was he raised one finger, as much as to say, I had but one eye, and I raised two fingers to signify that I could see out of my one eye as well as he could out of both of his. When he raised *three* fingers, as much as to say there were but three eyes between us, I doubled up my fist, and if he had not gone out of that room in a hurry I would have knocked him down."

By far the most satisfactory mode of securing accurate signs is to induce the Indians to tell stories, make speeches, or hold talks in gesture, with one of themselves as interpreter in his own oral language if the latter is understood by the observer, and if not, the words, not the signs, should be translated by an intermediary white interpreter. It will be easy afterward to dissect and separate the particular signs used. This mode will determine the genuine shade of meaning of each sign, and corresponds with the plan now adopted by the Bureau of Ethnology for the study of the aboriginal vocal languages, instead of that arising out of exclusively missionary purposes, which was to force a translation of the Bible from a tongue not adapted to its terms and ideas, and then to compile a grammar and dictionary from the artificial result. A little ingenuity will direct the more intelligent or complaisant gesturers to the expression of the thoughts, signs for which are specially sought; and full orderly descriptions of such tales and talks with or even without analysis and illustration are more desired than any other form of contribution. No such descriptions of any value have been found in print, and the best one thus far obtained through the correspondence of the present writer is given below, with the hope that emulation will be excited. It is the farewell address of Kin Chē-ĕss (Spectacles), medicine-man of the Wichitas, to Missionary A. J. Holt on his departure from the Wichita Agency, in the words of the latter.

A SPEECH IN SIGNS.

He placed one hand on my breast, the other on his own, then clasped his two hands together after the manner of our congratulations,—*We are friends.* He placed one hand on me, the other on himself, then placed the first two fingers of his right hand between his lips,—*We are brothers.* He placed his right hand over my heart, his left hand over his own heart, then linked the first fingers of his right and left hands,—*Our hearts are linked together.* He laid his right hand on me lightly, then put it to his mouth, with the knuckles lightly against his lips, and made the motion of flipping water from the right-hand forefinger, each flip casting the hand and arm from the mouth a foot or so, then bringing it back in the same position. (This repeated three or more times, signifying "talk" or talking.) He then made a motion with his right hand as if he were fanning his right ear; this repeated. He then extended his right hand with his index-finger pointing upward, his eyes also being turned upward,—*You told me of the Great Father.* Pointing to himself, he hugged both hands to his bosom, as if he were affectionately clasping something he loved, and then pointed upward in the way before described,—*I love him* (the Great Father). Laying his right hand on me, he clasped his hands to his bosom as before,—*I love you.* Placing his right hand on my shoulder, he threw it over his own right shoulder as if he were casting behind him a little chip, only when his hand was over his shoulder his index-finger was pointing behind him,—*You go away.* Pointing to his breast, he clinched the same hand as if it held a stick, and made a motion as if he were trying to strike something on the ground with the bottom of the stick held in an upright position,—*I stay*, or *I stay right here.*

Placing his right hand on me, he placed both his hands on his breast and breathed deeply two or three times, then using the index-finger and thumb of each hand as if he were holding a small pin, he placed the two hands in this position as if he were holding a thread in each hand and between the thumb and forefinger of each hand close together, and then let his hands recede from each other, still holding his fingers in the same position, as if he were letting a thread slip between them until his hands were two feet apart,—*You live long time.* Laying his right hand on his breast, then extending his forefinger of the same hand, holding it from him at half-arm's length, the finger pointing nearly upward, then moving his hand, with the finger thus extended, from side to side about as rapidly as a man steps in walking, each time letting his hand get farther from him for three or four times, then suddenly placing his left hand in a horizontal position with the fingers extended and together so that the palm was sidewise, he used the right-hand palm extended, fingers together, as a hatchet, and brought it down smartly, just missing the ends of the fingers of the left hand. Then placing his left hand, with the thumb and forefinger closed, to his heart, he brought his right hand, fingers in the same position, to his left, then, as if he were holding something between his thumb and forefinger, he moved his right hand away as if he were slowly casting a hair from him, his left hand remaining at his breast, and his eyes following his right,—*I go about a little while longer, but will be cut off shortly and my spirit will go away* (or will die). Placing the thumbs and forefingers again in such a position as if he held a small thread between the thumb and forefinger of each hand, and the hands touching each other, he drew his hands slowly from each other, as if he were stretching a piece of gum-elastic; then laying his right hand on me, he extended the left hand in a horizontal position, fingers extended and closed, and brought down his right hand with fingers extended and together, so as to just miss the tips of the fingers of his left hand; then placing his left forefinger and thumb against his heart, he acted as if he took a hair from the forefinger and thumb of his left hand with the forefinger and thumb of the right, and slowly cast it from him, only letting his left hand remain at his breast, and let the index-finger of the right hand point outward toward the distant horizon,—*After a long time you die.* When placing his left hand upon himself and his right hand upon me, he extended them upward over his head and clasped them there,—*We then meet in heaven.* Pointing upward, then to himself, then to me, he closed the third and little finger of his right hand, laying his thumb over them, then extending his first and second fingers about as far apart as the eyes, he brought his hand to his eyes, fingers pointing outward, and shot his hand outward,—*I see you up there.* Pointing to me, then giving the last above-described sign of "look," then pointing to himself, he made the sign as if stretching out a piece of gum-elastic between the fingers of his left and right hands, and then made the sign of "cut-off" before described, and then extended the palm of the right hand horizontally a foot from his waist, inside downward, then suddenly threw it half over and from him, as if you were to toss a chip from the back of the hand (this is the negative sign everywhere used among these Indians),—*I would see him a long time, which should never be cut off, i. e.*, always.

Pointing upward, then rubbing the back of his left hand lightly with the forefinger of his right, he again gave the negative sign,—*No Indian there* (in heaven). Pointing upward, then rubbing his forefinger over the back of my hand, he again made the negative sign,—*No white man there.* He made the same sign again, only he felt his hair with the forefinger and thumb of his right hand, rolling the hair several times between the fingers,—*No black man in heaven.* Then rubbing the back of his hand and making the negative sign, rubbing the back of my hand and making the negative sign, feeling of one of his hairs with the thumb and forefinger of his right hand, and making the negative sign, then using both hands as if he were reaching around a hogshead, he brought the forefinger of his right hand to the front in an upright position after their manner of counting, and said thereby,—*No Indian, no white man, no black man, all one.* Making the "hogshead" sign, and that for "look," he placed the forefinger of each hand side by side pointing upward,—*All look the same,* or alike. Running his hands over his wild Indian costume and over my clothes, he made the "hogshead" sign, and that for "same," and said thereby,—*All dress alike there.* Then making the "hogshead" sign, and that for "love" (hugging his hands), he extended both hands outward, palms turned downward, and made a sign exactly similar to the way ladies smooth a bed in making it; this is the sign for "happy,"—*All will be happy alike there.* He then made the sign for "talk," and for "Father," pointing to himself and to me,—*You pray for me.* He then made the sign for "*go away,*" pointing to me, he threw right hand over his right shoulder so his index-finger pointed behind him,—*You go away.* Calling his name he made the sign for "look" and the sign of negation after pointing to me,—*Kin Chē-ĕss see you no more.*

The following, which is presented as a better descriptive model, was obtained by Dr. W. J. Hoffman, of the Bureau of Ethnology, from Nátshes, the Pah-Ute chief connected with the delegation before mentioned, and refers to an expedition made by him by direction of his father, Winnemucca, Head Chief of the Pah-Utes, to the northern camp of his tribe, partly for the purpose of preventing the hostile outbreak of the Bannocks which occurred in 1878, and more particularly to prevent those Pah-Utes from being drawn into any difficulty with the authorities by being leagued with the Bannocks.

(1) Close the right hand, leaving the index extended, pointed westward at arm's length a little above the horizon, head thrown back with the eyes partly closed and following the direction,—*Away to the west*, (2) indicate a large circle on the ground with the forefinger of the right hand pointing downward,—*place* (locative), (3) the tips of the spread fingers of both hands placed against one another, pointing upward before the body, leaving a space of four or five inches between the wrists,—*house* (brush tent or wick′-i-up), (4) with the right hand closed, index extended or slightly bent, tap the breast several times,—*mine*. (5) Draw an imaginary line, with the right index toward the ground, from some distance in front of the body to a position nearer to it,—*from there I came*, (6) indicate a spot on the ground by quickly raising and depressing the right hand with the index pointing downward,—*to a stopping place*, (7) grasp the forelock with the right hand, palm to the forehead, and raise it about six inches, still holding the hair upward,—*the chief of the tribe* (Winnemucca), (8) touch the breast with the index,—*me*, (9) the right hand held forward from the hip at the level of the elbow, closed, palm downward, with the middle finger extended and quickly moved up and down a short distance,—*telegraphed*, (10) head inclined toward the right, at the same time making movement toward and from the ear with the extended index pointing towards it,—*I heard*, *i. e.*, understood.

(11) An imaginary line indicated with the extended and inverted index from a short distance before the body to a place on the right,—*I went*, (12) repeat gesture No. 6,—*a stopping place*, (13) inclining the head, with eyes closed, toward the right, bring the extended right hand, palm up, to within six inches of the right ear,—*where I slept*. (14) Place the spread and extended index and thumb of the right hand, palm downward, across the right side of the forehead,—*white man* (American), (15) elevating both hands before the breast, palms forward, thumbs touching, the little finger of the right hand closed,—*nine*, (16) touch the breast with the right forefinger suddenly,—*and myself*, (17) lowering the hand, and pointing downward and forward with the index still extended (the remaining fingers and thumb being loosely closed) indicate an imaginary line along the ground

toward the extreme right,—*went,* (18) extend the forefinger of the closed left hand, and place the separated fore and second fingers of the right astraddle the forefinger of the left, and make a series of arched or curved movements toward the right,—*rode horseback,* (19) keeping the hands in their relative position, place them a short distance below the right ear, the head being inclined toward that side,—*sleep,* (20) repeat the signs for *riding* (No. 18) and *sleeping* (No. 19) three times,—*four days and nights,* (21) make sign No. 18, and stopping suddenly point toward the east with the extended index-finger of the right (others being closed) and follow the course of the sun until it reaches the zenith,—*arrived at noon of the fifth day*

(22) Indicate a circle as in No. 2,—*a camp,* (23) the hands then placed together as in No. 3, and in this position, both moved in short irregular upward and downward jerks from side to side,—*many wick'-i-ups,* (24) then indicate the chief of the tribe as in No. 7,—meaning that *it was one of the camps of the chief of the tribe.* (25) Make a peculiar whistling sound of "phew" and draw the extended index of the right hand across the throat from left to right,—*Bannock,* (26) draw an imaginary line with the same extended index, pointing toward the ground, from the right to the body,—*came from the north,* (27) again make gesture No. 2,—*camp,* (28) and follow it twice by sign given as No. 18 (forward from the body, but a short distance),—*two rode.* (29) Rub the back of the right hand with the extended index of the left,—*Indian,* i. e., the narrator's own tribe, Pah-Ute, (30) elevate both hands side by side before the breast, palms forward, thumbs touching, then, after a short pause, close all the fingers and thumbs except the two outer fingers of the right hand,—*twelve,* (31) again place the hands side by side with fingers all spread or separated, and move them in a horizontal curve toward the right,—*went out of camp,* (32) and make the sign given as No. 25,—*Bannock,* (33) that of No. 2,—*camp,* (34) then join the hands as in No. 31, from the right towards the front,—*Pah-Utes returned,* (35) close the right hand, leaving the index only extended, move it forward and downward from the mouth three or four times, pointing forward, each time ending the movement at a different point,—*I talked to them,* (36) both hands pointing upward, fingers and thumbs separated, palms facing and about four inches apart, held in front of the body as far as possible in that posi-

tion,—*the men in council*, (37) point toward the east with the index apparently curving downward over the horizon, then gradually elevate it to an altitude of 45°,—*talked all night and until nine o'clock next morning*, (38) bring the closed hands, with forefingers extended, upward and forward from their respective sides, and place them side by side, palms forward, in front,—*my brother*, (39) followed by the gesture, No. 18, directed toward the left and front,—*rode*, (40) by No. 7,—*the head chief*, (41) and No. 2,—*camp*.

(42) Continue by placing the hands, slightly curved, palm to palm, holding them about six inches below the right ear, the head being inclined considerably in that direction,—*one sleep* (*night*), (43) make sign No. 14,—*white man*, (44) raise the left hand to the level of the elbow forward from the left hip, fingers pointing upward, thumb and forefinger closed,—*three*, (45) and in this position draw them toward the body and slightly to the right,—*came*, (46) then make gesture No. 42,—*sleep;* (47) point with the right index to the eastern horizon,—*in the morning*, (48) make sign No. 14,—*white man*, (49), hold the left hand nearly at arm's length before the body, back up, thumb and forefinger closed, the remaining fingers pointing downward,—*three*, (50) with the right index-finger make gesture No. 35, the movement being directed towards the left hand,—*talked to them*, (51) motion along the ground with the left hand, from the body toward the left and front, retaining the position of the fingers just stated (in No. 49),—*they went*, (52) tap toward the ground, as in gesture No. 6, with the left hand nearly at arm's length,—*to their camp*.

(53) Make gesture No. 18 toward the front,—*I rode*, (54) extend the right hand to the left and front, and tap towards the earth several times as in sign No. 6, having the fingers and thumb collected to a point,—*camp of the white men*. (55) Close both hands, with the forefingers of each partly extended and crooked, and place one on either side of the forehead, palms forward,—*cattle* (a steer), (56) hold the left hand loosely extended, back forward, about twenty inches before the breast, and strike the back of the partly extended right hand into the left,—*shot*, (57) make a short upward curved movement with both hands, their position unchanged, over and downward toward the right,—*fell over*, *killed*, (58) then hold the left hand a short distance before the body at the height of the elbow, palm downward,

fingers closed, with the thumb lying over the second joint of the fore-finger, extend the flattened right hand, edge down, before the body, just by the knuckles of the left, and draw the hand towards the body, repeating the movement,—*skinned,* (59) make the sign given in No. 25,—*Bannock,* (60) place both hands with spread fingers upward and palms forward, thumb to thumb, before the right shoulder, moving them with a tremulous motion toward the left and front,—*came in,* (61) make three short movements toward the ground in front, with the left hand, fingers loosely curved, and pointing downward,—*camp of the three white men,* (62) then with the right hand open and flattened, edge down, cut towards the body as well as to the right and left,—*cut up the meat,* (63) and make the pantomimic gesture of *handing it around to the visitors.*

(64) Make sign No. 35, the movement being directed to the left hand, as held in No. 49,—*told the white men,* (65) grasping the hair on the right side of the head with the left hand, and drawing the extended right hand with the edge towards and across the side of the head from behind forward,—*to scalp;* (66) close the right hand, leaving the index partly extended, and wave it several times quickly from side to side a short distance before the face, slightly shaking the head at the same time,—*no,* (67) make gesture No. 4,—*me,* (68) repeat No. 65,—*scalp,* (69) and raising the forelock high with the left hand, straighten the whole frame with a triumphant air,—*make me a great chief.* (70) Close the right hand with the index fully extended, place the tip to the mouth and direct it firmly forward and downward toward the ground,—*stop,* (71) then placing the hands, pointing upward, side by side, thumbs touching, and all the fingers separated, move them from near the breast outward toward the right, palms facing that direction at termination of movement,—*the Bannocks went to one side,* (72) with the right hand closed, index curved, palm downward, point toward the western horizon, and at arm's length dip the finger downward,—*after sunset,* (73) make the gesture given as No. 14,—*white men,* (74) pointing to the heart as in No. 4,—*and I,* (75) conclude by making gesture No 18 from near body toward the left, four times, at the end of each movement the hands remaining in the same position, thrown slightly upward,—*we four escaped on horseback.*

The above was paraphrased orally by the narrator as follows: Hearing of the trouble in the north, I started eastward from my camp in Western Nevada, when, upon arriving at Winnemucca Station, I received telegraphic orders from the head chief to go north to induce our bands in that region to escape the approaching difficulties with the Bannocks. I started for Camp McDermit, where I remained one night. Leaving next morning in company with nine others, we rode on for four days and a half. Soon after our arrival at the Pah-Ute camp, two Bannocks came in, when I sent twelve Pah-Utes to their camp to ask them all to come in to hold council. These messengers soon returned, when I collected all the Pah-Utes and talked to them all night regarding the dangers of an alliance with the Bannocks and of their continuance in that locality. Next morning I sent my brother to the chief, Winnemucca, with a report of proceedings.

On the following day three white men rode into camp, who had come up to aid in persuading the Pah-Utes to move away from the border. Next morning I consulted with them respecting future operations, after which they went away a short distance to their camp. I then followed them, where I shot and killed a steer, and while skinning it the Bannocks came in, when the meat was distributed. The Bannocks being disposed to become violent at any moment, the white men became alarmed, when I told them that rather than allow them to be scalped I would be scalped myself in defending them, for which action I would be considered as great a chief as Winnemucca by my people. When I told the Bannocks to cease threatening the white men they all moved to one side a short distance to hold a war council, and after the sun went down the white men and I mounted our horses and fled toward the south, whence we came.

Some of the above signs seem to require explanation. Nátshes was facing the west during the whole of this narration, and by the right he signified the north; this will explain the significance of his gesture to the right in Nos. 11 and 17, and to the left in No. 75.

No. 2 (repeated in Nos. 22, 27, 33, and 41), designates an Indian brush lodge, and although Nátshes has not occupied one for some years, the gesture illustrates the original conception in the round form of the foundation of poles, branches, and brush, the interlacing of which in the construction of the wick′-i-up has survived in gestures Nos. 3 and 23 (the latter referring to more than one, *i. e*, an encampment)

The sign for Bannock, No. 25 (also 32 and 59), has its origin from the tradition among the Pah-Utes that the Bannocks were in the habit of cutting the throats of their victims. This sign is made with the index instead of the similar gesture with the flat hand, which among several tribes denotes the Sioux, but the Pah-Utes examined had no specific sign for that body of Indians, not having been in sufficient contact with them.

"A stopping place," referred to in Nos. 6, 12, 52, and 54, represents the settlement, station, or camp of white men, and is contradistinguished by merely dotting toward the ground instead of indicating a circle.

It will also be seen that in several instances, after indicating the nationality, the fingers previously used in representing the number were repeated without its previously accompanying specific gesture, as in No. 61, where the three fingers of the left hand represented the men (white), and the three movements toward the ground signified the camp or tents of the three (white) men.

This also occurs in the gesture (Nos. 59, 60, and 71) employed for the Bannocks, which, having been once specified, is used subsequently without its specific preceding sign for the tribe represented.

The rapid connection of the signs Nos. 57 and 58, and of Nos. 74 and 75 indicates the conjunction, so that they are severally readily understood as "shot *and* killed," and "the white men *and* I." The same remark applies to Nos. 15 and 16, "the nine *and* I."

In the examination of the sign-language it is important to form a clear distinction between signs proper and symbols. All characters in Indian picture-writing have been loosely styled symbols, and as there is no logical distinction between the characters impressed with enduring form, and when merely outlined in the ambient air, all Indian gestures, motions, and attitudes might with equal appropriateness be called symbolic. While, however, all symbols come under the generic head of signs, very few signs are in accurate classification symbols. S. T. Coleridge has defined a symbol to be a sign included in the idea it represents. This may be intelligible if it is intended that an ordinary sign is extraneous to the concept, and, rather than directly suggested by it, is invented to express it by some representation or analogy, while a symbol may be evolved by a process of thought from the concept itself; but it is no very exhaustive or practically useful distinction. Symbols are less obvious and more artificial than mere signs, require convention, are not only abstract, but metaphysical, and often need explanation from history, religion, and customs. Our symbols of the ark, dove, olive branch, and rainbow would be wholly meaningless to people unfamiliar with the Mosaic or some similar cosmology, as would be the cross and the crescent to those ignorant of history. The last-named objects appeared in the lower class of *emblems* when used in designating the conflicting powers of Christendom and Islamism. Emblems do not necessarily require any analogy between the objects representing and those, or the qualities, represented, but may arise from pure accident. After a scurrilous jest the beggar's wallet became the emblem of the confederated nobles, the Gueux, of the Netherlands; and a sling, in the early minority of Louis XIV, was adopted from the refrain of a song by the Frondeur opponents of Mazarin. The several tribal signs for the Sioux, Arapaho, Cheyenne, &c., are their emblems precisely as the star-spangled flag is that of the United States, but there is nothing symbolic in any of them. So the signs for individual chiefs, when not merely translations of their names, are emblematic of their family totems or personal distinctions, and are no more symbols than are the distinctive shoulder-straps of army officers. The *crux ansata* and the circle formed by a snake biting its tail are symbols, but consensus as well as invention was necessary for their establishment, and our Indians have produced nothing so esoteric, nothing which they intended for hermeneutic as distinct from mnemonic purposes. Sign-language can undoubtedly be employed to express highly metaphysical ideas, indeed is so employed by educated deaf-mutes, but to do that in a system requires a development of the mode of expression consequent upon a similar development of the mental idiocrasy of the gesturers far beyond any yet found among historic tribes north of Mexico. A very few of their signs may at first appear to be symbolic, yet even those on closer examination will probably be relegated to the class of emblems, as was the case of that for "Partisan" given by the Prince of WIED. By that title he meant, as indeed was the common expression of the Canadian voyageurs, a leader of an occasional or volunteer war party, and the sign he reports as follows: "Make first the sign of the pipe, afterwards open the thumb and index-finger of the right hand, back of the hand outward, and move it forward and upward in a curve." This is explained by the author's account in a different connection, that to become recognized as a leader of such a war party as above mentioned, the first act among the tribes using the sign was the consecration, by fasting succeeded by feasting, of a medicine pipe without ornament, which the leader of the expedition afterward bore before him as his badge of authority, and it therefore naturally became an emblematic sign. There may be interest in noting that the "Calendar of the Dakota Nation" (Bulletin U. S. G. and G. Survey, vol iii, No 1), gives a figure (No. 43, A. D. 1842) showing "One Feather," a Sioux chief who raised in that year a large war party against the Crows, which fact is simply denoted by his holding out demonstratively an unornamented pipe. The point urged is that while any sign or emblem can be converted by convention into a symbol, or be explained as such by perverted ingenuity, it is futile to seek for symbolism in the stage of aboriginal development, and to interpret the conception of particular signs by that form of psychologic exuberance were to fall into mooning mysticism. This was shown by a correspondent of the present writer, who enthusiastically lauded the Dakota Calendar (edited by the latter, and a mere figuration of successive occurrences) as a numerical exposition of the great doctrines of the Sun religion in the equations of time, and proved to his own satisfaction that our Indians preserved hermeneutically the lost geometric cultus of pre-Cushite scientists. He might as well have deciphered it as the tabulated dynasties of the pre-Adamite kings.

A lesson was learned by the writer as to the abbreviation of signs, and the possibility of discovering the original meaning of those most obscure, from the attempts of a Cheyenne to convey the idea of *old man*. He held his right hand forward, bent at elbow, fingers and thumb closed sidewise. This not conveying any sense he found a long stick, bent his back, and supported his frame in a tottering step by the stick held, as was before only imagined. There at once was decrepit age dependent on a staff. The principle of abbreviation or reduction may be illustrated by supposing a person, under circumstances forbidding the use of the voice, seeking to call attention to a particular bird on a tree, and failing to do so by mere indication. Descriptive signs are resorted to, perhaps suggesting the bill and wings of the bird, its manner of clinging to the twig with its feet, its size by seeming to hold it between the hands, its color by pointing to objects of the same hue; perhaps by the action of shooting into a tree, picking up the supposed fallen game, and plucking feathers. These are continued until understood, and if one sign or group of signs proves to be successful that will be repeated on the next occasion by both persons engaged, and when becoming familiar between them and others will be more and more abbreviated. To this degree only, when the signs of the Indians have from ideographic form become demotic, are they conventional, and none of them are arbitrary, but in them, as in all his actions, man had at first a definite meaning or purpose, together with method in their after changes or modifications. The formation and reception of signs upon a generally understood principle, by which they may be comprehended when seen for the first time, has been before noticed as one of the causes of the report of a common code, as out of a variety of gestures, each appropriate to express a particular idea, an observer may readily have met the same one in several localities.

It were needless to suggest to any qualified observer that there is in the gesture-speech no organized sentence such as is integrated in the languages of civilization, and that he must not look for articles or particles or passive voice or case or grammatic gender, or even what we use as a substantive or a verb, as a subject or a predicate, or as qualifiers or inflexions. The sign radicals, without being specifically any of our parts of speech, may be all of them in turn. He will find no part of grammar beyond the pictorial grouping which may be classed under the scholastic head of syntax, but that exception is sufficiently important to make it desirable that specimens of narratives and speeches in the exact order of their gesticulation should be reported. The want before mentioned, of a sufficiently complete and exact collection of tales and talks in the sign-language of the Indians, leaves it impossible to dwell now upon their syntax, but the subject has received much discussion in connection with the order of deaf-mute signs as compared with oral speech, some notes of which, condensed from the speculations of VALADE and others, are as follows:

In mimic construction there are to be considered both the order in which the signs succeed one another and the relative positions in which they are made, the latter remaining longer in the memory than the former, and spoken language may sometimes in its early infancy have reproduced the ideas of a sign-picture without commencing from the same point. So the order, as in Greek and Latin, is very variable. In nations among whom the alphabet was introduced without the intermediary to any impressive degree of picture-writing, the order being, 1, language of signs, almost superseded by, 2, spoken language, and, 3, alphabetic writing, men would write in the order in which they had been accustomed to speak. But if at a time when spoken language was still rudimentary, intercourse being mainly carried on by signs, figurative writing was invented, the order of the figures will be the order of the signs, and the same order will pass into the spoken language. Hence LEIBNITZ says truly that "the writing of the Chinese might seem to have been invented by a deaf person." Their oral language has not known the phases which have given to the Indo-European tongues their formation and grammatical parts. In the latter, signs were conquered by speech, while in the former, speech received the yoke.

If the collocation of the figures of Indians taking the place of our sentences shall establish no rule of construction, it will at least show the natural order of ideas in the aboriginal mind and the several modes of inversion by which they pass from the known to the unknown, beginning with the dominant idea or that supposed to be best known. So far as studied by the present writer the Indian sign-utterance, as well as that natural to deaf-mutes, appears to retain the characteristic of pantomime in giving first the principal figure, and in adding the accessories successively, the ideographic expressions being in the ideological order.

As of sentences so of words, strictly known as such, there can be no accurate translation. So far from the signs representing words as logographs, they do not in their presentation of the ideas of actions, objects, and events, under physical forms, even suggest words, which must be skillfully fitted to them by the glossarist and laboriously derived from them by the philologer. The use of words in formulation, still more in terminology, is so wide a departure from primitive conditions as to be incompatible with the only primordial language yet discovered. No dictionary of signs will be exhaustive for the simple reason that the signs are exhaustless, nor will it be exact because there cannot be a correspondence between signs and words taken individually. Words and signs both change their meaning from the context. A single word may express a complex idea, to be fully rendered only by a group of signs, and, *vice versâ*, a single sign may suffice for a number of words. The list annexed to the present pamphlet is by no means intended for exact translation, but as a suggestion of headings or titles of signs arranged alphabetically for mere convenience.

It will be interesting to ascertain the varying extent of familiarity with sign-language among the members of the several tribes, how large a proportion possess any skill in it, the average amount of their vocabulary, the degree to which women become proficient, and the age at which children commence its practice. The statement is made by Titchkemátski that the Kaiowa and Comanche women know nothing of the sign-language, while the Cheyenne women are versed in it. As he is a Cheyenne, however, he may not have a large circle of feminine acquaintances beyond his own tribe, and his negative testimony is not valuable. A more general assertion is that the signs used by males and females are different, though mutually understood, and some minor points of observation may be indicated, such as whether the commencement of counting upon the fingers is upon those of the right or the left hand, and whether Indians take pains to look toward the south when suggesting the course of the sun, which would give the motion from left to right

CLASSIFICATION AND ANALYSIS.

An important division of the deaf-mute signs is into *natural* and *methodical*, the latter being sometimes called artificial and stigmatized as parasitical. But signs may be artificial—that is, natural, but improved and enriched by art—and even arbitrary, without being strictly what is termed methodical, the latter being part of the instruction of deaf-mutes, founded upon spoken languages, and adapted to the words and grammatical forms of those languages. This division is not appropriate to the signs of Indians, which are all natural in this sense, and in their beauty, grace, and impressiveness. In another meaning of "natural," given by deaf-mute authorities, it has little distinction from "innate," and still another, "conveying the meaning at first sight," is hardly definite.

The signs of our Indians may be divided, in accordance with the mode of their consideration, into innate (generally emotional) and invented; into developed and abridged; into radical and derivative; and into, 1. Indicative, as directly as possible of the object intended; 2. Imitative, representing it by configurative drawing; 3. Operative, referring to actions; and 4. Expressive, being chiefly facial. As they are rhetorically as well as directly figurative, they may be classified under the tropes of metaphor, synecdoche, metonymy, and catachresis, with as much or as little advantage as has been gained by the labeling in text-books of our figures of articulate speech.

The most useful division, however, for the analysis and report with which collectors are concerned is into *single* and *compound*, each including a number of subordinate groups, examples of which will be useful. Some of those here submitted are taken from the selected list before introduced to discriminate between the alleged universality of the signs themselves and of their use as an art, and the examples of deaf-mute signs have been extracted from those given for the same purpose by Mgr. D. De Haerne in his admirable analysis of those signs, which also has been used so far as applicable. Those will be equally illustrative, both the Indian and deaf-mute signs being but dialects of a common stock, and while all the examples might be taken from the collection of Indian signs already made, the main object of the present work is to verify and correct that collection rather than to publish more of it than necessary, with possible perpetuation of error in some details.

SINGLE SIGNS.

Single signs have been often styled "simple," which term is objectionable because liable to be confounded with the idea of "plain," in which sense nearly all Indian signs, being natural, are simple They are such as show only one phase or quality of the object signified. The following are the principal forms which they take:

1. *Indication or representation of the object to be described.* This is the Indicative division before mentioned. All the signs for "I, myself" given above, are examples, and another is the wetting of the tip of the finger by deaf-mutes to indicate humidity, the species being in the latter case used for the genus.

2. *Drawing the outlines* of the object, or more generally a part of the outlines. The Imitative or configurative division of signs reappears in this class and the one following. Example: The above sign for "dog," which conforms to the outline of its head and back.

3. *Imitation of the condition or of the action.*

(*a.*) Imitation of the condition or state of being. Under this form come nearly all the designations of size and measure. See some under "Quantity," above.

(*b.*) Imitation of the action, or of activity in connection with the object. Most of the ideas which we express by verbs come in this category, but in sign-language they are as properly substantives or adjectives. They may be Imitative when the action, as of "eating," is simulated in pantomime; or Operative, as when "walking" is actually performed by taking steps; or Expressive, as when "grief," "weeping," appears in facial play.

4. *The contact had with the object, or the manner of using it.* For "break" an imaginary stick may be snapped and the two parts looked at as if separated. See above signs for "destroyed." (*Dodge.*) A knife and most other utensils are expressed by their use.

5. *One part taken for the whole, or particular signs made to represent all the signs of an object.*

This class has reference to synecdoche. The Cheyenne sign for "old age" given above is an example.

6. *How an object is produced or prepared.*

Here is metonymy representing the cause for the effect. An example may be found among us when a still wine is indicated by the action of drawing a cork from a bottle, effervescent champagne by cutting the wires, and coffee by the imaginary grinding of the berry.

7. *The place where the object is to be found,* either according to its nature or as a general rule.

Here is again the application of metonymy. Example: "White," expressed by touching the teeth; "black," the hair (which nearly always has that color among Indians); "red," the lips Articles of clothing are similarly indicated.

8. *The effect, result, influence, and moral impression of the object.*

In this class are specially comprised the substantives, adjectives, and verbs which express the dispositions and impressions of the soul.

The Expressive gesture or sign dominates here, as might be supposed. It is generally the effect for the cause, by metonymy, which is expressed. Among the signs for "good" and "bad," above given, are several examples.

COMPOUND SIGNS.

Compound signs are those which portray several sides, features, or qualities of the object designed. They are generally more developed than those which are called single, although they also can be, and in fact often are, abridged in practice.

The various categories of compound signs may be reduced to certain heads, forming the following classes:

1. Objects that are represented by a *generical or radical indication, with one or more specific marks.* Example: The deaf-mute sign for "rich," which is the generic sign for "man" and the specific sign of activity in counting out money. Under this class are arranged—

(*a.*) The attributes, either adjective or participle, employed to indicate state or parentage, whether the generical sign is expressed or understood. The signs for "offspring" and "woman," given above, combined, mean "daughter."

(*b.*) The designation of most birds and many animals. Example: The deaf-mutes for "goose" make the generic sign for "bird," viz, an imitation of flying, and add that of a waddling walk.

(*c.*) The designation of flowers and plants. Example: The deaf-mutes gesture "rose" by the sign of "flower," growing from the fingers, and the action of smelling, then the sign for "red."

2. *Several parts or specific marks.* "Hail" is shown by the sign for "white," then its falling rapidly from above and striking head, arms, &c., or by signs for "rain" and "hard."

3. *Origin or source, and use of the object* (for the object itself, by metonymy). A pen would once have been understood by the sign for "goose," before mentioned, followed by the action of writing.

4. *Effects for causes* (also by metonymy). For "wind" blow with the mouth and make with the hands the motion of the wind in a determined direction.

5. *Form and use.*

The family of signs composing this category is very numerous. The *form* is generally traced with the forefinger of the right hand in space, or by the deaf-mutes sometimes upon a surface represented by the left hand open; but the latter device, *i. e.*, of using the left hand as a supposed drafting surface, has not been reported of the Indians. The *use*, or employment, is expressed by the position of the hands or arms, or by a pantomimic movement of the whole body. A good example is "hospital," composed of "house," "sick," and "many."

6. *Outline of the object and the place where it is found.* Example: The horns drawn from the head in one of the signs given above for "deer." (*Titchkemátski.*)

7. *Shape, and one or more specific marks.* Other signs given for "deer" may be instanced.

8. *Way of using and specific marks of the object.* "Chalk" would be distinguished from "pen," before given, by the sign of "white," followed by the action of writing.

9. *Shape, mode of using, and specific marks.* "Paper" would be shown by tracing its length and breadth, if necessary by the motion of folding, succeeded by that of writing, and, to make it still more distinct, by "white."

10. *End for which an object is used, or its make, and the place where it is found.* Example: "Sword," by drawing from a supposed sheath and striking; and "milk," by signs for "white," "milking," and "drinking."

11. *Place and specific mark.* The deaf-mute shows "spider" by opening

all the fingers of both hands, pointing with the left hand to a wall, then to a corner in the wall shown by the index of the right.

12. *Place, manner of using, or mode of arrangement.* The pantomime of putting on shoes or stockings by whites or moccasins by Indians indicates those articles.

13. *Negation of the reverse of what it is desired to describe.* Examples: "Fool—no," given above, would be "wise." "Good—no," would be "bad." This mode of expression is very frequent, and has led observers to report the absence of positive signs for the ideas negatived, with sometimes as little propriety as if when an ordinary speaker chose to use the negative form "not good," it should be inferred that he was ignorant of the word "bad."

14. *Attenuation or diminution of an object stronger or greater than that which it is desired to represent,* and the converse. *Damp* would be "wet—little"; *cool,* "cold—little"; *hot,* "warm—much." In this connection it may be noted that the degree of motion sometimes indicates a different shade of meaning, of which the graduation of the signs for "bad" and "contempt" (*Matthews*) is an instance, but is more frequently used for emphasis, as is the raising of the voice in speech or italicizing and capitalizing in print. The meaning of the same motion is often modified, individualized, or accentuated by associated facial changes and postures of the body not essential to the sign, which emotional changes and postures are at once the most difficult to describe and the most interesting when intelligently reported, not only because they infuse life into the skeleton sign, but because they may belong to the class of innate expressions. Facial variations are not confined to use in distinguishing synonyms, but amazing successes have been recorded in which long narratives have been communicated between deaf-mutes wholly by play of the features, the hands and arms being tied for the experiment.

There remains to be mentioned as worthy of attention the principle of *opposition,* as between the right and left hands, and between the thumb and forefinger and the little finger, which appears among Indians in some expressions for "above," "below," "forward," "back," but is not so common as among the methodical, distinguished from the natural, signs of deaf-mutes. This principle is illustrated by the following remarks of Col. DODGE, which also bear upon the subdivision last above mentioned: "Above" is indicated by holding the left hand horizontal, and in front of the body, fingers open, but joined together, palm upward. The right hand is then placed horizontal, fingers open but joined, palm downward, an inch or more above the left, and raised and lowered a few inches several times, the left hand being perfectly still. If the thing indicated as "above" is only a *little* above, this concludes the sign, but if it be *considerably* above, the right hand is raised higher and higher as the height to be expressed is greater, until, if *enormously* above, the Indian will raise his right hand as high as possible, and, fixing his eyes on the zenith, emit a duplicate grunt, the more prolonged as he desires to express the greater height. All this time the left hand is held perfectly motionless. "Below" is exactly the same, except that all movement is made by the left or lower hand, the right being held motionless, palm downward, and the eyes looking down.

The code of the Cistercian monks was based in large part on a system of opposition which would more likely be wrought out by an intentional process of invention than by spontaneous figuration, and is rather of mnemonic than suggestive value They made two fingers at the right side of the nose stand for "friend," and the same at the left side for "enemy," by some fanciful connection with right and wrong, and placed the little finger on the tip of the nose for "fool" merely because it had been decided to put the forefinger there for "wise man."

DETAILS OF DESCRIPTION AND ILLUSTRATION.

The signs of the Indians appear to consist of motions rather than positions—a fact enhancing the difficulty both of their description and illustration—and the motions are generally large and free, seldom minute. It seems also to be the general rule among Indians as among deaf-mutes that the point of the finger is used to trace outlines and the palm of the hand to describe surfaces. From an examination of the identical signs made for the same object by Indians of the same tribe and band to each other, they appear to make most gestures with little regard to the position of the fingers and to vary in such arrangement from individual taste. Some of the elaborate descriptions, giving with great detail the attitude of the fingers of any particular gesturer and the inches traced by his motions, are of as little necessity as would be a careful reproduction of the flourishes of tailed letters and the thickness of down-strokes in individual chirography when quoting a written word. The fingers must be in *some* position, but that is frequently accidental, not contributing to the general and essential effect, and there is a custom or "fashion" in which not only different tribes, but different persons in the same tribe gesture the same sign with different degrees of beauty, for there is calligraphy in sign-language, though no recognized orthography. It is nevertheless better to describe and illustrate with unnecessary minuteness than to fail in reporting a real differentiation. There are, also, in fact, many signs formed by mere positions of the fingers, some of which are abbreviations, but in others the arrangement of the fingers in itself forms a picture. An instance of the latter is one of the signs given for the "bear," viz, middle and third finger of right hand clasped down by the thumb, fore and little finger extended crooked downward. (*Titchkemátski.*) This reproduction of the animal's peculiar claws, with the hand in any position relative to the body, would suffice without the pantomime of scratching in the air, which is added only if it should not be at once comprehended. In order to provide for such cases of minute representation a sheet of "Types of Hand Positions" has been prepared, and if none of them exactly correspond to a sign observed, the one most nearly corresponding can be readily altered by a few strokes of pen or pencil. The sheet of "Outlines of Arm Positions," giving front and side figures with arms pendent, is also presented as a labor-saving device. The directions upon these sheets as illustrated by the sheet of "Examples," which concludes this pamphlet, are, it is hoped, sufficiently ample to show their proposed use, and copies of them, to any requisite number, will cheerfully be mailed, together with official stamps for return postage on contributions, by application to the address given below.

LIST OF SIGNS DESIRED.

The following is a condensed list, prepared for the use of observers, of the headings under which the gesture-signs of the North American Indians have been collated for comparison with each other and with those of deaf-mutes and of foreign tribes of men, and not intended to be translated into a mere vocabulary, the nature of the elementary principles governing the combinations in the two modes of expression being diverse. Many synonyms have been omitted which will readily fall into place when a sign for them may be noticed, and it is probable that many of them, depending upon the context and upon facial expression will be separately distinguished only with great difficulty. Even when the specific practice of the sign-language has been discontinued, the gesture formerly used for a sign as substitute for words may survive as a customary accompaniment to oratory or impassioned conversation, therefore should be noted. The asterisk prefixed to some of the words indicates those for which the signs or gestures made are specially desired—in some cases for their supposed intrinsic value, and in others on account of the incompleteness of their description as yet obtained, but it is not intended that signs corresponding with the words without an asterisk will not be welcomed. Observers should only regard this list as suggestive, and it is hoped, will add all signs that may be considered by them to be of interest. Those for many animals and utensils, weapons, articles of clothing, and similar common objects, have been omitted from the list because the number of them of a merely configurative or pantomimic character in the present collection was sufficient in comparison with their value, but when any distinct conception for them in signs is remarked it should be contributed.

Printed forms and outlines similar to those shown at the end of this pamphlet, prepared to diminish the labor of description and illustration, will be furnished on request mailed to

COL. GARRICK MALLERY, *U. S. A.*,
Bureau of Ethnology, Smithsonian Institution,
P. O. Box 585, *Washington, D. C.*

---

*Above.
Add, To; more.
Admiration.
Anger.
Arrow.
Arrow, To hit with an.
Autumn, fall.
Battle.
Bear.
Beaver.
*Before.
*Beginning; commencement.
*Behind.
*Below; under.

Big.
Bison, (buffalo.)
Black.
Blue.
Boat, canoe.
Bow, weapon.
Brave.
Break, broken.
Bring to me; or to us.
Broad.
Brother.
Capture, To.
Chief.
——, War.
Child; baby, infant.
——, offspring.
Clear.
Clothing; buffalo-robe or skin.
——, woolen blanket.
Cloud.
Cold; it is cold.
Come; arrive; coming.
—— come back.
—— come here.
Companion.
* Comparison; more, most.
* Contempt.
Content, satisfaction.
* Cross; sulky.
* Danger.
Daughter.
Day.
—— to-day.
—— to-morrow.
—— yesterday.
Dead; death.
Deer.
* Defiance.
* Destroyed, ruined.
* Different, contrasted.
Discontent, dissatisfaction.
* Disgust.
Dog.
Drink; drinking.
Earth, ground.
East.
End, done.
Enough.
Equal.
Exchange.
Fail, To.
Far.
Fat, of a person.
Fat, of meat.
Fear.
——, a coward; cowardice.
Female, applied to animals.
Fight.
Fire, flame.
Flat.
Flour.
Fly, To.
Fool, foolish.
Forest.
* Forever, always.
Forget; forgotten.
Found; discovered.
Friend.
Frost.
Full, as a box or sack.
* Future, to come (in time).
Gap; cañon.
* Generous.
Girl.
Give, to me or to us.
Glad; joy.
Go; go away.
God.
Good.
* Gone; departed.
* ——; lost, spent.
Grandmother.
Grass.
* Gray.
Grease.
Great.
Green.
* Grief, sorrow.
* Grow, To.
Gun.
——, To hit with a.
Gun shot.
Hair.
Halt!
* Halt; a stopping-place.
Hard.
* Hate.
He; another person; they.
Hear, heard.
Heavy.
* Help, To; to assist.
* Here.
Hide; to conceal; secret.
High; as a hill.
Hill.
* Honest.
* Horror.
* Humble, humility.
Hunting, for game.
Husband.
I; personal pronoun.
Ice.
* Imprudent, rash.
* In; within.
Indecision, doubt.
Kill, killing.
Kind.
Know, To.
—— I know.
—— I do not know.
Lance; spear.
Large, great in extent.
—— in quantity.
*Leaves, of a tree.
Lie, falsehood.
Lie, down.
Light, daylight.
—— in weight.
Lightning.
Listen, To.
Little; small in quantity.
——, in size.
Lodge; tepee; wigwam.
—— Entering a.
Long, in extent of surface.
—— in lapse of time.
Look! See!

Look, To.
Love, affection.
Male, applied to animals.
Man.
Many.
Marching, traveling.
* Medicine-man, Shaman.
Medicine in Indian sense.
Mine; my property.
Moon, month.
Morning.
Mother.
Mountain.
Much.
Near.
Negro.
Night.
No, negative.
None; I have none.
Nothing.
Now.
Number; quantity.
Obtain.
Old.
Opposite.
Out; outward; without (in position).
Paint.
Parent.
* Past; over (in time).
Patience.
Peace.
Pistol.
Poor, lean.
* ——, indigent.
Prairie.
Prayer.
Pretty; handsome.
* Pride.
Prisoner.
——, To take.
Property; possession; have; belong.
* Prudent, cautious.
Question; inquiry; what?
Rain.
Red.
Repeat, often.
Retreat; return through fear.
Ridge.
River.
Rocky, as a hill.
Run; running.
Same, similar.
Scalp.
Search, to seek for.
See, To; seeing.
Seen.
* Shame; ashamed.
* Short, in extent.
* Short, in time.
Sick, ill.
Sing.
Sister.
Sit down.
* Slave, servant.
Sleep.
Slow.
Small.
Snow.
Soft.
Son.
Sour.
Speak, To.
* Spring (season).
Steamboat.
* Stingy.
Stone.
Storm.
Strong, strength.
* Submission.
* Summer.
Sun.
Sunrise.
Sunset.
Surprise.
Surrender.
Surround.
Sweet.
Swift.
Talk, conversation.
* Time.
Taste.
Think.
Thunder.
Time of day; hour.
* —— a long time.
* —— a short time.
Tired, weary.
Told me, A person.
Tomahawk; ax.
Trade, barter, buy.
Travel, To.
* Tree.
True, It is.
Truth.
* Try, To; to attempt.
Understand.
Understand, Do not.
* Vain, vanity.
* Village, Indian.
* ——, White man's
War.
War, To declare.
Water.
* Well, in health.
When?
Whence?
Where?
White.
White man; American.
Wicked; bad heart.
Wide, in extent.
Wife.
* Wild.
Wind, air in motion.
* Winter.
* Wise; respected for wisdom.
* Wish; desire for.
Without; deprivation.
Woman; squaw.
Wonder.
Work, To; to perform.
Year.
* Yellow.
Yes, affirmation.
You.

OUTLINES FOR ARM POSITIONS IN GESTURE-LANGUAGE.

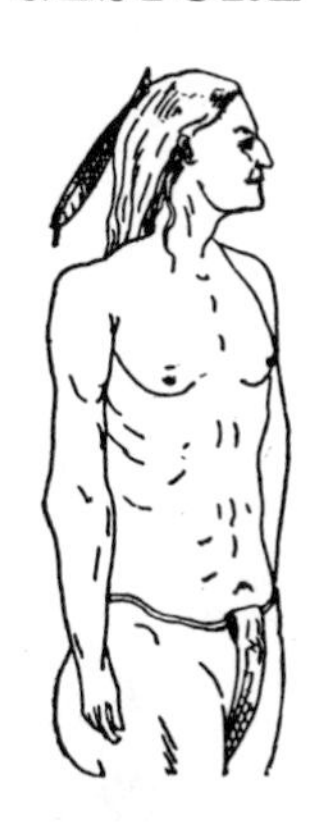

N. B.—The gestures, to be indicated by corrected positions of arms and by dotted lines showing the motion from the initial to the final positions (which are severally marked by an arrow-head and a cross—see sheet of EXAMPLES), will be always shown as they appear to an observer facing the gesturer, the front or side outline, or both, being used as most convenient. The special positions of hands and fingers will be designated by reference to the "TYPES OF HAND POSITIONS." For brevity in the written description, "hand" may be used for "right hand," when that one alone is employed in any particular gesture. In cases where the conception or origin of any sign is not obvious, if it can be ascertained or suggested, a note of that added to the description would be highly acceptable. Associated facial expression or bodily posture which may accentuate or qualify a gesture is necessarily left to the ingenuity of the contributor.

Word or Idea expressed by Sign:..

DESCRIPTION:

....

....

....

....

....

CONCEPTION OR ORIGIN:

....

Tribe:..

Locality:..

....

Observer.

TYPES OF HAND POSITIONS IN GESTURE-LANGUAGE.

A—Fist, palm outward, horizontal.

B—Fist, back outward, oblique upward.

C—Clinched, with thumb extended against forefinger, upright, edge outward.

D—Clinched, ball of thumb against middle of forefinger, oblique, upward, palm down.

E—Hooked, thumb against end of forefinger, upright, edge outward.

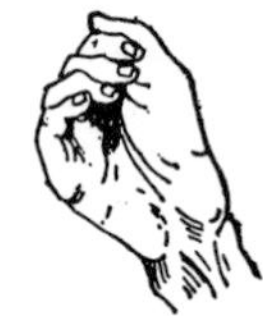

F—Hooked, thumb against side of forefinger, oblique, palm outward.

G—Fingers resting against ball of thumb, back upward.

H—Arched, thumb horizontal against end of forefinger, back upward.

I—Closed, except forefinger crooked against end of thumb, upright, palm outward.

J—Forefinger straight, upright, others closed, edge outward.

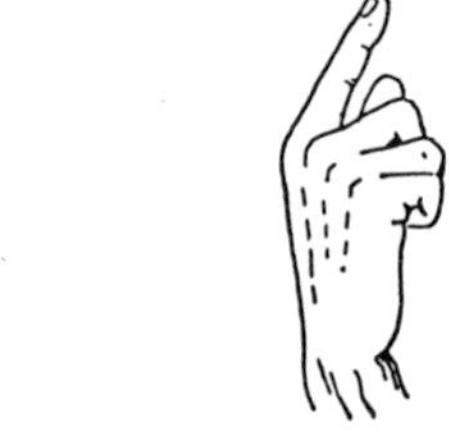

K—Forefinger obliquely extended upward, others closed, edge outward.

L—Thumb vertical, forefinger horizontal, others closed, edge outward.

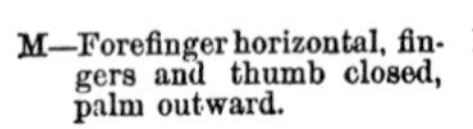

M—Forefinger horizontal, fingers and thumb closed, palm outward.

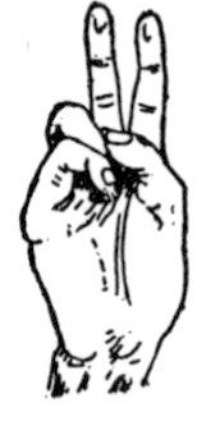

N—First and second fingers straight upward and separated, remaining fingers and thumb closed, palm outward.

O—Thumb, first and second fingers separated, straight upward, remaining fingers curved edge outward.

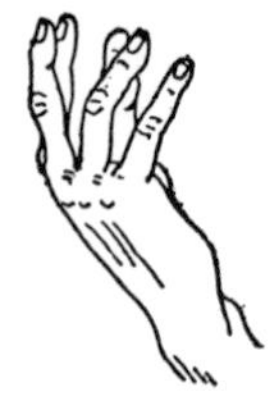

P—Fingers and thumb partially curved upward and separated, knuckles outward.

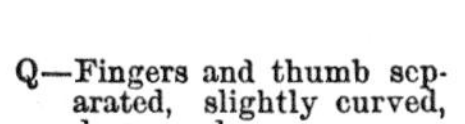

Q—Fingers and thumb separated, slightly curved, downward.

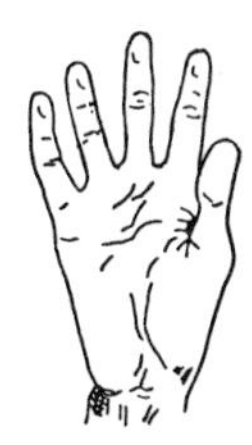

R—Fingers and thumb extended straight, separated, upward.

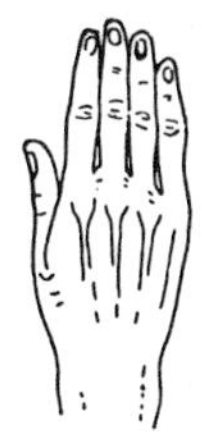

S—Hand and fingers upright, joined, back outward.

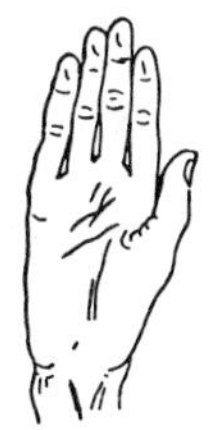

T—Hand and fingers upright, joined, palm outward.

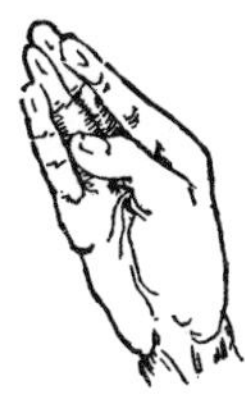

U—Fingers collected to a point, thumb resting in middle.

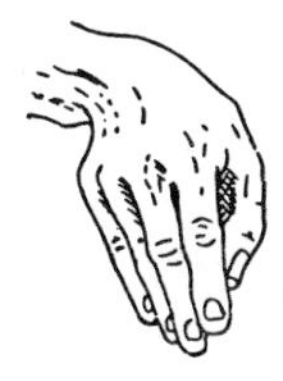

V—Arched, joined, thumb resting near end of forefinger, downward.

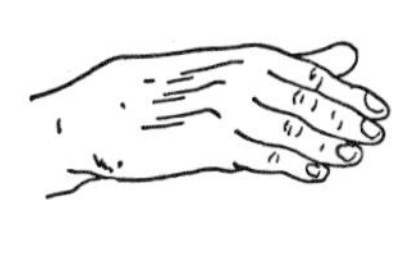

W—Hand horizontal, flat, palm downward.

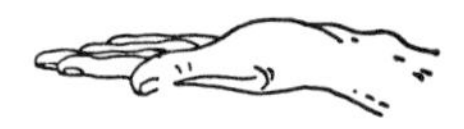

X—Hand horizontal, flat, palm upward.

Y—Naturally relaxed, normal; used when hand simply follows arm with no intentional disposition.

N. B.—The positions are given as they appear to an observer facing the gesturer, and are designed to show the relations of the fingers to the hand rather than the positions of the hand relative to the body, which must be shown by the outlines (see sheet of "OUTLINES OF ARM POSITIONS") or description. The right and left hands are figured above without discrimination, but in description or reference the right hand will be understood when the left is not specified. The hands as figured can also with proper intimation be applied with changes either upward, downward, or inclined to either side, so long as the relative positions of the fingers are retained, and when in that respect no one of the types exactly corresponds with a sign observed, modifications will be made by pen or pencil on that one of the types found most convenient, as indicated in the sheet of "EXAMPLES," and referred to by the letter of the alphabet under the type changed, with the addition of a numeral—*e. g.*, A 1, and if that type, *i. e.* A, were changed a second time by the observer (which change would necessarily be drawn on another sheet of types), it should be referred to as A 2.

EXAMPLES.

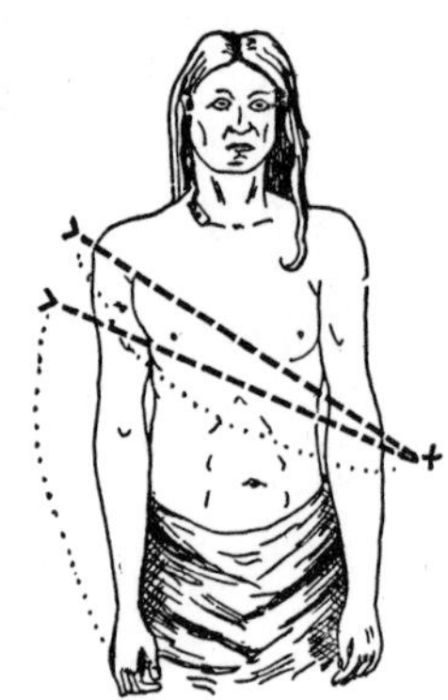

Word or idea expressed by sign: To cut, with an ax.

DESCRIPTION:

With the right hand flattened (X changed to right instead of left), palm upward, move it downward to the left side repeatedly from different elevations, ending each stroke at the same point.

Conception or origin: From the act of felling a tree.

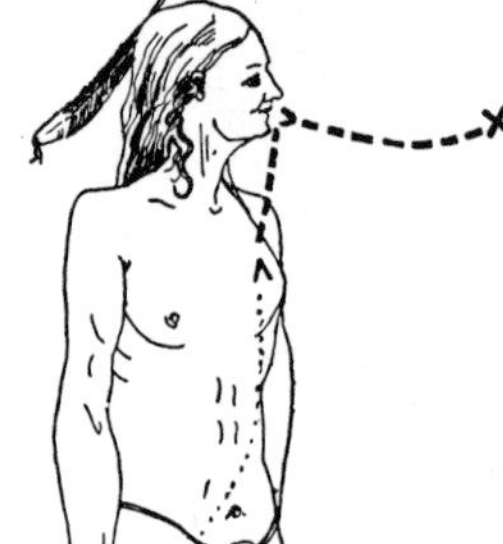

Word or idea expressed by sign: A lie.

DESCRIPTION:

Touch the left breast over the heart, and pass the hand forward from the mouth, the two first fingers only being extended and slightly separated (L, 1—with thumb resting on third finger).

Conception or origin: Double-tongued.

L, 1.

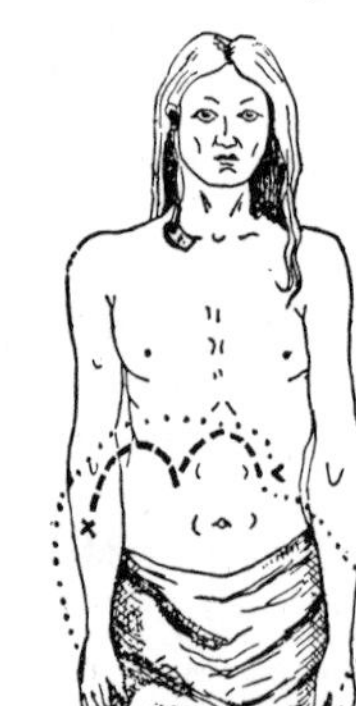

Word or idea expressed by sign: To ride.

DESCRIPTION:

Place the first two fingers of the right hand, thumb extended (N, 1) downward, astraddle the first two joined and straight fingers of the left (T, 1), sidewise, to the right, then make several short arched movements forward with hands so joined.

Conception or origin: The horse mounted and in motion.

N, 1.

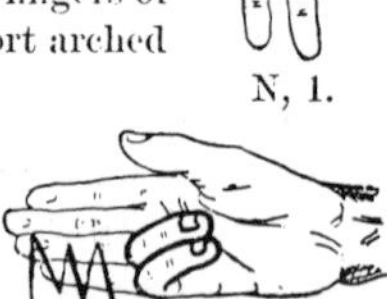

T, 1.

.... Dotted lines indicate movements to place the hand and arm in position to commence the sign and not forming part of it.

> Indicates commencement of movement in representing sign, or part of sign.

---------- Dashes indicate the course of hand employed in the sign.

X Represents the termination of movements.

⟶ Used in connection with dashes, shows the course of the latter when not otherwise clearly intelligible.

A COLLECTION

OF

GESTURE-SIGNS

AND SIGNALS

OF THE

NORTH AMERICAN INDIANS

WITH

SOME COMPARISONS

BY

GARRICK MALLERY

BREVET LIEUT. COL. AND FORMERLY ACTING CHIEF SIGNAL OFFICER, U. S. ARMY

A COLLECTION OF GESTURE-SIGNS OF THE NORTH AMERICAN INDIANS.

INTRODUCTORY LETTER.

SMITHSONIAN INSTITUTION,
BUREAU OF ETHNOLOGY,
Washington, D. C., July 31, 1880.

TO THE COLLABORATORS WITH THE BUREAU OF ETHNOLOGY IN THE STUDY OF SIGN LANGUAGE:

GENTLEMEN:

This paper contains the descriptions of the gesture-signs of the North American Indians which at the above date have been obtained by this Bureau. It will not be used for publication in its present shape, and will be distributed only to those correspondents who have contributed to its contents, and to others whose expected co-operation, the results of which are not yet received, is relied upon to add value to the final work. No discussion is now introduced. The descriptions alone, in a tentative arrangement, are presented for the purpose of the verification of observations, for verbal corrections of every kind, and for the study of all collaborators, as well as that of the editor, to secure accurate classification and comparison. Only such notes of resemblance or discordance between several of the Indian signs, and between some of them and those of deaf mutes, foreign tribes of men, and ideographic characters, are now printed as have already been attached to the same signs in the compilation for preliminary treatises already produced. It is convenient to retain those in the same connection. Many others of the same kind, remaining in MS. memoranda, are omitted, because their insertion will be more correctly made after the proper arrangement has been accomplished. Any such, occurring to collaborators, will, it is hoped, be suggested by them in the margin of the present paper where they may seem to be most appropriate.

The primary object of this paper is that every contributor to it may be enabled to revise his own contribution, which for the present is divided and arranged according to a scheme of linguistic families and subordinate languages or tribes, as set forth in the LIST OF AUTHORITIES AND COLLABORATORS, which also serves as an index to the VOCABULARY. It is supposed that this arrangement will prove the best to study the diversities and agreements of signs. For that important object it is more convenient that the names of the tribe or tribes among which the signs described have been observed should catch the eye in immediate connection with the signs, than that those of the observers only should follow. Some of the latter, indeed, having given both similar and differing signs for more than one tribe, the use of the contributor's name alone would create confusion. To print in every case the name of the contributor, and also the name of the tribe, would seriously burden the paper and be unnecessary to the student, the reference being readily made to each authority through the index.

No contribution has been printed which asserted that any described sign is used by "all Indians," for the reason that such statement is not admissible evidence unless the authority had personally examined "all Indians." If any credible correspondent had affirmatively stated that a certain identical, or substantially identical, sign had been found by him, actually used by Abenaki, Absaroka, Arikara, Assiniboins, etc., going through the whole list of tribes, or any definite portion of that list, it would have been so inserted under the several tribal heads. But the expression "all Indians," besides being insusceptible of methodical classification, involves hearsay, which is not the kind of authority desired in a serious study. Such loose talk long delayed the recognition of anthropology as a science. It is true that some general statements of this character are made by some old authors now quoted, but their descriptions are reprinted, as being all that can be used of the past, for whatever weight they may have, and they are kept separate from the linguistic classification.

Contributors will observe that there has been no attempt to change their phraseology even when it seemed to be defective. Besides the ordinary errors of the press, and those that may have crept into the copy by mistakes in reading or transcribing the written descriptions, some of the contributors will probably share the common experience of surprise at the extent to which details of expression and punctuation, when in the severe clearness of print, have altered the shade of meaning as intended to be conveyed in their MS. The wide margins and calendered paper will readily allow even of recomposition of sentences when desirable. For this purpose, as well as several others, this paper will be regarded by each correspondent as simply a proof-sheet sent directly to himself from the printer, and it will of course be understood that a correspondent who may make any kind of correction or note upon this paper will return it by mail (as book proof), so annotated, to the undersigned, thereby saving correspondence and securing accuracy. It is indeed requested that all copies shall be returned whether annotated or not, in order to prevent a professedly imperfect edition from falling into improper hands. It is much regretted that the illustrations and diagrammatic aids to the descriptions, furnished by most of the contributors, cannot be reproduced in this paper, so that their accuracy also might be determined, but the cost of such illustrations cannot be incurred at this time and for this purpose. The "Outlines for Arm Positions" and "Types of Hand Positions" were provided for from the appropriation for this Bureau, but its amount does not admit of such an undertaking as now in question. In this connection it may be mentioned that the descriptions frequently refer to illustrations furnished by the contributors or to the "Outlines" and "Types," and these references are retained in print. As all the contributors remember their own illustrations, etc.. the references will be intelligible to themselves, though unfortunately not always to others who might wish to compare them with their illustrations.

The ascertainment of the conceptions or origin of the several signs, embodying as they do, many sociologic, mythologic, and other ethnographic ideas, is of special importance. When those obtained through collaborators are printed in the VOCABULARY before the authority, they are to be understood to have been gathered from an Indian as being his own conception. When printed after the authority and within quotation marks, they are in the words of the collaborator as offered by himself. When printed after the authority and without quotation marks, they are suggested at this Bureau. All should be equally criticised and supplemented, and any error in printing the authority for the conceptions corrected. It has sometimes been impossible to decide whether the correspondent intended to give them as his own or as from an Indian. The importancee of an Indian's conception is so much greâter than any other that the fact should be made clear.

The margins will also allow of additions to all contributions, whether from intervening independent research or as suggested by any part of the material collected. This work being on the co-operative principle, it is not supposed that jealousies or questions of precedence will arise, and each contributor will be credited with the amount of capital advanced for the common stock. It is highly desirable that the signs as described by each should be compared by him with those of others, and notes of coincidence or discrepancy made. Perhaps, in some instances, the signs as described by one of the other contributors may be recognized as intended for the same sign for the same idea or object as that of the correspondent, and the former may prove to be the better description. The personal habitude of some individual in any tribe, and still more frequently the usage or "fashion" of different tribes, may, by a peculiar abbreviation or fanciful flourish, have induced a differentiation in description with no real distinction either in conception or essential formation. All collaborators will therefore be candid in admitting, should such cases occur, that their own descriptions are mere unessential variants from others printed, otherwise adhere to their own and explain the true distinction. When the descriptions show substantial identity, they will in the final publication be united, with a combined reference to all the authorities giving them, as they are in some cases of those taken at Washington in the present VOCABULARY.

It will probably be also noticed that a sign described will have the same actually substantive formation as some other in the VOCABULARY which is stated to be with a signification so markedly distinguished as to be insusceptible of classification as a synonym. It will then be important for each contributor of the rival signs to refresh his memory as to accuracy of description or significance, or both, and to announce his decision. No error is necessarily involved. It will be very remarkable if precisely the same sign does not prove to be used by different persons or bodies of people with wholly distinct significations, the graphic forms for objects and ideas being much more likely to be coincident than sound is for similar expressions, yet in all oral languages the same precise sound is used for utterly diverse meanings. The first conception of many objects must be the same. It has been found, indeed, that the homophony of words and the homomorphy of ideographic pictures is noticeable in opposite significations, the conceptions arising from the opposition itself. The differentiation in portraiture or accent is a subsequent and remedial step taken only after the confusion has been observed. Such confusion and contradiction would only be eliminated if the sign language were absolutely perfect as well as absolutely universal. Cast-iron inflexibility and adamantine endurance are certainly not found in any other mode of human utterance. It will be an abnormity in the processes of nature if signs do not have their births and deaths, their struggles for existence with survival of the fittest, as well as words, animals, and plants. For our purpose the inquiry is not what a sign might, could, would, or should be, or what is the best sign for a particular meaning, but what is any sign actually used for such meaning. If any one sign is honestly invented or adopted by any one man, whether Indian, African, Asiatic, or deaf-mute, it has its value. Its prevalence and special range present considerations of different and greater interest and requiring further evidence.

The editor takes occasion to declare that—for the good reason that his real study only now commences with the completion of the present paper which renders it practicable—he does not hold with tenacity any theory whatever, and particularly one which would deny that the Indian signs come from a common stock. On the contrary, it would be highly interesting to ascertain that the signs of this continent had a generic distinctiveness compared with those of other parts of the world. Such research would be similar to that into the Aryan and Semitic sources to which certain modern languages have been traced backwards from existing varieties, and if there appear to be existing varieties in signs their roots may still be found to be *sui generis*. It is, however, possible that the discrepancy between signs was formerly greater than at present. There is some evidence that where a sign language is now found among Indian tribes it has become more uniform than ever before, simply because many tribes have been for sometime past forced to dwell near together at peace. The use of signs, though maintained by linguistic diversities, is not coincident with any linguistic boundaries. The tendency is to their uniformity among groups of people who from any cause are brought into contact with each other while still speaking different languages. The longer and closer such contact, while no common tongue is adopted, the greater will be the uniformity of signs. A collection was obtained last spring at Washington from a united delegation of the Kaiowa, Comanche, Apache, and Wichita tribes, which was nearly uniform, but the individuals who gave the signs had actually lived together at or near Anadarko, Indian Territory, for a considerable time, and the resulting uniformity of their signs might either be considered as a jargon or as the natural tendency to a compromise for mutual understanding—the unification so often observed in oral speech, coming under many circumstances out of former differentiation. It may be found that other individuals of those same tribes who have from any cause not lived in the union explained may have signs for the same ideas different from those in the collection above mentioned; but this supposition should be disregarded, except to incite further inquiry, until such inquiry should collect specific facts to support the hypothesis. The whole of this controversy may be disposed of by insisting upon an objective instead of a subjective observation and study. Our duty is to collect the facts as they are, and so soon as possible, as every year will add to the confusion and difficulty. After the facts are established the theories will take care of themselves, and their final enunciation will be in the hands of men more competent than any of us, perhaps than any persons now living.

A warning seems necessary since the publication of an article in the number of "United Service" for July, 1880, in which the author takes the ground that the description of signs should be made according to a "mean" or average. There can be no philosophic consideration of signs according to a "mean" of observations. The final object is to ascertain the radical or essential part as distinct from any individual flourish or mannerism on the one hand, and from a conventional or accidental abbreviation on the other; but a mere average will not accomplish this object. If the hand, being in any position whatever, is, according to five observations, moved horizontally one foot to the right, and, according to five other observations, moved one foot horizontally to the left, the "mean" or resultant will be that it is stationary, which is not in any way corresponding with any of the ten observations. So if six observations give it a rapid motion of one foot to the right and five a rapid motion of the same distance to the left, the mean or resultant would be somewhat difficult to express, but perhaps would be a slow movement to the right for an inch or two, having certainly no resemblance either in essentials or accidents to any of the signs actually observed. In like manner the tail of the written letter "*y*" (which, regarding its mere formation, might be a graphic sign) may have, in the chirography of several persons, various degrees of slant, may be a straight line or looped, and may be curved on either side; but a "mean" taken from several manuscripts would leave the unfortunate letter without any tail whatever, or travestied as a "*u*" with an amorphous flourish. A definition of the radical form of the letter or sign by which it can be distinguished from any other letter or sign is a very different proceeding. Therefore, if a "mean" or resultant of any number of radically different signs to express the same object or idea, observed either among several individuals of the same tribe or among different tribes, is made to represent those signs, they are all mutilated or ignored as distinctive signs, though the result may possibly be made intelligible in practice, according to principles mentioned in the "Introduction to the Study of Sign Language" of the present writer; and still another view may be added, that because a sound of broken English may be understood by an intelligent Englishman it is no proof of that sound being an English word or a word of any language. The adoption of a "mean" may be practically useful in the formation of a mere interpreter's jargon, though no one can use it but himself or those who memorize it from him, but it elucidates no principle. It is also practically convenient for any one determined to argue for the uniformity and universality of sign-language as against the variety apparent in all the realms of nature. On the "mean" principle, he only needs to take his two-foot rule and arithmetical tables and make all signs his signs and his signs all signs. Of course they are uniform, because he has made them so after the brutal example of Procrustes.

In this connection it is proper to urge another warning, that a mere sign-talker is often a bad authority upon principles and theories. He may not be liable to the satirical compliment of Dickens's "brave courier," who "understood all languages indifferently ill"; but many men speak some one language fluently, and yet are wholly unable to explain or analyze its words and forms so as to teach any one else, or even to give an intelligent summary or classification of their own knowledge. What such a sign-talker has learned is by memorizing, as a child may learn English, and though both the sign-talker and the child may be able to give some separate items useful to a philologist or foreigner, such items are spoiled when colored by the attempt of ignorance to theorize. A German who has studied English to thorough mastery, except in the mere facility of speech, may in a discussion upon some of its principles be contradicted by any mere English speaker, who insists upon his superior knowledge because he actually speaks the language and his antagonist does not, but the student will probably be correct and the talker wrong. It is an old adage about oral speech that a man who understands but one language understands none. The science of a sign-talker possessed by a restrictive theory is like that of Mirabeau, who was greater as an orator than as a philologist, and who on a visit to England gravely argued that there was something seriously wrong in the British mind because the people would insist upon saying "give me some bread" instead of "donnez moi du pain," which was so much easier and more natural. A designedly ludicrous instance to the same effect was Hood's arraignment of the French because they called their mothers "mares" and their daughters "fillies." Not binding ourselves to theories, we should take with caution any statement from a person who, having memorized or hashed up any number of signs, large or small, has decided in his conceit that those he uses are the only genuine simon-pure, to be exclusively employed according to his direction, all others being counterfeits or blunders. His vocabulary has ceased to give the signs of any Indian or body of Indians whatever, but becomes the vocabulary of Dr. Jones or Lieutenant Smith, the proprietorship of which he fights for as did the original Dr. Townsend for his patent medicine. When a sign is contributed by one of the present collaborators, which such a sign-talker has not before seen or heard of, he will at once condemn it as bad, just as a United States Minister to Vienna, who had been nursed in the mongrel Dutch of Berks County, Pennsylvania, declared that the people of Germany spoke very bad German. The experience of the present editor is that the original authorities, or the best evidence, for Indian signs—*i. e.*, the Indians themselves—being still accessible, the collaborators in this work should not be content with secondary authority. White sign-talkers and interpreters may give some genuine signs, but they are very apt to interpolate their own inventions and deductions. By gathering the genuine signs alone we will be of use to scholars, and give our own studies proper direction, while the true article presented can always be adulterated into a composite jargon by those whose ambition is only to be sign-talkers instead of making an honest contribution to ethnologic and philologic science. The few direct contributions of interpreters to the present work are, it is believed, valuable, because they were made without expression of self-conceit or symptom of possession by a pet theory.

So far as only concerns the able gentlemen who have favored this Bureau with their contributions there is no need to continue these remarks. Suffice it to repeat with more emphasis, that their criticisms and suggestions are invited as to all matter herein contained, even to the details of grouping and title-words in the alphabetic arrangement, synonyms, and cross references. In the present private and tentative work many hundreds of separate slips of paper are for the first time connected together, thereby rendering perfect order unexpected. It may be mentioned that some of the title-words and phrases which have a quaint appearance are those used by the older printed authorities, for which it is not always safe to supply a synonym, and the signs of those same authorities being the most curtly and obscurely described of all in the collection, there is no alternative but to print them as they stand for such use as may be possible, which will chiefly be in their bearing upon the questions of persistency and universality. The present edition will allow the verbal expressions of the living and accessible to be revised and to be compared with, thus perhaps to correct the imperfections of descriptions made by the dead and inaccessible; but the language of the latter cannot now be changed. The arrangement of the VOCABULARY is more to group the concepts than the English title-words according to the synonyms of that language. A further step in the study will be to prepare a synoptic arrangement of the signs themselves—that is, of motions and positions of the same character apart from their individual significance in any oral speech.

The hearty thanks of this Bureau are rendered to all its collaborators, and will in future be presented in a manner more worthy of them. It remains to give to them an explanation of the mode in which a large collection of signs has been made in Washington. Fortunately for this undertaking, the policy of the government has brought here, during the last winter and spring, delegations, sometimes quite large, of most of the important tribes. Thus the most intelligent of the race from the most distant and farthest separated localities were here in considerable numbers for weeks, and indeed, in some cases, months, and, together with their interpreters and agents, were, by the considerate order of the honorable Secretary of the Interior, placed at the disposal of this Bureau for all purposes of gathering ethnologic information. The facilities thus obtained were much greater than could have been enjoyed by a large number of observers traveling for a long time over the continent for the same express purpose. The observations relating to signs were all made here by the same persons, according to a uniform method, in which the gestures were obtained directly from the Indians, and their meaning (often in itself clear from the context of signs before known) was translated sometimes through the medium of English or Spanish, or an aboriginal language known in common by some one or more of the Indians and by some one of the observers. When an interpreter was employed, he translated the words used by an Indian, and was not relied upon to explain the signs according to his own ideas. Such translations and a description of minute and rapidly-executed signs, dictated at the moment of their exhibition, were sometimes taken down by a phonographer, that there might be no lapse of memory in any particular, and in many cases the signs were made in successive motions before the camera, and prints secured as certain evidence of their accuracy. Not only were more than one hundred Indians thus examined individually, at leisure, but, on occasions, several of different tribes, who had never before met each other, were examined at the same time, both by inquiry of individuals whose answers were consulted upon by all the Indians present, and also by inducing several of the Indians to engage in talk and story-telling in signs between themselves. Thus it was possible to notice the difference in the signs made for the same objects and the degree of mutual comprehension notwithstanding such differences. Similar studies were made by taking the Indians to the National Deaf Mute College and bringing them in contact with the pupils.

By far the greater part of the actual work of the observation and record of the signs obtained at Washington has been ably performed by Dr. W. J. HOFFMAN, the assistant of the present editor. Dr. Hoffman acquired in the West, through his service as acting assistant surgeon, United States Army, at a large reservation, the indispensable advantage of becoming acquainted with the Indian character so as to conduct such researches as that in question, and in addition has the eye and pencil of an artist, so that he catches readily, describes with physiological accuracy, and reproduces in action and in permanent illustration all shades of gesture exhibited. It is therefore believed that the collection made here will be valuable for comparison with and to supplement those obtained during the same months in the field.

I remain, with renewed official and personal thanks and much regard,

Sincerely your friend,

GARRICK MALLERY.

NOTE SPECIALLY ADDRESSED TO CORRESPONDENTS IN FOREIGN COUNTRIES.

The present paper is a further step in the general line of research indicated in the "Introduction to the Study of Sign Language among the North American Indians," &c., in which the study of these signs was suggested as important to illustrate the gesture-speech of mankind. Its contents may be useful to collaborators in all parts of the world, both to facilitate description by annotated reference and in suggestion as regards modes of observation. It may also give assurance of thorough and painstaking work at this Bureau for the final collation, in the form of a vocabulary, of all authentic signs, ancient and modern, found in any part of the world, with their description, as also that of associated facial expression, set forth in language so clear that, with the assistance of copious illustrations, they can be reproduced by the reader. The success of this undertaking will depend upon the collaboration, now and before requested, of many persons of several classes. The present paper shows that arrangements have already been made probably sufficient to procure all the gesture-signs of the aboriginal tribes of this country which can still be rescued from oblivion. The conventional signs of deaf-mutes in institutions for their instruction are accessible to the present writer, who also has obtained a large number of the natural signs of deaf-mutes invented by them before systematic instruction, and used in intercourse with their families and friends. More of these would, however, be gladly received. Further assistance is urgently sought from philologists, travelers, and missionaries, whose attention has been directed to the several modes of expressing human thought.

The efforts at intercommunication of all savage and barbaric tribes, when brought into contact with other bodies of men not speaking an oral language common to both, and especially when uncivilized inhabitants of the same territory are separated by many linguistic divisions, should in theory resemble the devices of the North American Indians. They are not shown by published works to prevail in the Eastern hemisphere to the same extent and in the same manner as in North, and also, as believed from less complete observation, in South America. It is, however, probable that they exist in many localities, though not reported, and also that some of them survive after partial or even high civilization has been attained, and after changed environment has rendered their systematic employment unnecessary. Such signs may be, first, unconnected with existing oral language, and used in place of it; second, may be used to explain or accentuate the words of ordinary speech, or may consist of gestures, emotional or not, which are only noticed in oratory or impassioned conversation, such being, possibly, survivals of a former gesture-language.

All classes of gestures may be examined philologically to trace their possible connection with the radicals of language, syllabaries, and ideographic characters. Evidence has accumulated to show that the language of signs preceded in importance that of sounds, the latter remaining rudimentary long after gesture had become an art. The early connection between them was so intimate that gestures, in the wide sense of presenting ideas under physical forms, had a formative effect upon many words, thus showing that language originated partly, at least, from the sounds which naturally accompany certain gestures. It seems certain that the latter exhibit the earliest condition of the human mind, and that mainly through them was significance communicated to speech.

Even if the more material and substantive relations between signs and language cannot now be ascertained, we may at least expect, from the inquiries suggested, linguistic results in the analogy between their several developments. The mental processes are nearly the same in both cases, and the psychology of language may be studied in the older and lower means of communication as the physical and mental organization of man has been profitably compared with that of the lower animals. The examination of signs and of picture-writing, which is intimately associated with them, throws light upon the grammatic machinery of language, the syntactic principle, and the genesis of the sentence. Not until a large body of facts has been gathered by several classes of observers, and compared by competent scholars, can it be possible to ascertain with precision the principles of the primitive utterance of mankind. An exhaustive treatment of the subject will also bring to light religious, sociologic, and other ethnologic information of special interest. It is in this work that the Bureau of Ethnology of the Smithsonian Institution solicits the co-operation of learned men and observers in all lands, whose contributions, when received, will always be published with individual credit as well as responsibility.

G. M.

LIST OF AUTHORITIES AND COLLABORATORS.

---

1. A list prepared by WILLIAM DUNBAR, dated Natchez, June 30, 1800, collected from tribes then "west of the Mississippi," but probably not from those very far west of that river, published in the Transactions of the American Philosophical Society, vol. vi, as read January 16, 1801, and communicated by Thomas Jefferson, president of the society.

2. The one published in "An Account of an Expedition from Pittsburgh to the Rocky Mountains, performed in the years 1819–1820. By order of the Hon. J. C. Calhoun, Secretary of War, under the command of Maj. S. H. LONG, of the United States Topographical Engineers." Philadelphia, 1823. (Commonly called James' Long's Expedition.) This appears to have been collected chiefly by Mr. T. Say, from the Pani, and the Kansas, Otos, Missouris, Iowas, Omahas, and other southern branches of the great Dakota family.

3. The one collected by Prince MAXIMILIAN von WIED-NEUWIED in 1832–'34. His statement is "the Arikaras, Mandans, Minnitarris [Hidatsa], Crows [Absaroka], Cheyennes, Snakes [Shoshoni], and Blackfeet [Satsika] all understand certain signs, which, on the contrary, as we are told, are unintelligible to the Dakotas, Assiniboins, Ojibwas, Krihs [Crees], and other nations. The list gives examples of the sign language of the former." From the much greater proportion of time spent and information obtained by the author among the Mandans and Hidatsa then and now dwelling near Fort Berthold, on the Upper Missouri, it might be safe to consider that all the signs in his list were in fact procured from those tribes. But as the author does not say so, he is not made to say so in this work. If it shall prove that the signs now used by the Mandans and Hidatsa more closely resemble those on his list than do those of other tribes, the internal evidence will be verified. This list is not published in the English edition, but appears in the German, Coblenz, 1839, and in the French, Paris, 1840. Bibliographic reference is often made to this distinguished explorer as "Prince Maximilian," as if there were but one possessor of that christian name among princely families. For brevity the reference in this paper will be "*Wied.*"

No translation of this list into English appears to have been printed in any shape before that recently published by the present editor in the American Antiquarian, vol. ii, No. 3, while the German and French editions are costly and difficult of access, so the collection cannot readily be compared by observers with the signs now made by the same tribes. The translation now presented is based upon the German original, but in a few cases where the language was so curt as not to give a clear idea, was collated with the French edition of the succeeding year, which, from some internal evidence, appears to have been published with the assistance or supervision of the author. Many of the descriptions are, however, so brief and indefinite in both their German and French forms that they necessarily remain so in the present translation. The princely explorer, with the keen discrimination shown in all his work, doubtless observed what has escaped many recent reporters of aboriginal signs, that the latter depend much more upon motion than mere position—and are generally large and free—seldom minute. His object was to express the general effect of the motion rather than to describe it so as to allow of its accurate reproduction by a reader who had never seen it. For the latter purpose, now very desirable, a more elaborate description would have been necessary, and even that would not in all cases have been sufficient without pictorial illustration.

On account of the manifest importance of determining the prevalence and persistence of the signs as observed half a century ago, an exception is made to the general arrangement hereafter mentioned by introducing after the *Wied* signs, remarks of collaborators who have made special comparisons, and adding to the latter the respective names of those collaborators—as (*Matthews*)-(*Boteler*). It is hoped that the work of these gentlemen will be imitated not only regarding the *Wied* signs but many others.

4. That of Capt. R. F. BURTON, of signs which, it would be inferred, were collected in 1860–'61 from the tribes met or learned of by him on the overland stage route, including Southern Dakotas, Utes, Shoshoni, Arapahos, Crows, Pani, and Apaches. This is contained in "The City of the Saints," New York, 1862.

Information is recently received to the effect that this collection was not made by the distinguished English explorer from his personal observation, but was obtained by him from one man, a Morman bishop, who, it is feared, gave his own ideas of the usage of signs rather than their simple description.

5. A list read by Dr. D. G. MACGOWAN, at a meeting of the American Ethnological Society, Jan. 23, 1866, and published in the "Historical Magazine," vol. x, 1866, p. 86–87, purporting to be the signs of the Caddos, Wichitas, and Comanches.

6. A communication from Brevet Col. RICHARD I. DODGE, Lieutenant-Colonel Twenty-third Infantry, United States Army, author of "The Plains of the Great West and their Inhabitants," &c., relating to his large experience with the Indians of the prairies. Colonel Dodge, now on active duty, has been requested to assign his general descriptions to the tribe or tribes in which the signs were actually observed by him, and should such designation arrive, while the VOCABULARY is passing through the press, they will be classified accordingly. He is also preparing a larger contribution.

NOTICE

In the six collections above mentioned the generality of the statements as to locality of the observation and use of the signs, rendered it impossible to arrange them in the manner explained in the "Introductory Letter" hereto. They will therefore be referred to in the VOCABULARY by the names of the authors responsible for them. Those which now follow are arranged alphabetically by tribes, under headings of Linguistic Families, which are also given below in alphabetical order. Example: The first authority is under the heading ALGONKIAN, and, concerning only the Arapaho tribe, is referred to as (*Arapaho* I), Lieutenant LEMLY being the personal authority.

References to another title-word as explaining a part of a description or to supply any other portions of a compound sign will always be understood as being made to the description by the same authority of the sign under the other title-word. Example: In the sign for **Advance and Retreat** (*Mandan and Hidatsa* I) the reference to **Battle** is to that sign for **Battle** which is contributed by Dr. MATTHEWS, and is referred to under that title as (*Mandan and Hidatsa* I).

ALGONKIAN.

Arapaho I. A contribution from Lieut. H. R. LEMLY, Third United States Artillery, compiled from notes and observations taken by him in 1877 among the Northern Arapahos.

Cheyenne I. A list prepared in July, 1879, by Mr. FRANK H. CUSHING, of the Smithsonian Institution, from continued interviews with TITCHKEMÁTSKI (Cross Eyes), an intelligent Cheyenne, then employed at that Institution. It is expected that Mr. Cushing will make other contributions, especially from the Zuñi and other Pueblos, among whom he has been collecting material during the past year.

Cheyenne II. A special contribution with diagrams from Mr. BEN CLARK, scout and interpreter, of signs collected from the Cheyennes during his long residence among that tribe.

Ojibwa I. The small collection of J. G. KOHL, made about the middle of the present century, among the Ojibwas around Lake Superior. Published in his "Kitchigami. Wanderings around Lake Superior," London, 1860.

Ojibwa II. Notes from Very Rev. EDWARD JACKER, Pointe St. Ignace, Mich., respecting the Ojibwa.

Ojibwa III. A communication from Rev. JAMES A. GILFILLAN, White Earth, Minn., relating to signs observed among the Ojibwas during his long period of missionary duty, still continuing.

Ojibwa IV. A list from Mr. B. O. WILLIAMS, Sr., of Owosso, Mich., from recollection of signs observed among the Ojibwas of Michigan sixty years ago.

Sac, Fox, and Kickapoo I. A list from Rev. H. F. BUCKNER, D. D., of Eufaula, Ind. T., consisting chiefly of tribal signs observed by him among the Sac and Fox, Kickapoos, &c., during the early part of the year 1880.

DAKOTAN.

Absaroka I. A list of signs obtained from DEÉKITSHIS (Pretty Eagle), ETSHIDI-KAHOTSHKI (Long Elk), and PERITSHIKADIA (Old Crow), members of a delegation of Absaroka or Crow Indians from Montana Territory, who visited Washington, D. C., during the months of April and May, 1880.

Dakota I. A comprehensive list, arranged with great care and skill, from Dr. CHARLES E. MCCHESNEY, acting assistant surgeon, United States Army, of signs collected among the Dakotas (Sioux) near Fort Bennett, Dakota, during the last winter and spring. Dr. MCCHESNEY requests that recognition should be made of the valuable assistance rendered to him by Mr. WILLIAM FIELDEN, the interpreter at Cheyenne Agency, Dakota Territory.

Dakota II. A short list from Dr. BLAIR D. TAYLOR, assistant surgeon, United

States Army, from recollection of signs observed among the Sioux during his late service in the region inhabited by that tribe.

Dakota III. A special contribution from Capt. A. W. CORLISS, Eighth United States Infantry, of signs observed by him during his late service among the Sioux.

Dakota IV. A copious contribution with diagrams from Dr. WILLIAM H. CORBUSIER, assistant surgeon, United States Army, of signs obtained from the Ogalala Sioux at Pine Ridge Agency, Dakota Territory, during 1879–'80.

Dakota V. A report of Dr. W. J. HOFFMAN, from observations among the Teton Dakotas while acting assistant surgeon, United States Army, and stationed at Grand River Agency, Dakota, during 1872–'73.

Dakota VI. A list of signs obtained from PEZHĪ (Grass), chief of the Blackfoot Sioux; NAZULATANKA (Big Head), chief of the Upper Yanktonais; and TSHITOUAKIA (Thunder Hawk), chief of the Uncpapas, Teton Dakotas, located at Standing Rock, Dakota Territory, while at Washington, D. C., in June, 1880.

Dakota VII. A list of signs obtained from SHÚNGKA LÚTA (Red Dog), an Ogalala chief from the Red Cloud Agency, who visited Washington in company with a large delegation of Dakotas in June, 1880.

Hidatsa I. A list of signs obtained from TSHESHACHADACHISH (Lean Wolf), chief of the Hidatsa, located at Fort Berthold, Dakota Territory, while at Washington, D. C., with a delegation of Sioux Indians in June, 1880.

Mandan and Hidatsa I. A discriminating and illustrated contribution from Dr. WASHINGTON MATTHEWS, assistant surgeon, United States Army, author of "Ethnography and Philology of the Hidatsa Indians," &c., lately prepared from his notes and recollections of signs observed during his long service among the Mandan and Hidatsa Indians of the Upper Missouri.

Omaha I. A special list from Rev. J. OWEN DORSEY, missionary at Omaha Agency, Nebraska, from observations lately made by him at that agency.

Oto I. An elaborate list, with diagrams, from Dr. W. C. BOTELER, United States Indian service, collected from the Otos at the Oto Agency, Nebraska, during 1879–'80.

Oto and Missouri I. A similar contribution by the same author respecting the signs of the Otos and Missouris, of Nebraska, collected during the winter of 1879–'80, in the description of many of which he has been joined by Miss KATIE BARNES.

Ponka I. A short list from Rev. J. OWEN DORSEY, lately obtained by him from the Ponkas in Nebraska.

IROQUOIAN.

Iroquois I. A list of signs contributed by the Hon. HORATIO HALE, author of "Philology" of the Wilkes Exploring Expedition, &c., now residing at Clinton, Ontario, Canada, obtained in June, 1880, from SAKAYENKWARATON (Disappearing Mist), familiarly known as John Smoke Johnson, chief of the Canadian division of the Six Nations or Iroquois proper, now a very aged man, residing at Brantford, Canada.

Wyandot I. A list of signs from HÉNTO (Gray Eyes), chief of the Wyandots, who visited Washington, D. C., during the spring of 1880, in the interest of that tribe, now located in Indian Territory.

KAIOWAN.

Kaiowa I. A list of signs from SITTIMGEA (Stumbling Bear), a Kaiowa chief from Indian Territory, who visited Washington, D. C., in June, 1880.

KUTINEAN.

Kutine I. A letter from J. W. POWELL, Esq., Indian superintendent, British Columbia, relating to his observations among the Kutine and others.

PANIAN.

Arikara I. A list of signs obtained from KUANUCHKNAUIUCH (Son of the Star), chief of the Arikaras, located at Fort Berthold, Dakota Territory, while at Washington, D. C., with a delegation of Indians in June, 1880.

SAHAPTIAN.

Sahaptin I. A list contributed by Rev. G. L. DEFFENBAUGH, of Lapwai, Idaho, giving signs obtained at Kamiah, Idaho, chiefly from FELIX, chief of the Nez Percés, and used by the Sahaptin or Nez Percés.

SHOSHONIAN.

Comanche I. Notes from Rev. A. J. HOLT, Denison, Texas, respecting the Comanche signs, obtained at Anadarko, Indian Territory.

Comanche II. Information obtained at Washington, in February, 1880, from Maj. J. M. HAWORTH, Indian inspector, relating to signs used by the Comanches of Indian Territory.

Comanche III. A list of signs obtained from KOBI (Wild Horse), a Comanche chief from Indian Territory, who visited Washington, D. C., in June, 1880.

Pai-Ute I. Information obtained at Washington from NÁTSHES, a Pai-Ute chief, who was one of a delegation of that tribe to Washington in January, 1880.

Shoshoni and Banak I. A list of signs obtained from TENDOY (The Climber), TISIDIMIT, PETE, and UIAGAT, members of a delegation of Shoshoni and Banak chiefs from Idaho, who visited Washington, D. C., during the months of April and May, 1880.

Ute I. A list of signs obtained from ALEJANDRO, GALOTE, AUGUSTIN, and other chiefs, members of a delegation of Ute Indians of Colorado, who visited Washington, D. C., during the early months of the year 1880.

TINNEAN.

Apache I. A list of signs obtained from HUERITO (Little Blonde), AGUSTIN VIJEL, and SANTIAGO LARGO (James Long), members of a delegation of Apache chiefs from Tierra Amarilla, New Mexico, who visited Washington, D. C., in the months of March and April, 1880.

Apache II. A list of signs obtained from NAKANANITAIN (White Man), an Apache chief from Indian Territory, who visited Washington in June, 1880.

Apache III. A large collection made during the present summer by Dr. FRANCIS H. ATKINS, acting assistant surgeon, United States Army, from the Mescalero Apache, near South Fork, New Mexico. This MS. was received after the whole of the VOCABULARY had gone to the printer, and a large part actually printed, so it was not possible to insert all of the descriptions in the present edition. The interesting "Narrative" communicated by Dr. ATKINS is printed with similar matter following the VOCABULARY.

WICHITAN.

Wichita I. A list of signs from Rev. A. J. HOLT, missionary, obtained from KIN CHĒ-ĔSS (Spectacles), Medicine-man of the Wichitas, at the Wichita Agency, Indian Territory, in 1879.

Wichita II. A list of signs from TSODIÁKO (Shaved Head Boy), a Wichita chief, from Indian Territory, who visited Washington, D. C., in June, 1880.

ZUÑIAN.

Zuñi I. Some preliminary notes lately received from Rev. TAYLOR F. EALY, missionary among the Zuñi, upon the signs of that body of Indians.

Grateful acknowledgment must be made to Prof. E. A. FAY, of the National Deaf Mute College, through whose special attention a large number of the natural signs of deaf-mutes, remembered by them as having been invented and used before instruction in conventional signs, indeed before attending any school, was obtained, which are printed in this paper. The gentlemen who made the contributions in their own MS., and without prompting, are as follows: Messrs. M. BALLARD, R. M. ZEIGLER, J. CROSS, PHILIP J. HASENSTAB, —— LARSON. Their names will follow the several descriptions. Mr. BALLARD is now the teacher in the primary school of the college, and the other gentlemen were students during the last session.

Special thanks are also rendered to Prof. JAMES D. BUTLER, of Madison, Wis., for contribution of Italian gesture-signs, noted by him in 1843, and for many useful suggestions.

A small collection of AUSTRALIAN signs has been extracted from *The Aborigines of Victoria*, by R. BROUGH SMYTH, vol. ii, pp. 4–5, 308–9. London, 1878. Upon these the author makes the following curious remarks: "It is believed that they have several signs, known only to themselves, or to those among the whites who have had intercourse with them for lengthened periods, which convey information readily and accurately. Indeed, because of their use of signs, it is the firm belief of many (some uneducated and some educated) that the natives of Australia are acquainted with the secrets of Freemasonry."

VOCABULARY.

Abide. See **Stay.**

Above.

Place the right hand, by an ascending motion, upon the left hand, both extended, fingers joined and palms down. (*Arapaho* I.)

Point with fore finger of right hand raised from the side to the heavens above with extended arm quickly. (*Ojibwa* IV.)

Thumb and forefingers of both hands extended, pointing upright (other fingers closed) in front of body, level of breast, back of hand outward, and then held in this position, left hand outside of and higher than the right. The sign can also be made with one hand, by moving it after being held at the lower height to the higher one, and holding it there a short time. (*Dakota* I.) "Superior height—one person or thing above another."

With the back of the hand toward the right, the fingers bent at right angles with the palm and pointing toward the left, push the right hand from in front of the chin upward until it is a little higher than the head. (*Dakota* IV.)

Raise the hand very quickly above the head, palm to the front, and a little back of the head. (*Omaha* I.)

The right hand, with the index only extended, is elevated before the head. (*Comanche* I.)

The left flat hand is held in front of the body at the height of the elbow, palm down, the right similarly placed, over, and a little higher than the left. To express greater elevation the right hand is raised. (*Ute* I.)

——— A little.

Hold the left hand horizontal and in front of the body, fingers open but joined together, palm upward. The right hand is then placed horizontal, fingers open but joined, palm downward, an inch or more above the left, and raised or lowered a few inches several times, the left hand being perfectly still. (*Dodge*.)

Above, considerably.

Place the hands as in **Above, a little,** then raise the right hand higher and higher, as the height to be expressed is greater, until, if *enormously above,* the right hand is raised as high as possible; fixing the eyes toward the zenith, emit a duplicate grunt, the more prolonged the greater the height is expressed. The left hand must continue motionless during all this. (*Dodge.*)

Ache. See **Pain.**

Across.

Pass the hand, flattened and either partially or entirely extended, from the breast, forward, upward, and downward, forming an arch to the front. (*Absaroka* I; *Shoshoni and Banak* I.)

——— On the other side of.

Elevate the left fist palm down before the face, and pass the flat and extended right across the back of the left beginning at the thumb, sliding it down on the outer side so as to turn the tips of the fingers of the right nearly in toward the palm of the left. (*Ute* I.)

Add to; to put in some more; to add or put to. (Compare **Counting.**)

With right hand make downward motion as though to take up something; (2) move suddenly over to left as though depositing the some thing in it; to add one, the three first fingers are clasped down by the thumb. To add two, the little and third finger extended. To add three, the little and middle fingers extended. To add four, all the fingers extended. To add much, many, sign as before. (*Cheyenne* I.)

Hold the left hand in front of body scoop fashion with back of hand downward, then with the right hand held in the same scoop manner, with palm downward make a sort of diving motion downward and outward and lastly inward and upward to the left hand, as though gathering imaginary objects and putting them in the left hand. (*Dakota* I.) "Gathering and adding to."

With the fingers and thumb of the right hand, pretend to pick some imaginary object in the direction of the locality of the desired object, placing it near the body, and repeating the gesture several times. (*Dakota* VII.)

Bring the point of the extended forefingers together before the breast. (*Omaha* I.)

Left hand extended palm downward (**W**). Bring right hand directly over left hand downward, and let the right hand palm downward fall upon the back of the extended left hand. (*Comanche* I.) "Piling up furs."

Admiration, action of admiring; surprise; wonder. (Compare **Pretty, content.**)

Placing the hand upon the mouth, to show that language is inadequate to communicate their sensations. (*Long.*)

Hold the hollow hand for some time before the mouth. Perhaps the idea being that the mouth, widely open in amazement, is concealed beneath it, and it being improper to display emotion or admiration, the open mouth is concealed by the hand. (*Ojibwa* I.)

(1) Face turned to the right; (2) eyebrows elevated and contracted; (3) right hand lifted with fingers carelessly or loosely extended; (4) brought suddenly toward the mouth. To express *surprise* as distinct from *admiration* make the following sign: Eyebrows contracted and elevated, eyes indicating interest, right hand fingers outspread, elevated to side and front of eyes, gently oscillated. (*Cheyenne* I.)

Arms are crossed in front of body, the hands (**S**) pressing against the right and left breasts, which pressure is alternately relaxed and renewed. At the same time pleasure is expressed by facial emotion. (*Dakota* I.) "Almost involuntary on seeing an object or thing they admire."

Deaf-mute natural sign.—Draw one palm along upon the other; then press them against your breast, directly opposite the heart, making at the same time your face look like trying to kiss. (*Cross.*)

Adulation.

Italian sign.—The mouth kissing the hand—by which Job described a species of idolatry—is a species of adulation practiced by every cringing servant in Italy. (*Butler.*)

Advance and Retreat. (Compare **Battle.**)

With the upright hands about four inches apart, palms facing, fingers separated a little and semi-flexed, in front of the upper part of the chest, while all the fingers are in motion, move the left hand away from the right about four inches and bend it backward until its palm looks obliquely upward toward the right, and at the same time throw the right hand toward it and partly over it; then move the right hand away from the left and reverse the position of the hands. Repeat this maneuver several times. (*Dakota* IV.)

The hands held as in the sign for **Battle,** then their relations to one another remaining unchanged, they are moved alternately from side to side to represent the alternate advances and retreats of opposing forces. (*Mandan and Hidatsa* I.)

After. See **Before.** (*Cheyenne* II.)

Affection. See **Love.**

Affirmative. See **Yes.**

Aged. Old man.

Place the clinched right hand in front of the shoulder, a foot or so from it, palm to the left, then push it forward a few inches, drawing it back at a lower level. This is done three or four times, and the body is inclined to the front at the same time. (*Absaroka* I; *Shoshoni and Banak* I.) "Grasping a staff for assistance in locomotion."

Place the clinched fist in front of and nearly as high as the shoulder, then push it forward repeatedly, drawing it back at a lower level (*Dakota* VI.) "Imitates walking with a staff."

Hold the right fist in front of the right side nearly as high as the shoulder, move it forward and bring it back a little lower, repeating the motion several times. (*Kaiowa* I; *Comanche* III; *Apache* II; *Wichita* II.) "Walking with a staff."

Ahead. See **Before.**

Air. See **Wind.**

Alike. See **Same.**

Alive.

Right finger whirled upward. (*Macgowan.*)

The right hand, back upward, is to be at the height of the elbow and forward, the index extended and pointing forward, the other fingers closed, thumb against middle finger; then, while rotating the hand outward, move it to a position about four inches in front of the face, the back looking forward and the index pointing upward. (*Dakota* IV.)

——— Just alive, almost dead

The same motion as for **Alive,** but the index is to be bent and thumb placed against the palmar surface of the first joint of the index. (*Dakota* IV.)

All.

Move the right hand, palm downward, in a large circle, horizontally, two feet in front of the face, or move both hands in the same manner. (*Dakota* IV.)

Always; forever.

Pass the right hand, flat and extended, edgewise from the head outward toward the right, in two movements, the palm at last pointing to the right. The eyes are directed upward at the same time. (*Wyandot* I.)

Make the sign for **Time, a long,** and **long ago.** (*Comanche* III.) "The informant was requested to give the sign for the above word, but as none was known, he said the nearest approach to the idea would be expressed by making the signs meaning *long ago, a long time.*"

Same as the sign for **Earth, the.** (*Dakota* I.) "Without end."

American. See **White man.**

Among.

Bring the fingers and thumb of the left hand nearly together, so as to form an interrupted circle having a diameter of an inch or more, then introduce the extended index as far as the second joint. (*Kaiowa* I; *Comanche* III; *Apache* II; *Wichita* II.) "In the midst of others"

Anger. (Compare **Bad heart** and **Sad.**)

The fingers and thumb of the right hand with the ends together and near the breast, then turn the hand round two or three times so as to describe vertical circles, indicating that the heart is disturbed. (*Long.*)

Close the fist, place it against the forehead, and turn it to and fro in that position. (*Burton.*)

(1) Motion as if to touch the right breast with the right hand to express "self," "I;" (2) fingers partially closed, thumb resting on the tips and extending across the hand; (3) sudden motion forward and slightly to the right; (4) fingers sprung wide open; (5) rapidly shaken to and fro. (*Cheyenne* I.) "Shaking off, deprecating."

Close the right hand as if grasping a small object, hold it several inches before the forehead, and twist it around toward the left. (*Absaroka* I; *Shoshoni and Banak* I)

With the right hand, fist (**B** 2) raised to the upper part of the face in front, strike down on the left side of the body to the level of the stomach. **Very much angered,** repeat this movement several times, and express it by contraction of the muscles of the face. (*Dakota* I.)

The elbow in front of and as high as the shoulder, then strike with the fist across the face to the left, the face expressing impatience. (*Dakota* III.)

Touch the chest over the heart two or three times with the ends of the fingers of the right hand; then make the sign for bad. (*Dakota* IV.) "Heart bad."

The right arm is elevated and in type-position (**A**), brought to forehead above right eye and twisted spirally from right to left; motion ending with fists palm outward. (*Oto* I.)

Close the right hand as if grasping a small object, place it to the forehead palm down, then twist it forcibly, drawing the hand slightly to the front as if twisting off any projection. Another: The clinched right hand is twisted against the breastbone instead of the forehead. Another: When not very angry the index is slightly bent and twisted before the forehead. (*Ute* I.)

With the thumb and fingers of the right hand collected to a point, place the inner side of the hand over the heart, back up. (*Kaiowa* I; *Comanche* III; *Apache* II; *Wichita* II.)

Bring the tips of the fingers and thumb of the right hand to a point, and place them to the forehead just over the nasal eminence. (*Apache* I.)

Deaf-mute natural signs.—Make wrinkles in your face by frowning and shake your head. (*Cross.*)

Hide the face, next turn the head to another side, and then stretch down the open hand so as to indicate that the offender should be out of sight immediately. (*Hasenstab.*)

Move backward and suddenly raise the hands between the breast with a disagreeable expression of the face. (*Larson.*)

Move the lips as if speaking like a very angry man. (*Zeigler.*)

——— angry with you.

Hand closed, right index extended and points to heart; then sign for anger; then right index points to the individual intended. (*Oto* I.) "The mind turned away."

Antelope.

Pass the open right hand outward from the small of the back. (*Wied.*) This, as explained by Indians examined by the present editor, indicates the lighter coloration upon the animal's flanks. The Ute who could speak Spanish accompanied it with the word *blanco*, as if recognizing that it required explanation.

Extend and separate the forefingers and thumbs, nearly close all the other fingers, and place the hands with backs outward above and a little in front of the ears, about four inches from the head, and shake them back and forth. (*Dakota* IV.) " Antelope's horns."

With the index only extended hold the hand eighteen or twenty inches transversely in front of the head, index pointing to the left, then rub the sides of the body with the flat hands. (*Dakota* VI.) "The latter sign refers to the white sides of the animal; the former could not be explained."

Close the right hand, leaving the end of the index in the form of a hook and the thumb partly extended; then wave the hand quickly back and forth a short distance, opposite the temple. (*Hidatsa* I; *Arikara* I.) "Represents the pronged horn of the animal. This is the sign ordinarily used, but it was noticed that in conversing with one of the Dakotas (VI) the sign of the latter was used several times, to be more readily understood."

Place both hands, fingers fully extended and spread, close to the sides of the head. *Wied's* sign was readily understood as signifying the white flanks. (*Apache* I.)

Arrive. See **Come.**

Are you? See **Question.**

Arrow.

Pass the index-finger of the right hand several times across the left arm. (*Wied.*)

Expressed by notching it upon an imaginary bow and by snapping with the index and medius. (*Burton.*)

Forefinger of right hand extended, pointing upright in front of breast; back of hand out, then with the thumb and forefinger of left hand (other fingers loosely closed) rub up and down the extended forefinger of the right hand (straightening the arrow), thumb and forefinger on opposite sides, and then extend the left hand in front of body and draw the right back as though it was the arrow fixed in the string of the bow. (*Dakota* I.) "From the place of the arrow in drawing the bow."

Semi-flex the fingers and thumb of the left hand and place the hand in front of the chest with its palm inclining at an angle of 45° downward and backward and toward the right; draw the extended right index, its back forward from left to right downward and backward between the left index and thumb and along the palm, at the same time bringing the ends of the two latter together. (*Dakota* IV.) "Drawing an arrow out of the hand in which they hold their arrows."

Both arms are flexed as in position to shoot. The right arm is drawn up and flexed to an angle of 45°, with hand in position (**E** 1) modified, by palm facing the body. Left arm is extended *from* body about one foot and bent, with elbow horizontal at an angle of 80°. Left hand is in position (**G** 1) modified by reversing hand. The sign is completed by uniform movement, to and from, as of swinging-arrow, after assuming above positions. (*Oto* I.) "Clasping string and pointing arrow."

Hold the left hand as high as, and some distance in front of, the left breast, back forward, hand nearly half closed, then draw the extended index downward over the palm of the left, the whole distance extending about twenty inches. (*Kaiowa* I; *Comanche* III; *Apache* II; *Wichita* II.) "Drawing an arrow over the hand as in pulling the bow-string to shoot."

——— To hit with an.

Place the tips of the fingers downward upon the thumb, then snap them forward; then strike the hands together and elevate the index finger of the right hand. (*Wied.*) Probably when he says "strike the hands together," he wishes to describe my sign for **Shot.** When the person whom the prince saw making this sign raised the finger he may have done so to indicate a *man* shot or *one* shot. I do not think that the raising of the finger is an integral part of the sign. (*Matthews.*)

Pass the extended forefinger of the right hand (others closed) back of hand toward the right, between the fore and second fingers of the left hand, held about 18 inches in front of body, back of hand out, horizontal, all the fingers extended, and then close the two fingers of the left hand on the right index. (*Dakota* I.) "Arrow sticking in the target."

After making the sign for **Bow** strike the back of the right hand, its index extended, other fingers closed, against the palm of the opened left. (*Dakota* IV.)

Same as the sign for **Bow,** but with this addition: that after the finger has snapped from the thumb, the back of the right hand is struck against the palm of the left. (*Dakota* V.)

With the index only, extended and pointing upward, then elevate the left hand palm toward breast, first finger separated from the middle, pass the index of the right forcibly forward and through the space thus formed, striking the knuckles of the right against the palm of the left with a thud. (*Omaha* I.)

——— To kill with an.

Extend the left hand, closed, as if grasping a bow, drawing the right back toward the shoulder from the left, snap the first two fingers of the right forward from the thumb and throw the right over, toward the right and downward. (*Ute* I.)

——— To shoot with bow and.

The hands are placed as in the attitude of drawing the arrow in the bow (this is also the sign for the bow), and its departure is indicated by springing the fingers from the thumbs, as in the act of sprinkling water. (*Long.*)

Place the tips of the fingers downward upon the thumb, then snap them forward. (*Wied.*) I believe I have described this under the head of **Discharge of deadly missile.** I have always seen the same sign made for shooting both bullet and arrow. If it is necessary to distinguish the weapon, it can be done by appropriate signs in addition to this. (*Matthews.*) The left arm was then elevated, slightly bent at elbow, and extended from the body as in holding a bow. The right arm was then flexed, and the hand, in position (**G** 1), inverted as holding and pulling the string; the hands are thus uniformly swayed to and fro several times, as if holding the arrow *in situ* and taking an aim. If Wied's sign is complete, there is little resemblance to the sign among the Otos, save in the position of the right hand, which is similar, though inverted. In the Oto sign the distance between the hands would indicate length of the arrow, which is not shown in the latter. (*Boteler.*) "That which rests in the string and bow."

Make the sign for **Bow,** then right hand drawn suddenly back toward right shoulder. fingers of both hands snapped to indicate discharge of arrow and twang of string. (*Cheyenne* I.)

Extend the left arm with closed hand as if grasping the bow, draw the right hand back toward the right side of the face with the second finger resting against the thumb; then allow the finger to spring forward, move it in the direction of the object hit, and strike the backs of the fingers of the right hand against the palm of the left. (*Dakota* V.)

Same sign as *Dakota* V. (*Dakota* VII.)

Extend the left hand at a left oblique, the thumb and middle finger forming a circle, bring the right hand back to the right breast, and flip the fingers of both hands from the thumbs. (*Omaha* I.)

Arrow-head.

With the index finger of the right hand, touch the tip of the extended forefinger of the left hand several times. (*Wied.*)

Make the sign for **Arrow** and then place the right thumb, palm forward, on the last joint of the left index, its palm inward, the other fingers closed. (*Dakota* IV.)

Ashamed.

The extended right hand, palm inwards, is passed up and down two or three times, in front of the face and an inch or two from it, with a moderately rapid motion. (*Mandan and Hidatsa* I.)

——— I am.

Cross the hands in front of the face, palms backward, the right behind the left, to hide the face. (*Dakota* IV.)

Ashes.

Hands with fingers extended, brought together in front of the body, palmar surfaces of little fingers joined, and hands sloping obliquely upward with backs looking towards the sides, extended fingers pointing to the front, &c.; *i. e.*, hands held in imitation of a straight scoop, then without separating the hands carry them forward and downward with a quick upsetting motion. (*Dakota* I.) "Scooping up and throwing away ashes."

Asleep or **sleeping.**

Place the hands open over the face, close the eyes and gently press them down, at same time gently incline the head and body to the right, until attaining an angle of about 45°, or lower, remain in that position a few moments. (*Ojibwa* IV.)

Ass. See **Mule.**

Assent. See **Yes.**

Astonishment. See **Surprise.**

Attention.

Hold the right hand flat and extended, palm down, at arm's length and directed toward the person addressed, shaking it from side to side several times. (*Absaroka* I; *Shoshoni and Banak* I.)

—— To attract a person's attention previous to commencing conversation

The right hand (**T**) carried directly out in front of the body, with arm fully extended and there moved sidewise with rapid motions. (*Dakota* I.)

Aurora borealis.

First make the sign for the **Moon** in front of the body, at the level of the breast, and then on both sides of it at the same time make with both hands the sign for fire.

This requires explanation. The Sioux believe the northern lights to be the reflection of a fire built on either side of the moon to warm it. (*Dakota* I.)

Autumn.

With the thumb and forefinger of right hand describe a crescent (other fingers closed), back of closed fingers outward, hand carried in this position from above and to the left of the head in front of body toward the right and downward moderately rapid with a curved and small up and down waving motion made by turning the forearm and hand in imitation of a leaf carried from the tree by a slight breeze. (*Dakota* I.) "The falling of the leaves."

Hold the left hand upright in front of the left shoulder, fingers separated a little and slightly bent, the ends in a circle, and throw the right, its fingers separated a little and slightly bent, from immediately above the left several times downward sidewise on different sides, then, with its fingers pointing obliquely upward toward the left, carry the right hand spirally downward and obliquely toward the right, to imitate the falling of leaves. (*Dakota* IV.)

Elevate the left hand, pointing upward, before the face, palm to the right with fingers and thumb separated, then pass the right, with fingers in a similar position, upward past the left a short distance, then turn down two or more fingers of the right with the forefinger of the left, and throw the hands downward and forward with spread fingers pointing in the same direction. (*Kaiowa* I; *Comanche* III; *Apache* II; *Wichita* II.) "Growth of tree, and falling of leaves."

Awl.

The left forefinger is extended, and the right, also extended, is placed across it, and is then turned on its axis, so as to imitate the action of the awl in making a hole. (*Long.*)

(1) Thumb and three fingers of right hand clasped as though grasping handle of an awl, and the index finger extended to represent the point; (2) thrust against the palm or some portion of the left hand. (*Cheyenne* I.)

Ax. See **Tomahawk.**

——— Cutting with an. See **Cutting.**

Baby. See **Child.**

Bacon.

Separate the thumb and fingers of the right hand, the former under, the latter over the extended left hand, palms down, as if feeling its thickness. (*Arapaho* I.)

Left hand with thumb and fingers extended, joined, horizontal edge of hand downward is held in front of the left breast back outward, right hand with fingers extended and joined, thumb extended and forked is passed over the left from above with ends of fingers downward, and then the palmar surfaces of the right fingers are passed along the backs, and the thumb along the palmar surface of the left hand. (*Dakota* I.) "Thin" is also denoted by this sign. "Fat like a side of bacon."

Bad, mean.

Make the sign for **Good** and then that of **Not.** (*Long.*)

Close the hand, and open it whilst passing it downward. (*Wied.*) This is the same as my description, but differently worded; possibly notes a less forcible form. I say, however, that the hand is moved forward. The precise direction in which the hand is moved is not, I think, essential. (*Matthews.*)

Scatter the dexter fingers outward, as if sporting away water from them. (*Burton.*)

(1) Right hand partially elevated, fingers closed, thumb clasping the tips; (2) sudden motion downward and outward accompanied by equally sudden opening of fingers and snapping from the thumb, of the fingers. (*Cheyenne* I.)

Right hand closed (**B**) carried forward in front of the body toward the right and downward. during which the hand is opened, fingers downward, as if dropping out the contents. (*Dakota* I.) "Not worth keeping."

Half close the fingers of the right hand, crook the thumb over the fore and middle fingers and move the hand, back upward, a foot or so toward the object referred to, and suddenly let the fingers fly open. (*Dakota* IV.) "Scattered around, therefore bad."

Close the fingers of the right hand, resting the tips against the thumb, then throw the hand downward and outward toward the right to arm's length, and spring open the fingers. (*Dakota* VI, VII.)

Hands open, palms turned in; move one hand towards, and the other from, the body; then *vice versa.* (*Omaha* I.)

Throw the clinched right hand forward, downward, and outward, and when near at arm's length, suddenly snap the fingers from the thumb as if sprinkling water. (*Wyandot* I.) "To throw away contemptuously; not worth keeping."

Raise hand in front of breast, fingers hooked, thumb resting against second finger, palm downward (**G** 1), then with a nervous movement throw the hand downward to the right and a little behind the body, with an expression of disgust on the face. During motion of hand the fingers are gradually extended as though throwing something out of hand, and in final position the fingers and thumb are straight and separated, palm backward (**R** 1), with fingers pointing downward, palm backward. (*Sahaptin* I.) "Away with it."

Same motion of arm and hand as in **All right; Good.** But in the first position fingers are closed, and as the hand moves to the right they are thrown open, until in final position all are extended as in final for **All right; Good.** (*Sahaptin* I.)

Deaf-mute natural sign.—Use the sign for **Handsome,** at the same time shake the head as if to say **No.** (*Ziegler.*)

The Neapolitans, to express contempt, blow towards the person or thing referred to. The deaf-mutes preserve the connection of "bad" and "taste" by brushing from the side of the mouth. This may be compared with the deaf-mute sign of flipping an imaginary object between thumb nail and forefinger, denoting something small or contemptible. The motion of snapping the finger either on or from the thumb in disdain is not only of large modern prevalence in civilization, but is at least as ancient as the contemporary statute of Sardanapalus at Anchiale.

——— Very.

Sign for **Very** as in **Very good**, and sign for **Bad.** (*Sahaptin* I.)

Bad heart; Wicked. (Compare **Anger.**)

Sign for **Bad,** then folded right hand struck two or three times suddenly against the heart to imitate palpitation. (*Cheyenne* I.)

Tap several times the region over the heart with the right hand, horizontal, fingers extended, pointing toward the right, back outward, and then make the sign for **Anger.** (*Dakota* I.) "My heart is bad."

The sign most commonly used for this idea is made by the hand being closed and held near the breast, with the backs toward the breast, then as the arm is suddenly extended the hand is opened and the fingers separated from each other. (*Mandan and Hidatsa* I.)

Place the fingers of the flat right hand over the heart, then make the sign for **bad.** (*Dakota* VI, VII.)

Ball.

Middle fingers and thumbs of both hands brought together to represent a circle. (*Cheyenne* I.)

——— For gun. See **Bullet.**

Barter. See **Trade.**

Basin.

Same as **Kettle,** except final motion of setting on the fire, which is omitted. (*Dakota* I.) "From its shape."

Basket.

Interlock the separated fingers of the hands in front of body, backs outward, hands horizontal, in imitation of the interlacing of basket-work. (*Dakota* I.) "From the interwoven splinters of a basket."

Battle. (Compare **Fight** and **Kill.**)

The clinched hands are held about as high as the neck and five or six inches asunder, then waved two or three times laterally to show the advances and retreats of the combatants; after which the fingers of each hand are suffered to spring from the thumb towards each other, as in the act of sprinkling water, to represent the flight of missiles. (*Long.*)

To show that fighting is actually taking place, make the gesture of **Kill;** tap the lips with the palm like an Oriental woman when "keening," screaming the while O-a! O-a! to imitate the war song. (*Burton.*)

(1) Fists of both hands closed and brought to the level of the chin, near together and knuckles facing each other; (2) moved suddenly forward and backward with a sort of churning, grinding motion. (*Cheyenne* I.)

Both hands (**A** 1) brought to the median line of body on a level with the breast and close together; describe with both hands at the same time a series of circular movements of small circumference. (*Dakota* I.) "Two opposing forces of Indians at battle in a circular manner, after their usual custom of surrounding their enemy."

Two or more men fighting. Move the upright fists alternately several times back and forth about eight inches in front of the upper part of the chest, palms inward, and about four inches apart. (*Dakota* IV.)

The clinched fists are held before the chest, backs outward (the forearm neither prone nor supine), and passed straight up and down rapidly a distance of six to twelve inches, alternately in opposite directions to one another, either a short distance apart or with the joints touching. This indicates any angry contest or struggle as well as a pitched battle fought with weapons. (*Mandan and Hidatsa* I.)

The arms are equally flexed, as in position of defense. Hands are loosely clenched and fists rotated in palmar proximity three times; the sign is completed by retaining the arms and fists in posture for defense about three seconds. Hands are position (**B**) doubled or (**B B**). (*Oto* I.) "To ward off."

Both hands at height of breast, palms facing, the left forward from the left shoulder, the right outward and forward from the right, fingers pointing up and spread, move them alternately towards and from one another. (*Ute* I.) "Mingling of men in strife."

Place both hands on a level in front of the chest, half closed, fingers pointing downward, the backs of the right-hand fingers pointing forward, those of the left facing them, held about six inches apart, the hands are then quickly moved toward and from one another several times. (*Apache* I.)

Deaf-mute natural sign.—Vibrate your fingers, implying how many soldiers; then copy the manner of shooting a gun. (*Cross.*)

Battle, White man's.

Both hands clinched with ball of thumb on the second and third fingers, clinched forefingers of each hand touching. Then push the hands from you, letting the fingers all fly outward at the same time, as if you were trying to throw water off your fingers. This repeated twice or more, according to the severity of the engagement. (*Comanche* I.) "Soldiers standing in line delivering their fire."

—— Charge and counter-charge.

The hands are held as in the sign for **Battle**, and are then simultaneously moved from and toward one another. (*Mandan and Hidatsa* I.)

Beads, glass.

Stroke the fingers of the right hand over the left upper arm. (*Wied.*) Do not remember. A sign for necklace is sometimes made by extending the thumb and index finger and placing them against the throat. (*Matthews.*)

Tips of the joined thumb and forefinger left hand (others closed), wet by touching to the tongue, are passed down in front of the body, as though picking up loose beads from the lap, and then hand held in front of the left breast, horizontal, back outward, tips of joined thumb and finger toward the right, and then the right hand, with thumb and forefinger extended, crooked and joined at tips (others closed), as though holding a sinew, is passed toward the left, to and over the left hand as though passing the sinew through the beads. (*Dakota* I.) "From the use of beads in embroidery."

Bear, animal.

Pass the hand before the face to mean ugliness, at the same time grinning and extending the fingers like claws. (*Burton.*)

(1) Middle and third finger of right hand clasped down by the thumb, forefinger and little finger extended, crooked downward; (2) the motion of scratching made in the air. (*Cheyenne* I.)

Fingers of both hands closed, except the little finger, which is extended and pointing straight towards the front, thumbs resting on the backs of the second phalanges of all the fingers, hands horizontal, backs upward, are held in front of their respective sides near the body, and then moved directly forward with short, sharp jerking motions. (*Dakota* I.) "From the motion of the bear in running."

—— Grizzly.

Right hand flat and extended, held at height of shoulder, palm forward, then bring the palm to the mouth, lick it with the tongue, and return it to first position. (*Omaha* I.) "Showing blood on the paw."

Seize a short piece of wood, say about two feet long, wave in the right hand, and strike a blow at an imaginary person. (*Omaha* I.)

Seize a short thing about six inches long, hold it as dagger, pretend to thrust it downward under the breast bone repeatedly, and each time farther, grunting or gasping in doing so; withdraw the stick, holding it up, and, showing the blood, point to the breast with the left forefinger, meaning to say *so do thou when you meet the bear.* (*Omaha* I.)

Pretend to stab yourself with an arrow in various parts of the body, then point towards the body with the left-hand forefinger. (*Omaha* I.)

Arms are flexed and hands clasped about center of breast; then slowly fall with arms pendulous and both hands in position of (**Q**). The sign is completed by slowly lifting the hands and arms several times in imitation of the animal's locomotion.

Movement and appearance of animal's front feet. (*Oto* I.)

Hold the closed right hand at the height of the elbow before the right side, palm downward, extend and curve the thumb and little finger so that their tips are nearly directed toward one another before the knuckles of the closed fingers; then push the hand forward several times. (*Kaiowa* I, *Comanche* III, *Apache* II, *Wichita* II.) "Paw and long claws."

Hold both closed hands before the body, palms down, and about eight inches apart; reach forward a short distance, relaxing the fingers as if grasping something with them, and draw them back again as the hands are withdrawn to their former position. (*Ute* I.) "Scratching, and grasping with the claws."

The right hand thrown in the position as for **Horse,** then extend both hands with fingers extended and curved, separated, palms down, and push them forward several times making a short arch. (*Apache* I.) "The animal that scratches with long claws."

Beard, whiskers.

Place the back of the wrist under the chin, spread the fingers slightly, allowing them to extend downward and forward. (*Dakota* VI, VII.)

Beaver.

With the back of the open right hand, strike the palm of the left several times. (*Wied.*) I have seen this. It represents the beaver striking the water with his flat tail. (*Matthews.*) The arms are semi-flexed and approach the body with the hands opened, palms down, the right over and above the left. The right hand finally sweeps back and is held extended, flatly open in position of animal's tail. There is no similarity in the execution or conception of this sign and that of Wied;

the conception in the sign among the Otos is clear, but that for the latter obscure. The height and broad flat tail of the animal. (*Boteler.*)

Describe a parenthesis ( ) with the thumb and index of both hands, and then with the dexter index imitate the wagging of the tail. (*Burton.*)

(1) Right hand flattened to form an extension of forearm, palm downward at an angle of 45° to the body; (2) suddenly slapped down two or three times. (*Cheyenne* I.) "From the manner in which the beaver slaps water or mud."

Both hands in front of body, fingers extended, horizontal flat, palms downward (**W**), ends of fingers pointing obliquely forward, pat the back of first one hand and then the other several times, not too rapidly. (*Dakota* I.) "These Indians believe that beavers are able to converse with each other and do so while building their dams and lodges. The sign, however, would seem to come from the noise made by the beavers in patting down their dams."

Hold the left hand, palm looking obliquely, forward, outward and upward, a foot in front of the lower part of the chest, and strike its palm several times with the back of the open right hand. (*Dakota* IV.)

Both arms are semi-extended and the hands in type-position (**W**) are held right above left to represent size of animal. Hands in this position execute uniformly a creeping movement forwards, then right hand sweeps to back of body and is extended from cocygeal region to represent the broad, flat tail. (*Oto* I.) "Creeping animal with flat tail."

Hold both extended flat hands, palms down, side by side, then extend the first three fingers, separated, the little fingers and thumbs closed and the hands retaining the same relative position. (*Apache* I.) "Flat tail, and claws."

——— Trap. See **Trap.**

Before.

Bring the hand close to the right breast (**M**) with palm to left, forefinger pointed outward; bring left hand (**M** palm inward) in front of and few inches from the breast, move right hand to the front and at same time move the left toward the breast slightly. **After** or **afterward** is done by having the hands in the same position (**M**) except to have palm of right hand down. The left is kept stationary and the right is drawn back. (*Cheyenne* II.) "The left hand representing an imaginary line, the action of the right makes it the front or before."

Right hand moved quickly forward from side back of hand in front, the hand drawn back less quickly, again projected forward rapidly, with slight inclination of head and body forward. (*Ojibwa* IV.)

Left hand held forefinger straight, upright, palm of hand outward (other fingers closed) in front of body about 18 inches, and then the right hand held in the same manner close to the body on the same level. (*Dakota* I.)

Close the fingers of the right hand, thumb crossing the middle finger, index extended, palm forward, thrust the hand forward forcibly. (*Omaha* I.)

——— Going before or passing another person.

Move both fingers up and down and away from the body on the same line, but allowing the right to gain on and finally pass the left. (*Dakota* I.) "From the idea of being or going before or ahead of another person in walking or anything else. A derivative of superior."

——— In place; ahead; superior.

The hands are held horizontally, in front, backs upward, parallel about an inch apart, the forefingers only extended. The right hand is then advanced before the left, usually about a finger's length; but the distance may be increased or diminished to express degree. This sign is to be considered in connection with those for *behind* or *inferior*, and *beside* or *equal*. They all refer to the relative position of objects in space but to different degrees of quality. **Comparison** is expressed by these signs. Changes of position—as horses in a race—are shown by changes in the position of the hands. (*Mandan and Hidatsa* I.)

——— In time.

Same sign as for **Time, long,** and **short,** but in referring to length of time, as to the previous day, or previous month, etc., the hands are drawn apart carelessly about ten or twelve inches, quickly, the absence of any haste indicating a longer time, and consequently applicable to matters referring to a longer period. (*Kaiowa* I; *Comanche* III; *Apache* II; *Wichita* II.)

Beg, Beggar.

First make the sign for **Lodge, entering a,** and then the sign for **Give to me.** (*Dakota* I.)

First make the sign for **Give to me,** and then the sign for **Man** or **Woman,** as the case may be. **Lodge, entering a,** can be added or not. (*Dakota* I.) "From an Indian going from lodge to lodge asking for things."

Begone. See **Go.**

Beginning. Commencement.

The right hand horizontal, with fingers arched (**H**) back of hand upward, of fingers obliquely downward and outward, is pushed out in front of the right breast and drawn back to original position two or three times. (*Dakota* I.) "Going to do it. Going to commence anything. Going ahead, pushing things."

Behind.

Bring both hands in front of the breast (**M** palms down), the forefingers near together pointing to front; move the right suddenly to rear few inches, the left remains still. In describing a person being left farther and farther to the rear, keep moving the right a couple inches or so at a move until you make three or four moves back. (*Cheyenne* II.)

Same motion as **Before** repeated by swinging the hand backwards from thigh, with quickened motion as the hand went back. (*Ojibwa* IV.)

Forefinger of the right hand, straight, upright (others closed) palm of hand outward is drawn from in front of and on a level with the shoulder, behind the body on as near a straight line as possible and with forefinger pointing upright. (*Dakota* I.) "Reverse of going before. He is still falling behind me."

With its index extended, pointing backward, over the right shoulder, back upward, push the right hand backward about four inches from just in front of the shoulder. (*Dakota* IV.)

——— Inferior. (Compare **Ahead.**)

Place the hands in the same position as for **ahead,** except that the right hand is behind the left, *i. e.*, nearer the body. (*Mandan and Hi datsa* I.)

Place the spread right hand behind the body, moving the hand slightly a little up and down, and back and forth. (*Omaha* I.)

The left arm is flexed and hand upright with hooked index (as in **I** 1—modified by the hand being held edgewise) is brought before the face. The right fist is then brought to the same position and the palms face each other. The left hand remains before the face, but the right fist is swept back past the right side with the index finger extended as in type **J**. (*Oto* I). "What I have gone by."

Belong. See **Possession.**

Below, under.

Made like the sign for **above,** with this exception—that all movement is made by the left or lower hand, the right being held motionless, palm downward, and the eyes looking down. (*Dodge.*)

Place the right hand, by a descending motion, just under the left hand, both extended, fingers joined and palms down. (*Arapaho* I.)

Same motion as **Above** with hand starting from above level of elbow, finger pointing down, rest gently closed. (*Ojibwa* IV.)

Same as **Above,** only draw the right hand a considerable distance below the left. (*Dakota* I.)

Sign can also be made with one hand by the reverse of the movements in **Above.** (*Dakota* I.) "Refer to the person below or under by moving that finger and hand which represents the under person, animal, or thing. *Inferior height*, variant of superior."

Both hands are extended open as in type **W.** The hands are then approximated edgewise and the left is superimposed, and swept palm over back of right. Both hands now diverge and arms are extended from the right and left sides—palms down. The right hand is now brought forward as in type (**I** 1)—modified by index being more extended, and describes a quadrant's arc towards the ground. (*Oto* I.) "The vast depths into which all suddenly fall."

Place the flat left hand in front of the body, palm down, fingers directed toward the right; place the flat right hand, palm down, below the left, the greater the distance the hands are moved apart the greater the depths to which allusion is made. (*Ute* I.)

Belt.

Motion of putting a belt around the body with thumb and forefinger of both hands on their respective sides of body—thumb and forefinger U-shaped, pointing inward and carried around the sides of body to the front. other fingers of both hands closed. (*Dakota* I.) "From the placing of the belt."

Beside, equal. See **Same.**

Big. In the sense of **Broad** or **Flat.**

Bring both hands in front of breast, palms down, then extend them forward to length of arms and sweep around outwards, on a level. For small extent, the hands indicate a less extended arc or circle. (*Ojibwa* IV.)

The right hand is held horizontal, extended and flat, pointing forward. A slight arched curve from right to left is made at whatever height the speaker wishes to indicate. (*Apache* I.)

Deaf-mute natural sign.—Place the hands near each other and move them apart. (*Ballard.*)

——— In the sense of **High** as a hill.

An ascending motion of the extended hand, fingers joined, palm down, outlining and indicating relative height. (*Arapaho* I.)

(1) Both hands unclosed, brought to the front of chest, palms facing each other, right hand above left and elevated to level of chin; (2) right hand carried upwards in proportion to the degree of height to be expressed (*Cheyenne* I.)

Raise the extended hand, fingers joined, palm down and horizontal, to the proper or greatest possible elevation. (*Arapaho* I.)

With the forefinger of the right hand straight, upright (**J**), carry it straight upward in front of the body above the head as high as the extended arm will reach. This would indicate a mountain, and lesser degrees of arm extension would denote lesser heights. (*Dakota* I.) "Great high—a mountain."

Deaf-mute natural signs.—**High**—Move the hand upward. (*Ballard.*)

Raise the hand above the head. (*Larson.*)

—— In the sense of **Large around.**

(1) Motion for **width** except that the elbows are not pressed against the sides, but the arms and hands are extended and give a curved form as though clasping some large object; (2) motion made to increase or enlarge the circle in proportion to the size to be expressed. (*Cheyenne* I.)

Indicate the general outlines, and in the direction thereof extend both hands, palms towards each other, to the required or greatest possible distance apart. (*Arapaho* I.)

Palms facing, and moved apart farther and farther by jerks. (*Omaha* I.)

—— Large.

The opened upright hands, palms facing, fingers relaxed and slightly separated, being at the height of the breast and about two feet apart, separate them to nearly arm's length. (*Dakota* IV.)

Deaf-mute, natural signs.—Place the hands near each other palm toward palm, and move them apart, and at the same time distending the cheeks. (*Ballard.*)

Point your finger at your chest and face expanded. (*Cross.*)

Move both open hands from each other with the palms looking toward each other, and, at the same time, cause the cheeks to look big and round by blowing against them. (*Hasenstab.*)

Stretch out the hands. (*Larson.*)

—— In the sense of **Long.**

(1) Hands brought upward and to the front of the body, forefinger and thumbs of both brought together as though grasping the ends of a

string; (2) suddenly drawn apart in proportion to the length to be expressed. (*Cheyenne* I.)

Deaf-mute natural signs.—Direct the forefinger downward and extend it forward toward a distant point. (*Ballard.*)

Point to the land, if the land is intended, with the finger; then move horizontally the outstretched hand along the line, which describes an ellipse. (*Cross.*)

Having held the hands, the palms touching each other, separate them by moving them in a straight line, showing how far apart. (*Cross.*)

Stretch out the arm, at the same time place the forefinger on the shoulder. (*Zeigler.*)

——— In the sense of **Long**, in extent.

Push the opened right hand, palm toward the left, from the lower part of the chest upward to about a foot above the head. (*Dakota* IV.)

Deaf-mute natural signs.—By moving the hands apart. (*Ballard.*)

Stretch out both arms in a straight line. (*Larson.*)

——— In the sense of **Wide.**

(1) Elbows close to side, forearm and open flattened hands extended, palms facing and nearly touching each other; (2) separated in proportion to the degree of width to be represented. (*Cheyenne* I.)

(1) Hands extended to front of body, open to fullest extent, but fingers and thumbs contiguous, palms up, little fingers of each hand touching; (2) gradually moved apart in proportion to the size of the surface to be represented. (*Cheyenne* I.)

Bring both hands to the front of the body, on the same level and close together, fingers and thumbs extended, palms downward (**W**), fingers pointing forward, with both hands on the same level, make a sidewise movement with the left to nearly its arm's full extent, but only move the right a few inches. (*Dakota* I.)

This sign resembles that for **Big**, in the sense of **Flat.** (*Dakota* I.) "A broad, level piece of land."

Both hands brought together in front of breast, with a curved motion, hands horizontal, flat, palm downward (**W**). Ends of fingers pointing toward the front are then carried out sidewise, with a curved downward motion to their natural positions. (*Dakota* I.) "Covering a large surface."

Place both flat hands side by side before the breast, palms down, then pass them horizontally outward toward their respective sides. (*Dakota* VI, VII.)

Deaf-mute natural sign.—Put the palms of the hands near each other and then move them far apart. (*Ballard.*)

Bird.

The hands are flapped near the shoulders. If specification be required, the cry is imitated or some peculiarity is introduced. (*Burton.*)

(1) Both hands open to fullest extent, palms downward, brought to level of and against the chest, forefinger pointing away from chest; (2) pushed forward and downward and withdrawn with a curve motion, as the motion of wings in flying. (*Cheyenne* I.)

The front outline will represent the sign. Both arms are elevated and flexed at right angles to the shoulders, the arms are then made to diverge, and in position **W** made to flap or waver to and from the body after the manner of the wings of a bird. (*Oto and Missouri* I.) "That has wings or flies by them."

Bison, buffalo.

The two forefingers are placed near the ears, projecting, so as to represent the horns of the animal. (*Long.*)

Raise the forefingers crooked inward, in the semblance of horns, on both sides of the head. (*Burton.*)

Lower the head and project the forefingers from their respective sides of the head, like horns. (*Arapaho* I.)

Both hands elevated to or toward the sides of the head, forefinger of each crooked obliquely forward to represent horns. (*Cheyenne* I.)

Both hands tightly closed, leaving the forefingers only extended and curved; place the fists before the chest, about ten inches apart, palms facing, and forefingers above. (*Absaroka* I; *Shoshoni and Banak* I.) "Horns."

Both hands raised to the sides of the head in front of ears, back of hands looking forward, forefingers crooked backward, rest of fingers closed, thumb resting on second finger. (*Dakota* I.) "Horns of the buffalo."

Both hands closed except forefinger, and then applied to each temple, with the latter pointing a little forward. (*Dakota* II.) "Horns of animal."

Semi-flex the forefingers, loosely close the others, thumbs either under the middle fingers or against them, and place the hands upright, backs outward, one on each side of the head, near it or against it, above and just in front of the ears, then move the hands suddenly forward a couple of inches. (*Dakota* IV.) "The horns of the buffalo."

All the digits of both hands are completely flexed except the forefingers, and these are slightly curved; each hand, back outwards and held obliquely outwards and upwards, is brought in contact with the temple of the same side at the wrist. (*Mandan and Hidatsa* I.) "Thus the horns of the animal are indicated."

Same sign as for **Cattle.** (*Oto* I.)

Both hands, flattened, slightly arched, and flexed at the wrists, are held as high as the ears and over the shoulders pointing backwards, then incline the head to the front and quickly throw the hands upward a short distance. (*Kaiowa* I; *Comanche* III; *Apache* II; *Wichita* II.) "The humped shoulders and motion of the animal."

Bison, Generic.

Close both hands loosely, extending and crooking the forefingers, place them several inches from either temple. (*Ute* I.)

——— Buffalo calf.

Place the fists upright, backs outward, with the thumbs strongly extended above the ears near the head, and shake them back and forth several times to imitate the shaking of a calf's ears. (*Dakota* IV.)

Same sign as for **Buffalo cow,** but made as low as the waist, and the hands are moved slightly forward and backward. (*Kaiowa* I; *Comanche* III; *Apache* II; *Wichita* II.)

——— Female.

Curve the two forefingers, place them on the sides of the head, and move them several times. (*Wied.*) I have given you a similar sign for the Bison without regard to sex, except that I do not mention any motion of the hands after they are placed in the position of horns, and I do not remember seeing such motion. (*Matthews.*)

Same sign as made by *Shoshoni* and *Crows.* When this sign is made before the person it signifies both sexes, without special reference to either. (*Kaiowa* I; *Comanche* III; *Apache* II; *Wichita* II.)

——— Male.

Place the tightly-closed hands on both sides of the head with the fingers forward. (*Wied.*) If he does not mean to describe the same sign (minus the motion) as in *Bison, female,* I know not what he does mean. I have seen but one sign for Buffalo (which I am certain was generic), and I opine that one sign was all the Prince saw. The movements he speaks of in *Bison, female,* may have been unnecessary or accidental. Additional signs are used to indicate sex when necessary. (*Matthews.*) It is conjectured that *Wied's* sign for the *Male buffalo* indicates the short, stubby horns, and that for the *Female,* the ears seen moving, not being covered by the shock mane of the male.

Bite, To.

Thumb of right hand extended, fore and second fingers also extended, joined, and slightly arched, thumb separated from fingers (other fingers closed), hand horizontal or obliquely upwards, edge of hand downward, the arm extended to nearly full capacity in front of right breast, quickly draw the fingers against the thumb several times, at the same time draw the arm in toward the body. (*Dakota* I.) "From the snapping bite of dogs. The two jaws in motion."

Blanket. See **Clothing**.

Blind.

First touch the closed eyes with the tips of the extended fore and second fingers of right hand, back upward (other fingers closed), then turn the hand and make the sign for **Seeing** as contained in **Look,** and then the sign for **None** or **I have none** as contained in **None.** (*Dakota* I.) "I have no sight."

Blood.

Right hand with thumb extended and pointing upward, fingers extended and separated as much as possible and pointing obliquely upward and downward, is brought along the body in front to the mouth, where it is carried forward for a few inches with a downward curved motion, back of hand toward the right. (*Dakota* I.) "From a buffalo when seriously wounded standing and spirting blood from the mouth or nose."

Both hands with fingers collected at the palm points are brought to the temples. The fingers are then suddenly flipped outward. The right arm is then elevated and the index-finger and thumb brought to the nose and made to drop in jerks to the ground. (*Oto and Missouri* I.) "That which pulsates in the temples and drops from the nose."

Close the right hand, leaving the index and second fingers only extended, then draw the palmar surface across the lips from left to right. (*Kaiowa* I; *Comanche* III; *Apache* II; *Wichita* II.)

Boat, row.

Make with both hands raised to the level of the shoulders the forward and backward motions as though rowing a boat. (*Dakota* I.)

Bring the hands together, hollowed, little fingers joining, the thumbs farther apart, to represent the body of a boat, held before the breast. (*Omaha* I.)

——— Canoe.

Both hands at the same side of body, one above the other, make the movement of using a paddle first at one side of the body and then at the other. (*Dakota* I.) "From the manner of using the oars and paddle."

Make the motion of paddling. (*Iroquois* I.)

Body.

The hands with the fingers pointed to the lower part of the body are then drawn upwards. (*Dunbar*.)

Both hands, backs outward, carelessly lowered to front and sides of thighs, forefingers extended, tips slightly touching the thighs, then draw rapidly upward along the sides of the body, and out toward the shoulders. (*Cheyenne* I.)

Boiling.

Same as the sign for **fire,** heat being the idea. (*Dakota* I.)

Bold. Imprudent; rash.

Bold is included in *Brave*. Rashness or imprudence cannot be separated from *Brave*, as the distinction is too fine a one for the Indian's comprehension. (*Dakota* I.)

Book.

Place the right palm upon the left palm, and then open both before the face. (*Burton*.)

(1) Both hands brought to a reading distance and made to represent a book open, or much as it would be while holding it in the hand and reading, little fingers not touching; (2) motion made with the face and eyes as though intently reading. (*Cheyenne* I.)

Make with the right hand at the right side of the body the movements of using the pen in writing, and then carry the hand to the front of the breasts, where the left hand is held horizontal with fingers extended, pointing outward, palm upward (**X**), and close the right hand with palm down, fingers extended, &c. (**W**), on the left, as though closing an open book. First part of the sign denotes *writing*. (*Dakota* I.) "From the act of closing an open book."

Both arms are elevated and semi-extended, with the hands open and parallel before the face; palms upward; the eyes directed intently therein. The right hand is then raised and turned into a modified position (**H**), and seemingly writes in the open palm of the left. The hands are then reapproximated in position (**X**) before the face and approach and diverge as in the opening and closing of a book's leaves. The thickness is indicated by holding the right hand at the desired distance above the left, open. (*Oto and Missouri* I.) "Something written that opens and closes from which we read."

Bone.

Make the sign for the animal to which the bone belongs, and then touch the particular part or place in the body to which the bone belongs. (*Dakota* I.) "Locating the bone."

Born, To be.

Place the left hand in front of the body, a little to the right, the palm downward and slightly arched; pass the extended right hand downward, forward, and upward, forming a short curve underneath the left. (*Dakota* V.) "This is based upon the curve followed by the head of the child during birth, and is used *generically*." The sign, with additions, means "father," "mother," "grandparent," but its expurgated form among the French deaf-mutes means "parentage" generically, for which term there is a special sign reported from our Indians. See **Parentage** (*Dodge*).

Pass the flat right hand downward, forward, and upward in front of the lower portion of the abdomen and pubis. (*Dakota* VI.)

Pass the right hand, naturally relaxed, downward from the lower part of the chest, forward from the pubis, and upward a short distance, forming a curve with the convexity downward. (*Kaiowa* I; *Comanche* III; *Ute* I; *Apache* II; *Wichita* II.)

Hold the extended left hand transversely in front of the lower portion of the abdomen, about eight inches in front of it, and slightly arched, then pass the flat right hand downward between the body and the left hand, forward and slightly upward beyond it. (*Apache* I.) "The left hand represents the pubic arch; the right, the curve of Carus."

Bottle.

Denote the size of the bottle on the upheld left forearm by drawing across the edge of the right hand with fingers extended and palm downward, and then make the sign for **Drinking.** (*Dakota* I.) "From drinking out of a bottle."

The left arm is semi-extended before the body, and the hand is held up with the fingers collected up and down as in (**E**), more horizontal; the index and thumb make a circle. A small bottle is represented by the right index extended sawing on the knuckle-joint of the left index. (*Oto and Missouri* I.) "The shape and grasp of the bottle."

Bow, weapon.

The left hand being a little extended, the right hand touches it and makes the motion of drawing the cord of the bow. (*Dunbar.*)

The hands are placed as in the attitude of drawing the arrow in the bow. (*Long.*)

Draw the right arm back completely, as if drawing the bow-string, while the left arm is extended with clinched hand. (*Wied.*) There is a similarity of conception in the sign given by me (*Oto and Missouri* I) and *Wied's*, but in execution the latter represents almost identically the the Oto sign for "To shoot an **Arrow.**" (*Boteler.*)

Make the movement of bending it. (*Burton.*)

(1) Left hand slightly closed as though grasping bow, and middle finger of right hand within three or four inches of the left hand and apparently grasping a string; (2) drawn two or three times in imitation of drawing a bow-string. (*Cheyenne* I.)

Draw back the right hand from the left, held in front of the body, as though drawing back the arrow fixed in the string of the bow. (*Dakota* I.) "From the drawing of the bow."

Incline the head a little toward the right; raise the nearly closed left hand, back outward to the height of the shoulder, nearly at arm's length forward, and while pushing the left hand ahead and a little toward the left, draw the nearly closed right hand, its back outward from near it backward and toward right, as if drawing the bow. (*Dakota* IV.)

The left fist, back forward, thumb upward, is held a foot or two in front of the chest as if grasping a bow. The right hand, with thumb upward, the finger tips forward and two or more of them in contact with the thumb (position of holding the string), is held a few inches behind the left hand. Then a slight motion backwards of the right hand may or may not be made. (*Mandan and Hidatsa* I.)

The left arm was then slightly bent and extended antero laterally from the body. In this position it is made to resemble the shape of a bow. The position of the hand is variable and unimportant, but it is generally clinched as in (**D**), probably to represent the holding of the bow-string, as it is always secured at the end of the bow. The right hand is sometimes approximated, as in setting an arrow. (*Oto and Missouri* I.) "The shape of the bow and its use."

——— To draw the.

The hands are held as in the sign for **Bow,** then the right hand is moved backwards a foot or more as in the act of drawing a bow. (*Hidatsa* I.)

——— To shoot or kill with the.

Hold the left fist, back outward, nearly at arm's length forward, point toward it with the right index, palm inward, from in front of the right shoulder; then drop the left hand and after hooking the right thumb over the fingers of right hand, suddenly let them fly forward. (*Dakota* IV.)

Bowl.

Same as the sign for **Kettle,** except the latter part of the sign indicating placing it on the fire. (*Dakota* I.)

The distinction is not a fine one between *Kettle*, *Bucket*, *Cup*, *Bowl*, *Basin*, &c., and either of them may be indicated in the same sign, but the connection in which the sign is used is generally sufficient to denote the particular article to which reference is made. (*Dakota* I.) "From its use."

Box.

Designate size of box on the upheld left arm by drawing the edge of the right hand over it with extended fingers, and then both arms extended in front of body, hands horizontal, fingers extended, backs outward, to the sides, are held on the level of the breast. (*Dakota* I.) "Size of the box."

Boy. See **Child.**

Brave.

Close the fists, place the left near the breast, and move the right over the left toward the left side. (*Wied.*) A motion something like this, which I do not now distinctly recall—a sort of wrenching motion with the fists in front of the chest—I have seen used for **Strong.** If *Wied's* sign-maker's hand first struck the region over the heart (as he may have done) he would then have indicated a "strong heart," which is the equivalent for **Brave**. (*Matthews.*)

Clinch the right fist, and place it to the breast. (*Absaroka* I; *Shoshoni and Banak* I.)

Both hands fist; the left (**A**) moved up close to the body to the height of the chin, at the same time the right is brought up and thrown over the left (as it is moved up) with an outward and downward curved motion; the left hand is carried upward and downward and the movement of the right reversed two or three times. *Endurance* is expressed by this sign, and it is connected with the sun-dance trials of the young man in testing his bravery and powers of endurance before admission to the ranks of the warriors. (*Dakota* I.)

Push the two fists forward about a foot, at the height of the breast, the right about two inches behind the left, palms inward. (*Dakota* IV.) "The hands push all before them."

The right arm is flexed and elevated; the hand is then placed over left eye much in type-position (**B**). It is then twisted from left to right across forehead until at right side of head when the motion stops with fists palm outwards. That is readily unwound, *As good will.* (*Oto* I.)

Strike the breast gently with the palmar side of the right fist. (*Wyandot* I.)

Strike the clinched fist forcibly toward the ground in front of and near the breast. (*Arikara* I.)

——— He is the bravest of all.

Raise right hand, fingers extended, palm downward (**W** 1), swing it around "over all," the point to the man (fig. 1), raise left fist (**A** 1, changed to left and palm inward) to a point in front of and near the body, close fingers of right hand and place the fist (**A** 2, palm inward) between left fist and body and then with violent movement throw it over left fist, as though breaking something, and stop at a point in front of and little below left fist, and lastly point upward with right hand, (fig. 2). (*Sahaptin* I.) "Of all here he is strongest."

The right fist, palm downward, is struck against the breast several times, and the index is then quickly elevated before the face, pointing upward. (*Apache* I.)

Only by showing willingness to fight. (*Apache* III.)

Deaf-mute natural sign.—To run forward with a bold expression of the countenance. (*Larson.*)

Pantomime.—Not to run back but to run forward. (*Zeigler.*)

Brave.

"Hold firmly closed left hand about eight inches in front of centre of body, left forearm horizontal, pointing to right and front, back of left hand vertical, and on line of prolongation of forearm; bringing the firmly closed right hand some six inches above and a little in front of left hand, back to right and front, and on line of forearm; strike downward with right hand, mostly by elbow action, the second joints of right hand passing close to and about on a line with knuckles of left hand.

"I believe there is no gesture in the Indian sign-language that is as flexible and possesses as much strength and character in its meaning as this, for, when added to other signs, it intensifies their description wonderfully; adds, in fact, the superlative to every idea; gives heroic character to bravery; arrant cowardice to timidity; makes an ordinary meal a feast, and of a fast, starvation; pleasure becomes bliss, and care most bitter sorrow. Pointing to a man and making this sign would convey to an Indian's mind the idea that he was brave, fearless; and this, to them, is the highest, most perfect, virtue, and creates not only respect, but positive reverence."

The foregoing is quoted from an article in the *United Service* for July, 1880, written by an author who, according to his strong expressions, is obviously afflicted with a theory of a stereotype and universal sign language among the Indians, which is supported by his avowedly taking "means" of signs. His descriptions are therefore liable to some of the cautionary remarks in the Introductory Letter to this pamphlet, but as it is highly interesting to ascertain the use of an intensive or superlative sign, the quotation is printed to attract the attention of correspondents. If they meet with, or can recall the precise sign as above described, and in the same exact sense, they will please state the tribe and all particulars. Something like this sign, with some resemblance to its use as an intensive, appears elsewhere in this paper, but not with the signification of "brave," under which head it was given by the author and therefore is so placed here.

Bread.

Combine signs for **Flour** and **Fire.** (*Arapaho* I.)

Both hands in front of body, fists, backs obliquely downward, push up and down several times as though kneading the dough, and then at a higher elevation both hands, one above the other about 8 inches, horizontal, left hand below, back downward (**X**) right hand above, back upward (**W**) change the positions of the hands (**W** taking the place occupied by **X**) two or three times. (*Dakota* I.) "From the packing and shaping of the loaf."

The right hand being about a foot in front of the chest, opened and relaxed, pointing obliquely forward toward the left, quickly throw the left hand, palm downward, and lay it across the right, then rapidly separate the hands, and turning the palm of the left upward and the right downward, lay the right across the left. Repeat this maneuver several times. (*Dakota* IV.) "Forming a piece of dough into a cake or loaf."

——— Hard; crackers, &c.

After making the first part of the above sign, carry the left hand in front of the left breast with fingers extended, joined, flat, horizontal ends of fingers forward, and the edge of the right hand with fingers extended is drawn across the palm of the left at proper distances crosswise. (*Dakota* I.)

Break.

Both hands brought one above the other around to front of body, closed as though grasping small stick, and suddenly turned in opposite directions to imitate breaking. (*Cheyenne* I.)

The extended forefinger of both hands (**J** 1) meeting at their tips in front of body, palms inward, and then separate the fingers by an outward movement as though breaking anything in two. (*Dakota* I.) "From the breaking of a twig."

The fists being near together, directed forward and backs upward, turn the outer sides downward as if breaking a stick. (*Dakota* IV.)

With both hands placed side by side, thumbs touching, throw them outward and downward as if breaking a stick. (*Dakota* VI, VII.)

Both fists (backs upward, knuckles forward, thumbs one or two inches apart) are held at a common level six or more inches in front of the chest, then simultaneously the forearms are semi-rotated so as to bring the thumbs uppermost. Other methods of breaking have other signs to represent them. (*Mandan and Hidatsa* I.)

The arms are uniformly flexed and the hands in type-position (**B**) approximated before the body. The hands are thus palms downwards.

The sign is completed by turning them over towards the side as in regularly bending or breaking a stick. Sign ends with palms up. (*Oto* I.) "Something torn apart."

Both hands closed, palms down, thumbs touching, then throw both downward and outward, toward their respective sides as if breaking a stick. If this gesture is accompanied by a movement of the body forward, and facial accompaniment of exertion, it represents greater destruction by breaking or the increased requirement of strength to break the object referred to. (*Ute* I; *Apache* I.)

Seize an imaginary object, hold it in two fists (**A** palm down), turn fists outward and palms up (broken). (*Apache* III.)

Deaf-mute natural signs.—Clinch the hands and turn them as in the act of breaking a stick. (*Ballard.*)

Use the shut hands as if to break a stick. (*Hasenstab.*)

Nod the head slowly with the upper teeth rested on the lower lip and the eyes opened widely to express astonishment, and, at the same time, use the shut hand with the forefinger up as if to give a warning, and then use the shut hands as if to break something. (*Hasenstab.*)

Place the fists together and suddenly raise them apart, forming an angle (either right or obtuse.) (*Larson.*)

To use both shut hands as if to break a stick. (*Zeigler.*)

Broken.

Twist off left forefinger with right forefinger and thumb. (*Apache* III.

Breech-cloth.

Pass the flat hand from between the legs upward toward the belly. (*Wied.*) This sign is still in use among these Indians. (*Matthews.*) Metaphorically speaking there is identity with Wied's sign and mine. (*Oto and Missouri* I.) In the latter, the sign is given for the thing signified. By the Prince of Wied the conception is the article itself. By the Oto, that which invariably accompanies and supports it. (*Boteler.*)

Draw the opened right hand, palm upward, between the legs from behind forward. (*Dakota* IV.)

Both arms were flexed and somewhat raised at the sides; the hands, then in position (**K**), inverted, are brought to the prominences of the hips. The hands are then brought around to the pubis in an approaching line and there twisted over each other, as in tying a knot. The movement in this sign represents the tying of the cord that passes around over the hips to support the breech-cloth. (*Oto and Missouri* I.) "That which is tied on with a cord at the hips."

Bridle.

The extended forefinger of both hands (all others closed) (**M**), meeting at their tips are placed backs against and covering the mouth (this is the bit) and then separated, drawn upward along their respective sides of the face in front of the ears until they meet at the back part of the top of the head. (*Dakota* I.) "From the wearing of the bridle."

Both hands, with extended indices as in (**K**), meet on the top of the head and describe a line down the sides of the face, then meet in the center of the mouth and diverge. (*Oto and Missouri* I.) "The position of the same on animal's head."

Bring to me. See **Give.**

Broad, wide. See **Big** in the sense of **Wide.**

Brother; brother and sister. See **Relationship.**

Broom.

Both hands joined, in front of the stomach, backs outward, fingers extended and pointing obliquely downward, make a pushing forward and backward movement from the wrists as though brushing or pushing dirt before the broom. (*Dakota* I.) "From the manner of using their primitive brush broom."

Bull.

First make at the sides of the head the horns of the animal by crooking backward the index-fingers, and then make the sign for **Male** applied to animals, which is the same as applied to human beings. (*Dakota* I.) "Denoting the sex of the animal."

Bullet.

Make the sign of the exploding of the powder, then grasp the forefinger of the hand with the remaining fingers and thumb, so that the tip of it will so extend beyond them as to represent the ball. (*Long.*)

(1) Make the sign for **Ball;** (2) forefinger and thumb of right hand used as in the act of grasping some small object in front of the face, and between it and left hand, which represents a half circle by means of the middle finger and thumb; (3) forefinger of right hand suddenly extended and pushed through the imaginary circle of which the middle finger and thumb of the left hand forms one-half. (*Cheyenne* I.)

Left hand hooked (**G**), nearly closed; *i. e.*, leaving a small circular opening, thumb resting on the backs of the 2d and 3d phalanges of all the fingers, is held horizontal in front of the left breast, with back toward the left (this represents the barrel of a gun), then the index and thumb of the right extended and nearly joined at tips as though holding a round bullet between them (this is the bullet), is carried to the left hand and the fingers relaxed as though placing the bullet in the circular opening, and then strike down on the left hand, with the edge of the right fist (**A** 1) as though driving the bullet home. This sign is that for loading a gun. (*Dakota* I.)

Burn.

——— To destroy.

Move the right hand (**F**) in front of the body as though throwing something forward, and then make the sign for **Fire.** (*Dakota* I.)

——— Of the body.

Touch the burnt part with the right index and then make the sign for **Fire.** (*Dakota* I.) "From fire."

Bury, To.

Both flat and extended hands held at arms'-length before the abdomen, edges down, and about an inch or two apart, thumbs touching; both hands are then thrown outward toward their respective sides repeatedly, as if throwing away from their upper surfaces some light substance. Then reverse the motion from without inward. (*Shoshoni* and *Banak* I.) "Throwing up earth from a ditch, and returning it."

Buy. See **Trade.**

Camp.

Clinch both hands, hold them upward before the breasts, thumbs touching, then pass them forward, each describing a semicircle, so that the outer edges touch at the termination of the gesture. (*Arikara* I.)

Cannon.

Thumb of the right hand under the fingers; *i. e.*, in the palm of the hand, otherwise fist (**B**); snap out the fingers and thumb in front of the right breast, and then make the sign for **Big.** (*Dakota* I.) "The big explosion From the smoke made on the discharge of a cannon."

Cap, percussion. See **Gun.**

Capture, To. See **Prisoner, to take.**

Cards, playing.

(1) Left hand fully opened, slant, and held up to the level and in front of the shoulder; (2) sudden pecking motion made with the partially-closed fingers of the right hand five or six times towards the palm of the open left hand, as though throwing cards. (*Cheyenne* I.)

Go through the motion of dealing from a pack of cards, and throwing one each to imaginary players, right, front, and left. (*Dakota* VI, VII.)

Cattle.

First make the sign for **Buffalo** and then the sign for **Spotted.** (*Dakota* IV.) "Spotted buffalo."

Both arms are elevated to head flexed. The hands, both, are brought to sides of head at parietal ridges in position (**K**), and thus held a minute; the head is then rotated to the several sides. (*Oto* I.) "Animal with horns."

Make the sign for **Buffalo,** then extend the left forefinger and draw the extended index across it repeatedly at different places. (*Absaroka* I; *Shoshoni* and *Banak* I.) "Literally, spotted buffalo."

——— To round up.

Place the closed hands horizontally before the chest, leaving the forefingers curved, then make a beckoning motion with both simultaneously, the right moving over the left, as if hooking at imaginary stalks of grain. (*Kaiowa* I; *Comanche* III; *Apache* II; *Wichita* II.) "Cattle. (horns) brought together."

Chair.

Right forearm extended from the elbow at right angles (**L**), with the hand fist (**A**), with back downward. (*Dakota* I.) This sign also denotes sit down. "From its use."

Challenge.

Florentine sign.—A fist clinched, with the thumb thrust out under the forefinger. (*Butler.*)

Chicken cock, domestic.

Bring the thumb and fingers of the right hand together, and, holding the hand moderately elevated, move it across, imitating the motion of the head of a cock in walking. (*Dunbar.*)

(1) Sign for **Bird;** (2) tips of fingers and thumb of right hand closed and pointing downward; (3) motion of scratching and pecking imitated. (*Cheyenne* I.)

Chief, a.

The forefinger of the right hand extended, pass it perpendicularly downward, then turn it upward, and raise it in a right line as high as the head. (*Long.*) "Rising above others."

Raise the index finger of the right hand, holding it straight upward, then turn it in a circle and bring it straight down, a little toward the earth. (*Wied.*) The right hand is raised, and in position (**J**) describes a semicircle as in beginning the act of throwing. The arm is elevated perfectly erect aside of the head, the palm of the index and hand should be outwards. There is an evident similarity in both execution and conception of this sign and *Wied's;* the little variation may be the result of different interpretation. The idea of superiority is most prominent in both. "A prominent one before whom all succumb." (*Boteler.*) *Wied's* air-picture reminds of the royal scepter with its sphere.

Raise the forefinger, pointed upwards, in a vertical direction, and then reverse both finger and motion; the greater the elevation the "bigger" the chief. (*Arapaho* I.)

(1) Sign for **Man;** (2) middle, third, little finger, and thumb slightly closed together, forefinger pointing forward and downward; (3) curved motion made forward, outward, and downward. (*Cheyenne* I.) "He who stands still and commands," as shown by similarity of signs to **Sit here** or **Stand here,** "the man who stands still and directs."

The extended forefinger of the right hand (**J**), of which the other fingers are closed, is raised to the right side of the head and above it as far as the arm can be extended, and then the hand is brought down in front of the body with the wrist bent, the back of hand in front and the extended forefinger pointing downward. (*Dakota* I.) "Raised above others."

Move the upright and extended right index, palm forward, from the shoulder upward as high as the top of the head, turn it through a curve, and move it forward six inches, and then downward, its palm backward, to the height of the shoulder. (*Dakota* IV.) "Above all others. He looks over or after us."

Elevate the extended index before the shoulder, pass it upward as high as the head, forming a short curve to the front, and downward again slightly to the front. (*Dakota* VI, VII; *Hidatsa* I; *Arikara* I.)

Right hand closed, forefinger pointing up, raise the hand from the waist in front of the body till it passes above the head. (*Omaha* I.)

Another sign: Bring the closed right hand, forefinger pointing up, on a level with the face; then bring the palm of the left hand with force against the right forefinger; next send up the right hand above the head, leaving the left as it is. (*Omaha* I.)

The right arm is extended by side of head, with the hand in position (**J**). The arm and hand then descend, the finger describing a semicircle with the arm as a radius. The sign stops with arm hanging at full length. (*Oto* I.) "The arm of authority before whom all must fall."

Both hands elevated to a position in front of and high as the shoulders, palms facing, fingers and thumbs spread and slightly curved; the hands are then drawn outward a short distance towards their respective sides and gently elevated as high as the top of the head. (*Wyandot* I.) "One who is elevated by others."

Elevate the closed hand—index only extended and pointing upward—to the front of the right side of the face or neck or shoulder, pass it quickly upward, and when as high as the top of the head, direct it forward and downward again toward the ground. (*Kaiowa* I; *Comanche* III; *Apache* II; *Wichita* II.)

Close the right hand, index raised, extended, and placed before the breast, then move it forward from the mouth, pointing forward, until at arm's length. (*Ute* I.)

——— Head chief of tribe.

Place both flat hands before the body, palms down, and pass them horizontally outward toward their respective sides, then make the sign for **Chief.** (*Arikara* I.) "Chief of the wide region and those upon it."

After pointing out the man, point to the ground all fingers closed except first (**J** 1, pointing downward instead of upward), then point upward with same hand (**J** 2), then move hand to a point in front of body (Fig. 2), fingers extended palm downward (**W** 1) and move around in circle—over all. (*Sahaptin* I.) "In this place he is head over all."

Grasp the forelock with the right hand, palm backward, pass the hand upward about six inches and hold it in that position a moment. (*Pai-Ute* I.)

Elevate the extended index vertically above and in front of the head, holding the left hand, forefinger pointing upward, from one to two feet below and underneath the right, the position of the left, either elevated or depressed, also denoting the relative position of the second individual to that of the chief. (*Apache* I.)

——— War. Head of a war party; Partisan.

First make the sign of the **Pipe;** then open the thumb and index-finger of the right hand, back of the hand outward, moving it forward and upward in a curve. (*Wied.*) By the title of "Partisan" the author meant, as indeed was the common expression of the Canadian voyageurs, a leader of an occasional or volunteer war party. The sign is explained by his account in a different connection, that to become recognized as a leader of such a war party, the first act among the tribes using the sign was the consecration, by fasting succeeded by feasting, of a medicine pipe without ornament, which the leader of the expedition afterward bore before him as his badge of authority, and it therefore naturally became an emblematic sign. There may be interest in noting that the "Calandar of the Dakota Nation" (Bulletin U. S. G. and G. Survey, vol. iii., No. 1), gives a figure (No. 43, A. D. 1842), showing "One Feather," a Sioux chief, who raised in that year a large war party against the Crows, which fact is simply denoted by his holding out, demonstratively, an unornamented pipe.

Combine signs for **Chief** and **Battle.** (*Arapaho* I.)

First make the sign for **Battle** and then that for **Chief.** (*Dakota* I.) "First in battle."

——— Of a band.

Point the extended index forward and upward before the chest, then place the spread fingers of the left hand around the index, but at a short distance behind it, all pointing the same direction. Ahead of the remainder. (*Arikara* I.)

Grasp the forelock with the right hand, palm backward, and pretend to lay the hair down over the right side of the head by passing the hand in that direction. (*Pai-Ute* I.)

The French deaf-mute sign for *Order*, *Command*, may be compared with several of the above signs. In it the index tip first touches the lower tip, then is raised above the head and brought down with violence. (*L'enseignment primaire des sourds-muets; par M. Pélissier. Paris*, 1856.)

Child; Baby; Infant; Offspring.

Bring the fingers and thumb of the right hand and place them against the lips, then draw them away and bring the right hand against the forearm of the left as if holding an infant. Should the child be male, prefix the sign of a man before this sign, and if a female, do so by the sign of the woman. (*Dunbar.*)

If an Indian wishes to tell you that an individual present is his offspring, he points to the person, and then with the finger still extended, passes it forward from his loins in a line curving downward, then slightly upward. (*Long.*)

Push the index-finger rapidly into the air then draw the hand back downward. (*Wied.*) The right arm is brought up and flexed toward the body. The open hand then describes a semicircle downward and outward similar to the curve of Carus, and stops with the hand erect. The palm is first toward the body, but at the completion of the sign, outward. There is no similarity between this sign and *Wied's* save in the method of indicating its age or size. The upturned palm as indicating species will not obtain among the Otos. (*Boteler.*) "That has been born or come forth." The distance from the ground when the motion ceases indicates the height of the child referred to. Indians often distinguish the height of human beings by the hand placed at the proper elevation, back downward, and that of inanimate objects or animals not human, by the hand held back upward.

A son or daughter is expressed by making with the hand a movement denoting issue from the loins; if the child be small, a bit of the index held between the antagonized thumb and medius is shown. (*Burton.*)

Caresses, by a man; by a woman, by a motion representing parturition. (*Macgowan.*)

Pass the hand downward from the abdomen, between the legs, indicating that it came that way. (*Dodge.*)

Right hand fingers somewhat curved and separated and held forward and higher than the wrist, palm down, moved in a short arch from side to side at the height representing the child indicated, and as if rubbing the top of the head. (*Apache* I.)

——— Small.

Place the right hand a couple of feet from the ground, or lower, back forward and fingers pointing upward, then close both hands and place them in front of the navel, backs outward, the right fist about three inches above the left, and while in this position extend and flex the hands at the wrist once or twice. (*Dakota* IV.) "In arm and small."

——— One able to walk.

Right hand extended in front of the body on level with the breast, back of hand out, fingers joined and pointing obliquely toward the left, turn the hand over with palm looking downward (**W**), and carry the hand downward as though laying its palm on the top of a child's head, the distance of the hand from the ground indicating the height and approximately the age of the child. (*Dakota* I.) "Indicating a child's age by its height."

Deaf-mute natural sign for child, not a baby in arms.—Put the hand when naturally stretched out down to the knee. (*Larson.*)

The Cistercian monks, vowed to silence, and the Egyptian hieroglyphers, notably in the designation of Horus, their dawn-god, used the finger in or on the lips for "child." It has been conjectured in the last instance that the gesture implied, not the mode of taking nourishment, but inability to speak—*in-fans.* This conjecture, however, was only made to explain the blunder of the Greeks, who saw in the hand placed connected with the mouth in the hieroglyph of Horus (the) son, "Hor-(p)-chrot," the gesture familiar to themselves of a finger on the lips to express "silence," and so mistaking both the name and the characterization, invented the God of Silence, Harpokrates. A careful examination of all the linear hieroglyphs given by Champollion (Dictionnaire Égyptien), shows that the finger or hand to the mouth of an adult (whose posture is always distinct from that of a child) is always in connection with the positive ideas of voice, mouth, speech, writing, eating, drinking, &c., and never with the negative idea of silence. The special character for "child" always has the above-mentioned part of the sign with reference to nourishment from the breast.

——— Baby, Infant.

Place the forefinger in the mouth, *i. e.*, a nursing child. (*Arapaho* I.)

(1) Sign for woman made from shoulder downward; (2) lowered in accordance with size and height of child. If a baby, both hands open and palms slightly curled up brought up to front of breast, slight upward and sidewise rocking motion imitated. (*Cheyenne* I.)

Lay the back of the right hand in the palm of the left crosswise on the left side of the breast, and make the up- and down movements as though holding and dandling an infant. (*Dakota* I.) "Sex of the child can be designated by its appropriate sign."

Move the opened right hand, palm backward, fingers pointing downward, from the lower part of the chest downward until it is in front of the lower part of the abdomen, here turn the palm downward and move the hand forward about eighteen inches; then raise the left elbow and fix it about six inches forward from the side, the wrist in front of it and three or four inches higher, holding the hand slightly flexed, its palm backward and fingers pointing upward, and lay the back of the opened right hand on the left forearm near the elbow-joint. (*Dakota* IV.) "Giving birth to, and holding in the arm."

The right arm is flexed and hand brought against abdomen about the umbilicus. The hand is in position (**S** 1); modified by being inverted. The hand with palm towards the body then roughly describes "the curve of Carus" or course of the fœtus in delivery. The sign is completed at end of curve by the hand being held erect, palm outward, back towards the body. (*Oto* I.) "That which hath come forth."

Right and left hands carried to the left breast as if holding a very small infant there. (*Comanche* I.)

Deaf-mute natural signs.—By sucking the finger and placing the hands a small distance apart to denote the size of the child. (*Ballard.*)

Dandle an imaginary baby in the arms. (*Larson.*)

To move the arms as if to dandle a baby in the arms. (*Zeigler.*)

——— Boy.

Hold up the index-finger. (*Dodge.*) "This is used when indicating *my boy*, as when given alone it would also signify *man*."

Right hand tightly closed; then place it before the body, extend the index pointing upward. (*Absaroka* I; *Shoshoni and Banak* I.)

Place the hand in first position for **Man,** then open all of the fingers and move the hand to the right about a foot to the height of the boy referred to, the hand to be horizontal, pointing forward, and its back upward. (*Dakota* IV.) "Male one."

——— Boy, girl.

The most natural signs descriptive of the sexual organs. (*Comanche* I.)

——— Girl.

Join the two outstretched thumbs and forefingers and place them before the crotch. Given when conversing with a person little acquainted with signs. (*Dodge.*)

Sign for **Squaw.** To indicate stature, hold the hand, palm down and fingers extended and joined, at the proper height. (*Arapaho* I.)

Pass the flat extended hands, fingers joined, down the sides of the head as far as the shoulders, when they are drawn forward and outward a short distance, ending with the tips pointing towards one another and palms down. Then hold the left hand and arm transversely before the body, pointing to the right, and pass the right index downward along the abdomen, passing it underneath the left hand, then outward and upward, holding the index as high as the face. (*Absaroka* I; *Shoshoni and Banak* I.) ' "Woman born."

Make the sign for **Woman** and designate age approximately by the distance the right hand is held from the ground, *i. e.* the child's height. The longer hair denoting the sex and the height age. (*Dakota* I.)

First make the sign for **Woman;** then move the hand, back forward, down to the height of the girl referred to, turning the fingers upward and slightly flexing them and gathering their ends (thumb included) into a circle about two inches in diameter. (*Dakota* IV.) "The women wear the hair behind the ears and plaited."

Right hand carried to the navel, then with extended palm, fingers together and pointing downward, move the hand downward to the groins, and then outward, palm still downward and fingers together. (*Comanche* I.)

Extend the left forearm at an angle of 45°, forward and upward from the elbow, place the extended flat right hand with the outer edge against the middle of the forearm, then draw the left towards the breast, the right retaining its relative position while doing so. (*Ute* I.)

Both hands arched or curved, palms facing and about four inches apart; then place the right hand, fingers extended but joined, to the left breast. (*Apache* I.)

——— Offspring.

Pass the hand, fingers extended downward and joined, palm toward the body, downward, close to and in front of the body, changing the direction outward between the thighs; literally, "out of the loins," or else implying the act of parturition. (*Arapaho* I.)

Is "denoted by a slightly varied dumb show of issuance from the loins," the line traced showing a close diagnosis of parturition. (*Dakota* I.)

The right hand, back forward, in the position of an index-hand pointing downward, is held before the abdomen and then moved downward and forward in a curve. (*Mandan and Hidatsa* I.)

Place the left elbow against the side of the chest and bring the hand up to within about eight or ten inches of the face, then lay the right flat hand edgewise transversely into the angle formed by the left arm and forearm. (*Wyandot* I.)

Children. Young men and women.

Both hands closed and held at the height of the shoulders before the body, forefingers straight and extended, pointing upward; move them up and down alternately and repeatedly. (*Absaroka* I; *Shoshoni and Banak* I.) "The individuals; represented by the sign for man."

Chinaman.

Place the tips of the right hand, thumb, and fingers together, then from the center or top of the head make a spiral movement downward and in front of the shoulder reaching as far as the hip. (*Absaroka* I; *Shoshoni and Banak* I.) "Represents the queue."

Cinnabar. See **Color, Vermilion.**

Clean.

With the thumb and forefinger of both hands, an arc of a circle (semicircle), rest of fingers closed, and then the hands are carried downward at the sides of the face in front of body below the breasts, tips of fingers and thumb looking inward, and complete by making the sign for **handsome.** (*Dakota* I.) "From 'handsome, pretty.' Clean, pretty face would seem to be intended."

Clear. (Compare **Light.**)

The hands are uplifted and spread both ways from the head. (*Dunbar.*)

Both hands with palms downward, fingers extended, pointing straight to the front (**W**), are brought together in front of the body on a level with the stomach, and then moved sidewise from each other on the same level for a few inches. (*Dakota* I.) "Resembles somewhat the sign for **broad,** and also for **flat, level.** The separation of two bodies, the heavier falling to the bottom as in the clearing of water."

Deaf-mute natural sign.—Look at the sky and arch the arms and hands towards it, and move them apart to indicate the absence of obstruction by clouds. (*Ballard.*)

Clock or watch.

Make the sign for **Sun** to the right of the body height of head, and then close all the fingers except the index which points upright, carry the hand obliquely downward toward the right, describing with the index a circle decreasing in size as the hand is carried downward. (*Dakota* I.) "Telling time by the sun's movements."

The right hand with the index hooked is made to describe the arc of the horizon before the forehead. The left arm is then semi-extended, fingers collected, but the index and thumb crooked to form a circle. The right index in position (**K**) now describes a circle over the left index and thumb as held above; then a second circle to indicate hour-marks is made in dots, as it were, then the arc of the horizon is divided off, as it were, in points. (*Oto and Missouri* I.) "Something circular that marks the divisions of daily time."

Clothing. Blanket, robe.

Pass both fists, crossing, in front of the breasts, as if wrapping one's self up. (*Wied.*) Sign still in use, but nowadays for blanket as well as robe. (*Matthews.*)

——— Robe, red.

First indicate the wrapping about the shoulders, then rub the right cheek to indicate the red color. (*Wied.*) Here he means blanket, not buffalo-robe, which shows that in his day the same radical sign was used for both. (See sign for **Blanket.**) Buffalo robes were never painted of a uniform color, except when rubbed with white or yellow earth, never certainly green or blue throughout; but red, green, and blue are favorite colors for Mackinaw blankets. The signs for the colors are the essential points to be noticed in these descriptions. A color may be indicated by rubbing any object that possesses it, or pretending to rub such an object. (*Matthews.*)

——— Robe, green.

Indicate the wrapping about the shoulders, and with the back of the left hand make the gesture of stroking grass upon the earth. (*Wied.*) Same remarks are applicable to this as to **Robe, Red,** *q. v.* (*Matthews.*)

Put them on in pantomime. (*Burton.*)

Pantomimic show of enveloping oneself in a blanket as worn by Indians. (*Arapaho* I.)

Both hands closed, as though loosely clasping the edge of a blanket, and brought up in front of the middle of the breast, the left hand over the right, as though folding the blanket around the shoulders. (*Cheyenne* I.)

Touch the article in question; in its absence, a pantomimic show of where it should be if present. (*Arapaho* I.)

Is denoted by crossing both arms in front of the body on a level with the breast, and close to the body, with the hands grasping a fold of the blanket—the same as a shawl would be worn. (*Dakota* I.) "From its use as a covering."

From an upright position, just above the corresponding shoulder, palm forward, move each hand across the chest, and, gradually rotating the hand until its palm is backward, place it against the opposite shoulder, crossing the fore-arms on the chest, then rub the back of the left hand with the ends of the fingers of the right. (*Dakota* IV.) "Wrapping a blanket around the shoulders."

Use both hands as if drawing a blanket around the body and shoulders, crossing the forearms over the breast in doing so. (*Dakota* VI, VII.)

The arms are flexed and hands in type-position (**S**) crossed on the front of breast. Then withdrawn open downwards and outwards over limbs. Sometimes both hands in above positions are made simply to touch successively limbs, body, and arms. (*Oto* I.) "That which wraps or incloses me."

The left palm is laid upon the chest, then both hands are opened and thrown back on a level with the shoulders. The hands are now gradually crossed on the breast, being closed fist-like as they come together. (*Oto and Missouri* I.) "The manner of folding something over the body."

Fold arms across the breast, signifying drawing the blanket about the shoulders. (*Sahaptin* I.)

Elevate both hands clinched (**A**) on a level with the shoulders and then jerk them across to the opposite shoulder after the manner of wrapping up in a blanket when it is cold. For a buffalo robe the sign is the same, only a previous sign indicating the robe is used. (*Comanche* I.)

——— Of skin, or a buffalo robe.

The hands are placed near the shoulders, as if holding the ends of the robe, and then crossed, as if drawing the robe tight around the shoulders. (*Long.*)

Combine signs for **Clothing**, **Woolen Blanket**, and **Buffalo.** (*Arapaho* I.)

Sign for **Skin** followed by sign for **Blanket.** (*Cheyenne* I.)

First make the sign for **Buffalo,** and then the sign for **Blanket.** (*Dakota* I.) "From its use as a covering."

Both hands with thumbs and forefingers extended (**K** 1), back of hands outward, are brought to the sides of the head and carried downward slightly in front of the sides of the body to the level of the lower ribs where the hands are turned so the thumbs and forefingers point downward. Pants, coat, and every other garment must be pointed to the position where worn. (*Dakota* I.) "The entire covering of a person."

Make the sign for wrapping a blanket around the shoulders, and then to indicate the hair on the robe, hold the left forearm horizontal and directed obliquely forward and toward the right, and move the right hand along it from the elbow to the wrist, the back of the right hand against the forearm, its fingers and thumb somewhat curved and separated, pointing upward (as in **P**). (*Dakota* IV.)

——— Coat.

Separate the thumb and index-finger of each hand, and press them downward over the sides of the body. (*Wied.*) I have described a sign much the same for *shirt*. Perhaps he regarded the Indian hunting-shirt as a coat, since it is used as an outside garment. The motion he describes depicts the pulling down of a shirt over the head, not the putting on of a coat of European fashion. Coats must have been rare among the Upper Missouri Indians in 1834. He says that the thumb and forefinger are separated (but being thus mentioned together he evidently saw them approximated). I have described them as in contact. Perhaps the sign is variable to this trifling degree even when made by the same person, the mere indication of the pulling down of the shirt being the essential point. (*Matthews.*)

With the fingers held as for **Dress, Tunic,** but with the thumbs pointing, first carry the right hand along the left arm from the wrist to the shoulder and the left hand along the right arm. (*Dakota* IV.) "Covering the arms."

The left arm is partially elevated and semi-extended in a passive manner from the body. The right hand is then brought over the extended left in type-position (**W**). The back of hand more arched, and is drawn up over the left arm and *vice versa*. The fingers of both hands are then twisted over each other at points from the neck down the chest as in buttoning. (*Oto and Missouri* I.) "Something that covers our arms and is buttoned around the body."

Deaf mute natural signs.—Indicated by moving the hands along the arms up toward the shoulders. (*Ballard.*)

Take hold of the front lappel of the coat with one hand to make a movement of it back and front, and point to it with the other, nodding the head as if to indicate goods of the same kind. (*Hasenstab.*)

——— Dress, Tunic.

With the forefingers extended and pointing inward, backs forward, and thumbs extended and pointing backward, the other fingers closed, move the hands from the front of the shoulders downward to the waist. (*Dakota* IV.) "The motion of passing a garment over the head and covering the body from the shoulders down."

——— Shirt, hunting.

The forefinger and thumb so opposed as to form a curve are passed near the surface of the body, from the forehead to the abdomen. (*Long.*)

Forefingers pointing towards the hips, brought up along the sides and above the shoulders and pointed backwards. This sign may be reversed. (*Cheyenne* I.)

Same as the sign for **Clothing.** (*Dakota* I.)

The tips of the thumbs of each hand are opposed to one or more of the corresponding fingers and the hands are then passed rapidly downward in front from the top of the head to below the stomach. (*Mandan and Hidatsa* I.)

——— Trowsers.

With the fingers held as for **Tunic,** carry the hands from the knees upward to the waist. (*Dakota* IV.)

Cloud.

Begin with the sign of water, then raise the two hands as high as the forehead and, placing them with an inclination of 15°, let them gently cross one another. (*Dunbar.*)

(1) Both hands partially closed, palms facing and near each other, brought up to level with or slightly above but in front of the head; (2) suddenly separated sidewise, describing a curve like a scallop; this scallop motion is repeated for "many clouds." (*Cheyenne* I.)

Both arms fully extended at the sides of the body with hands horizontal, straight out, palms downward (**W**), are brought together with a curved motion in front of, but higher than, the head, so that the tips of the fingers meet backs of hands upward (**W**). (*Dakota* I.) Resembles the sign for **Night** or **Darkness.** "The coming together of clouds. Darkness in the heavens."

Cloudiness, dampness.

May be signified by making the sign for **Smoke,** the hand ascending, then descending, by constantly revolving motion. (*Arapaho* I.)

Club.

Point to a piece of wood with the right index (**M**), and then strike the palm of the flat, horizontal left hand (**X**), held in front of the body, fingers pointing toward the right, with the edge of the fingers of the right crosswise. (*Dakota* I.) "From striking with a club."

Coal.

The left hand is carried down and held at the level of and in front of the left breast from a little higher elevation, with palm inward, fingers joined extended, slightly arched, and pointing towards the right, hand horizontal (bank or bluff), and then the right hand in the same position, except ends of extended fingers pointing toward the left, is carried out in front of the body, and its back struck several times against the palm of the left (hard), and then the right index is carried from left to right along the middle of the palm of the yet stationary left (this indicates the vein of coal,) and then the left hand is dropped down in front. The thickness of the vein itself and its depth below the surface can also be approximately indicated in this sign. No written language could convey to the mind a more graphic picture than does this to the person who has ever ascended the Missouri, and seen the veins of coal cropping out of the river bluffs and banks. (*Dakota* I.) "A vein of coal cropping out of a bank."

Coffee.

Left-hand fist (**A** 1) held to the left side of body in front of stomach, the right hand brought over it, end of index resting on end of thumb (other fingers closed), back of hand upward; then the right is turned in a circular manner as though turning the handle of an upright coffee-mill. The sign for *Kettle* can be made to indicate the boiling of the coffee or the sign for *Drinking* to indicate that it is made. (*Dakota* I.) "From the grinding of the coffee."

The arms are semi-extended and hands (as in type-posture (**C** 1) modified by being horizontal) made to rub circularly, the palms slightly separated This is followed by signs for **Water, Fire,** and **Drink.** (*Oto* I.) "Something to be ground, then subjected to fire and water, and drunk."

Coin.

A shaking of fingers and thumbs. (*Macgowan.*)

The fingers of the right hand closed, leaving the thumb and index curved, with tips joining, thus forming a circle. The hand is then held outward toward the right side, showing the circle to the observer. (*Absaroka* I; *Shoshoni and Banak* I.) "Round, like silver or gold coin."

Join the tips of the thumb and forefinger of the same hand, the interior outline approximating a circle. (*Arapaho* I.)

The arm is raised to the head, and the right index finger hooked describes the hat-mark of the forehead in sign for **White man** or American. The arms then diverge wave like from the sides of the body. The left hand is then brought before the body. The index and thumb form a circle in imitation of the outline of the silver dollar, as in (**G**). The extended right index finger is then drawn across the circle of the left hand twice. (*Oto and Missouri.*) "The round, marked currency of the white man."

Coitus.

The extended index of the right hand (the hand usually back downwards, other fingers flexed) is grasped by left hand, palm upwards. In this position the index is usually moved back and forth, *i. e.*, in and out, once or oftener. (*Mandan and Hidatsa* I.)

Australian sign.—Fingers of both hands closely interlocked, backs up and horizontal. This is used ceremonially in offering their women to a visitor as a rite of hospitality. (*Smyth.*)

Cold.

The same as for **Air,** but when applied to a person the right hand is shut and held up nearly opposite the shoulder, and put into a tremulous motion. (*Dunbar.*)

The arms with clinched hands held up before the breast, thrown into a tremulous motion, as if shivering with cold. (*Long.*)

(1) Both hands, palms facing breast, awkwardly closed, as though numb with cold, and brought to a level with the shoulders which (2) shrugged; (3) hands slightly motioned downward, forward, and sidewise, to imitate violent trembling. (*Cheyenne* I.)

Hold the clinched right hand (or both) in front of the shoulder, then cause the fist to tremble as if shivering from cold. (*Dakota* VII.)

With both fists clinched and held drawn up near to the shoulders, imitate trembling, as from cold. (*Dakota* VI; *Hidatsa* I; *Arikara* I.)

Both hands clinched, cross the forearms before the breast with a trembling motion. (*Hidatsa* I; *Arikara* I.)

Both arms are uniformly flexed and made to approach the chest. The hands are (in type-posture (**B** 1) modified by palms facing chest) then made to tremble before the body. (*Oto* I.) "That makes me shake or tremble."

Elevate both hands, clinched, to the shoulders; then let them shiver a little back and forth. This sign, varied as follows, indicates chill and fever. After the cold sign is used, place both hands clinched (**B**) beside the temples, and let the fingers fly outward, and clinch them again, and let them fly outward again. This indicates fever in the head. (*Comanche* I.)

Same sign as for **Winter.** (*Apache* I.)

—— It is.

Wrap up, shudder, and look disagreeable. (*Burton.*)

Cross both hands (**B**) at the wrists a few inches in front of the body on a level with the breast, and then make a shivering motion with both hands and arms. (*Dakota* I.) "Resembles somewhat the sign for **Blanket; covering.** The idea of shivering with cold."

Deaf-mute natural signs.—Shudder with the shoulders. (*Ballard.*)

Shiver, with your fists near the breast. (*Cross.*)

Make a rapid movement of the clinched hands from and toward each other with the eyes lifted toward the person spoken to, and then point in the direction from which the wind is blowing, so as to indicate that the weather is cold. (*Hasenstab.*)

To shiver, moving the closed fists. (*Larson.*)

Close the fists and draw the arms toward the body with a motion of shivering. (*Zeigler.*)

Color.

First and second fingers of the right hand extended, thumb resting on the third finger which with the little finger is closed, are brought to the front of the body and to the left side over the left hand, which is held slightly oblique (**B**) on the left side of the body about a foot to the front of the left breast, and then the two extended fingers of the right hand are rubbed over the back of the left hand.

This is generic. The particular color must be designated after making this sign in each instance by touching something of that color. (*Dakota* I.) "Mixing the paint on the hand to see its color before using; much the same as painters try a color on a board, or anything with a proper back ground to display it."

—— Black, White, Red, Blue, Yellow, etc.

With arm elevated, semi-flexed, the hand in position (**K**) pointed to these different colors, whether represented in the vault of the firmament or articles of dress. White is sometimes indicated by pointing to the sun, and black by pointing to sun then executing sign for **No** or the **Sun's Setting.** (*Oto* I.) "As objects appear."

—— Black.

Rub the hair on the right side of the head with the flat hand. (*Wied.*) The hair of the Indians being nearly always black, that object is naturally selected as indication of that color.

First make the sign for **Color,** and then touch something black. (*Dakota* I.) "Designating a particular color."

Rub along the back of the left hand with the palm of the fingers of the right, back of left up palm of right down. Or point at a black object. (*Dakota* IV.)

Pass the fingers of the right hand gently over the hair on the right side of the head. (*Dakota* V.)

Deaf-mute natural sign.—Show the end of the nail of one finger of the one hand with the forefinger of the other hand pointing at it, on account of the color of the dirt under the nail. (*Larson.*)

——— Blue.

With two fingers of the right hand rub the back of the left. (*Wied.*) It is conjectured that the veins on the back of the hand are indicated.

First make the sign for **Color,** and then touch something blue. (*Dakota* I.) "Designating color."

——— Gray.

First make the sign for **Color,** and then touch something of gray color with the right index. (*Dakota* I.) "Designating color."

——— Green.

With the back of the left hand make the gesture of stroking grass upon the earth. (*Wied.*)

First make the sign for **Color,** and then touch something of green color. (*Dakota* I.) "Designating color."

Point at a green object and then rub the radial edge of the left hand with the fingers of the right. (*Dakota* IV.) "Colors are usually designated by pointing at or rubbing some object of the color referred to."

Deaf-mute natural sign.—Point the finger to grass. (*Larson.*)

——— Red.

Rub the right cheek to indicate the red color. (*Wied.*) The red refers to the paint habitually used on the checks, not to the natural skin. The Indians know better than to designate between each other their natural color as red, and have been known to give the designation *red man* to the visiting Caucasian, whose blistered skin often better deserves the epithet, which they only apply to themselves in converse with the conquering race that insisted upon it. The author mentions in another connection that the Mandans used red on the cheek more than on other parts of the body on which parts other colors were generally displayed.

Make the sign for **Color,** and then touch any red object with the right index. (*Dakota* I.)

Hold the left hand as for **Spotted,** and rub its upper edge with the ends of the fingers of the right hand. (*Dakota* IV.)

Rub the cheek with the fingers. (*Dakota* VI.)

Rub the right cheek with the palmar surface of the extended fingers of the right hand. (*Kaiowa* I; *Comanche* III; *Apache* II; *Wichita* II.) "From the custom of coloring the cheeks red."

Rub the cheek with the palmar surface of the extended and joined fingers of the right hand. Sometimes both hands are used in communication with Americans. (*Ute* I.)

——— Vermillion, cinnabar.

Rub the right cheek with the fingers of the right hand. (*Wied.*) Still in use. (*Matthews.*)

——— White.

With the underside of the fingers of the right hand rub gently upon that part of the left hand which corresponds with the knitting of the bones of the forefinger and thumb. (*Dunbar.*)

First make the sign for **Color,** and then touch anything white with the right index. (*Dakota* I.)

Extend the left hand, palm up, flat, with fingers spread before the body, and draw the index of the right from the tip of the middle finger of the left back across the palm to the wrist. Light color compared with the darkness of the skin generally. (*Ute* I.)

Deaf-mute natural sign.—Point the finger to the human skin. (*Larson.*)

——— Yellow.

Sign for **Color,** and then touch something yellow. (*Dakota* I.) "Designating color."

Comb.

Curve the spread fingers of the right hand into a half circle, then pass them over the hair of the right side from above downward, as if combing one's self. The hand is then brought forward again, still in the same condition, to indicate the object. (*Dakota* V.)

Combat. See **Battle.**

Come, To; to arrive.

The forefinger moved from right to left with an interrupted motion as if imitating the alternate movement of stepping. (*Dunbar.*)

Elevate the index finger near the face, extend the hand, and return it with a number of gentle jerks. (*Wied.*) The simple idea of "come" is expressed by a straight and unvarying motion of the finger, as you imitate in your remark. Prince W. gives here a variation which signifies coming from a distance or making a journey with halts. He says "gentle jerks;" I say "wavering motion." We both mean the same thing. The interruption of motion may indicate nightly camps. (*Matthews.*) The right arm extended with the hand in type-position (**K** 1), index a little more opened, pointing to the individual, then describing an arc towards the body with *slight jerking*, the index sharply hooked. There appears to be reasonable similarity to *Wied.* It is probable that, were the latter description more explicit, with some allowance for misinterpretation, the identity of the conceptions and similarity of signs would be more plain. (*Boteler.*) "To approach or draw near unto in both."

(1) Forefinger of right hand pointed to person addressed, the finger tips pointing upwards and palm inward; (2) hand drawn toward face. In rapid communication, the pointing with the forefinger is superseded by a motion of the whole hand towards the object addressed. The back of the hand is kept towards the person, the finger tips upward, palm inward, and motion made towards the speaker's face with the first two or three fingers of his right hand. (*Cheyenne* I.)

Hold the left hand a foot in front of the chest, its back forward, fingers pointing a little upward toward the right, and bring the palm of the right hand backward against it smartly, the fingers of the right pointing a little upward toward the left. (*Dakota* IV.)

Elevate the right hand, back forward, quickly elevate the index and throw it back into its place again. (*Dakota* VI; *Hidatsa* I; *Arikara* I.)

The right arm is extended forwards nearly or quite to full length, the right hand erected by full flexion of the wrist, making the back of the hand look forwards, the index only is extended. Then the hand is drawn inwards close to the chest. This is used for *Come* in both an indicative and imperative sense. (*Mandan and Hidatsa* I.)

Place the closed hand, flexed at the wrist and pointing upward, palm toward the face at arm's length to the front and right of the body, elevate the index, and bring it slowly in jerks toward the body. (*Kaiowa* I; *Comanche* III; *Apache* II; *Wichita* II.)

——— (In the imperative.)

Right hand extended length of arm, palm upward (**Y**), bring it to you. The same sign is used by whites in the States. (*Comanche* I.)

——— I or we.

Place both hands palm to palm, and pass them in gentle and interrupted arched movements to the front, resembling the motion of walking. (*Pai-Ute* I.)

——— Back.

Beckon in the European way, and draw the forefinger toward yourself. (*Burton.*)

(1) Sign for **Go away**; (2) sign for **Come**; that is, gone away and come back. (*Cheyenne* I.)

The sign can be made by a simple motion of the right hand perpendicular, held at the arm's extended capacity at the side of the body and drawn back toward the left in front of the body, provided the person being called back is near enough to see it. (*Dakota* I.) "Calling a person back."

Deaf-mute natural sign.—Stretch out the hand to denote going away, and then move it towards one's self. (*Ballard.*)

——— Here.

The hands stretched outward with the palm under, and brought back with a curve motion downward and inclining to the body. (*Dunbar.*)

Beckon with the forefinger as is done in Europe, not as is done in the East. (*Burton.*)

(1) The sign for **Come** is first made; (2) drooped fingers slightly extended and reversed from position assumed in making the sign for "come;" (3) suddenly dropped to indicate desired position. (*Cheyenne* I.)

Place the right index upright, back forward, at height of shoulder, at arm's length, other fingers closed, thumb against middle finger, then bring the hand near to the shoulder. (*Dakota* IV.)

The right arm was extended and the hand in type-position (**K** 1), modified by being held back outward and downward, palm upward and inward, was made to point with index to the object. The arm is then gradually and uniformly flexed toward the body; the hand in above position approaching the body, palm upward, in a semicircle or by jerks, the index sharply hooked. (*Oto and Missouri* I.) "Draw near me."

Deaf-mute natural signs.—Moving the hand toward one's self. (*Ballard.*)

Use the open hand as if to beckon to somebody to come, and, at the same time, nod the head. (*Hasenstab.*)

——— Come to this place, To.

Make the sign for **Come,** and as the hand is brought in front of the body throw the palm against the horizontal palm of the left hand. (*Kaiowa* I; *Comanche* III; *Apache* II; *Wichita* II.)

Coming (participle).

Right hand and arm extended, hand clinched, index-finger pointing outward. Then bring the hand slowly to the body, the index-finger meanwhile moving backward and forth, as if it alone were motioning some one to come. (*Comanche* I.)

——— Arrival from a great distance.

The hand is placed as in **Come,** but as far to the front as possible, and then drawn slowly toward the body, sometimes with a laterally tortuous motion. (*Mandan and Hidatsa* I.)

——— Of a person; to arrive; soon to be here.

Place the forefinger in a vertical position, with the arm extended towards the point from which the person came, or is to come, then bring it gradually near the body, but not in contact with it, or, if he continued on, carry it in the direction he passed. (*Long.*)

Clap the hands, elevating the index-finger of the right hand. (*Wied.*)

Is always used in connection with the object acting; for instance, approaching objects are pointed out, described, enumerated, and sign for "come" made from the direction of the approaching object to the front of the person speaking. (*Cheyenne* I.)

——— Of a person.

Right arm fully extended to the right side of the body (which must be so placed that the hand in this position will point in the direction in which the person is coming), forefinger extended, straight, upright, resting on the thumb extended along it (other fingers closed), back of hand outward: bring the hand to the body (breast) in this position by a series of jerking movements (nearly stopping its motion), in imitation of the stepping of the coming person. *From the walking movements of the approaching person. Come.*—In the sense of he *Has* come; he *Has* arrived; he *Is* here. On the completion of the above sign, clap the hands once at the left side of the body, indicating the person has come to his journey's end. (*Dakota* I.)

Hold the right hand at arm's length, a little higher than the shoulder, palm turned toward the face, and index elevated; then bring it backward in an interrupted motion to near the breast, where the left palm is held edgewise, pointing forward, when the ball of the right hand is brought against it with a slap. (*Dakota* VII.)

——— Toward you.

Right hand flat and extended, held edgewise, thrust the hand forward in a curve either upward or downward. (*Omaha* I.)

Deaf-mute natural sign.—Stretch up the open hand over the shoulder or the hand, indicating the height of the person coming, next point to him, and then use the open hand as if to beckon to come, and at the same time nod the head. (*Hasenstab.*)

Companion; in company. (Compare **Relationship.**)

The two forefingers are extended and placed together, with their backs upward. This sign is also used for **Husband.** (*Long.*)

Two forefingers held motionless together, touching throughout their length in front of breast, backs upward. (*Cheyenne* I.)

The forefinger of each hand extended, pointing straight to the front and *joined* (all other fingers of both hands closed), hands horizontal, backs upward, on level of the stomach, and close to the body, are carried forward for about eighteen inches with a curved upward movement, so that when the sign is completed the fingers are on a level with the upper part of the breast, pointing obliquely upward. (*Dakota* I.) "Inseparable, united, equal."

The arms are flexed before the body and the hands, in type position (**S**), approximated palms in contact, before the chest. The arms are then semi-extended, and the indices in type position (**J**), horizontal, are crossed, the hands, thus "*in situ*," describe a wave-like motion forward, as in going on one's way together with one to whom we are attached. (*Oto and Missouri* I.) "A friend who is dear and accompanies me."

Both hands closed (**M**) and brought within two inches of each other, index-fingers pointing outward, then let both hands move outward a foot or so. (*Comanche* I.)

——— Traveling. See **Friend.**

——— For life. See **Husband** and **Wife**, the **Same**, **Equal.**

Comparison; More, Most.

In comparison the signs for **Little** and **Big** are used as representing "more," "most." (*Dakota* I.)

Complaisance. (Compare **Glad.**)

Compulsion.

Italian sign.—The thumb is held under the chin to indicate being laid under necessity. (*Butler.*)

Contempt. Insult. (Compare **Disgust.**)

Close or shut the right hand and hold it drawn toward the chest and on a level with it, with the palm up and the shut fingers and thumb up; and the expression of contempt is given by extending out the hand and arm directly in from the body, at the same time opening the thumb and fingers wide and apart, so that at the termination of the motion the arm is nearly extended and the thumb and fingers all radiating out as if it were from the center of the hand, and the palm of the hand still pointing upward. (*Ojibwa* III.)

Indicate by turning from the object, move both hands and arms to right or left, with palms outward, as if in the act of pushing away the person or object; stand erect or lean back a little. (*Ojibwa* IV.)

My observations agree with Dr. Matthews, connection with "bad," (see p. 26, Introduction to the Study of Sign-Language, by G. Mallery), and also with Gilfillan (*loco citato*), as expressing the highest degree of "contempt;" at the same time the contempt is expressed by facial emotions. (*Dakota* I.) "Derivative of bad."

This is expressed by the sign for **Bad,** but the motion is more forcibly made. If the person at whose expense the sign is made is present, the hand is moved toward him and the face is sometimes averted from him. (*Mandan and Hidatsa* I.)

Italian sign.—Thrust out the forefinger and the little finger, calling the gesture by a name very similar in sound to "fig." "To turn up the nose at" (English phrase) is translated into the Italian gesture by doing the thing it describes, and possibly owes its origin to the same gesture. (*Butler.*)

Content. See **Glad.**

Corn.

Same as the sign for the **Arikara Indians,** which see. *Planting corn* is made with the right hand nearly as (**U**), pointing toward the ground at the right side of the body, and moved along as though dropping the grains of corn into a hill, and then the hand is turned so that the fingers point upward, which probably indicates that the corn has been planted. I am not aware that the sign resembles any other made by the Sioux, but the hand is invariably turned upward at the time indicated above. (*Dakota* I.) "From the planting of corn."

Same sign as for **Arikara.** (*Kaiowa* I; *Comanche* III; *Apache* II; *Wichita* II.)

Collect the fingers and thumb of the right hand to a point, pass the tips upward from the height of the pubis as high as the head, then pretend to grasp an imaginary object, holding the radial side of the hand downward toward the left, then throw the hand forcibly on its back over toward the right. Represents the stalk and the breaking off of the ear. (*Ute* I.)

——— Standing.

Close the right hand, extend the index, holding it upward, back of hand near the ground, and gradually and interruptedly elevated to the height of the head. (*Ute* I.)

Correct. See **True.**

Counting, or numeration.

The fingers and thumbs expanded count ten. In order to proceed with the enumeration by tens the hands must be clinched, and if again expanded it counts twenty, and so on, the hands being clinched between every ten. In order to indicate the digits, clinch the hands and extend the little finger of the left hand for one, extend also the ring finger for two, and so on, the thumb for five; these must remain extended whilst the thumb of the right is extended for six, &c. Any number within five, above any number of tens, is indicated by clinching the left hand and crossing the right over it, with the requisite number of fingers extended. For the number of sixteen, exhibit the sign of ten and then extend four fingers and the two thumbs in the order of enumeration; for seventeen, proceed by extending the forefinger of the right hand, and so on to twenty. In this manner any sum can be denoted, always holding the backs of the hands upward. When enumerating a small number where a considerable exertion of the memory is requisite, the Indians extend the left hand with the palm upward, whilst with the index of the right the fingers are successively bent into the palm, beginning as before with the little finger, and the greater difficulty in recalling to mind the numbers or events the more apparent resistance is offered to the inflexion of the finger. (*Long.*)

Elevate the index-finger and move it forward to indicate one, twice for two, etc. When counting on the fingers begin at the left hand. (*Wied.*)

Show the required number of fingers; the system of tens obtains. (*Arapaho* I.)

Left hand held up to, on a level with, and in front of the shoulder, right hand partially closed, forefinger slightly extended and (2) touched successively to the fingers of the left hand from little finger to forefinger. This is the abstract idea of counting. (*Cheyenne* I.)

Left hand (**P**) extended in front of body, then with the right hand, of which the thumb and forefinger are extended (**K** 1), tap with the forefinger of the right hand each finger and thumb of the left hand; closing the finger at the time of tapping, it indicating that it has been counted. Where multiples of ten are to be used, one or both hands, as may be required, are held in front of the body nearly together, with fingers extended, palm outward, hands upright, and the fingers closed and opened as often as may be necessary.

Your remarks on page 23 of "Introduction to the study of sign language," after Kohl's sign for "Quantity, many, much," are correct. That observer has without doubt confounded these signs, as I have seen many Indians, belonging to different tribes, using the sign he gives for quantity in counting, and if there is any one universal sign it is this one for counting. (*Dakota* 1.)

Deaf-mute natural signs.—Move alternately each forefinger on the tips of all the fingers. (*Larson.*)

Touch the fingers one after another with one finger. (*Cross.*)

Country, Land.

Point toward the ground a short distance before the body, slightly stooping, and directing the eyes to the same point. (*Absaroka* I; *Shoshoni and Banak* I.) "When the country is at a distance, the fist is thrown toward the ground, outer edge down, at arm's length, in the direction of the location of the region."

——— My.

First make the sign for **Country,** followed by that for **Mine.** (*Absaroka* I; *Shoshoni and Banak* I.) "When possession is elsewhere the arm is extended in that direction, which, with the sign for **Possession,** signifies *My country*, the sign for the latter being dropped."

Courtship.

Place the closed right hand near the hip, the index extended and pointing forward, the thumb extended toward the left and upward at right angles to the axis of the index, then move the hand forward to arm's length, rotating the hand and forearm during motion. (*Kaiowa* I; *Comanche* III; *Apache* II; *Wichita* II.)

Cow.

The two forefingers brought up to the side of the head and extended outwards so as to represent the position of the horns. (*Dunbar.*)

(1) Sign for **Bison;** (2) motion of milking made with both hands. (*Cheyenne* I.)

Hold the crooked right index at the right side of the head to represent the horns, and then make the sign for **Female,** applied to animals. (*Dakota* I.) "Horns and sex."

Coward, cowardice. See **Fear.**

Crane.

Open both hands, move them from the sides of the chest outward and backward (as if swimming, but near the body); then close the right hand, leaving the forefinger extended and slightly curved; pass it from before the chin, upward, forward, and slightly downward, forming an arc to indicate the long neck of the bird. (*Dakota* V.)

Crazy or demented.

Raise right hand, with fingers partly distended, above the head, the hand drooping, and make quick circles close around top of the head from right to left, and a shake of the head. (*Ojibwa* IV.)

Move the opened right hand through a circle above and around the head. (*Dakota* IV.) "Head turned."

Cross, Sulky.

Place clinched right hand (**C** with thumb close to forehead), turn the hand till palm faces outward, then move it to the front slightly. (*Cheyenne* II.)

Same as the sign for **Anger**, not made however with as much force and omitting any decided facial expression. (*Dakota* I.) "Derivative of anger."

Crow.

Fingers and thumb of right hand brought to a point and motion of pecking slightly made. (*Cheyenne* I.)

Cutting anything in pieces.

Draw the right hand, palm backward, fingers pointed obliquely upward toward the left, four or five times across the advanced left hand, forearm, and arm, each time at a different place. (*Dakota* IV.)

——— With an ax.

With the right hand flattened (**X** changed to right instead of left), palm upward, move it downward toward the left side repeatedly from different elevations, ending each stroke at the same point. (*Dakota* I, V.) "From the act of felling a tree."

Repeat the sign for **Ax** several times, making the cuts from different points, but terminating about the same place each time. (*Dakota* VI; *Hidatsa* II; *Arikara* I.)

——— With a knife.

Left hand (**M** 1) extended in front left side of body on level breast, and the right hand (**S**) brought crosswise over the extended fingers just below the knuckles. (*Dakota* I.) "Cutting anything in two with a knife."

(1) Left hand open, flattened and held out, fingers of right hand also open and flattened and placed above left hand, little finger toward palm of left hand; (2) suddenly drawn once or twice across the palm of the left hand to imitate the act of cutting. (*Cheyenne* I.)

With the right hand, or index only, imitate cutting the left forefinger as if it were a stick. (*Dakota* VII.)

The hands are held as in sign for **Knife**, and then the right hand is moved upon the left crosswise to represent the act of cutting. (*Mandan and Hidatsa* I.)

Deaf-mute natural sign.—Place the forefinger of one hand upon that of the other, and slide it along in imitation of the action. (*Ballard.*)

Use the shut hands as if to cut a stick with a knife. (*Hasenstab.*)

Dance, Calumet.

The hand extended with the edge upward, and with the arm waved sideways, with a motion like that of a swing. (*Long.*)

Dance, To. Dancing.

Rise on toes and fall two or three times, the hands and arms by the side naturally. (*Ojibwa* IV.)

Fingers and thumb of the right hand hooked (**E**), hand horizontal, back outward at the right side of the body is carried up and down several times with a moderately quick movement, ending each at the same point, in imitation of the up and down motion of the body in Indian dancing, consisting in keeping time with the legs to the tom-tom. (*Dakota* I.) "From a method of dancing."

Danger.

Crouch the body slightly, bend the knees forward, bend the head forward, raise the shoulders a trifle, extend both hands in front naturally, palms down, bend elbow at nearly right angles, right hand in advance of left, move both slowly down a few inches and hold still. (*Ojibwa* IV.)

Right hand with first and second fingers extended (others closed, thumb resting on third finger (**N** 1), except that the hand is horizontal, back upward), directly forward in front of the right shoulder and then drawn back at the same time the body is thrown back. *Cautious, pru-*

dent are indicated in this sign. The idea being that the person shall be cautious in his movements as there may be danger ahead. (*Dakota* I.)

Bring the body quickly to an erect posture, at the same time gazing intently toward the expected source of danger with a look of fright. (*Wyandot* I.)

Place hand in front of breast, fingers hooked as though holding a knife, back outward (**F** 1, horizontal and back outward), then make motion as though cutting out the heart, first with the downward movement turning back of fist upward, then with the quick movement upward throwing back outward again. (*Sahaptin* I.) "Cutting out the heart."

Bring the right hand from the right side and back of the body as if grasping a twig, bring the hand before the breast, make the sign for **Battle,** for **No,** and for **Go.** (*Apache* I.) "Information of the coming of that which will cause a fight, or questionable security, and a desire to go rather than encounter it."

——— Dangerous.

The left fist placed horizontally before the lower end of the sternum, the right forearm passing before it pointing to the front and left, so that the fist is about four inches before the left. Then raise the left fist slightly and throw it forward and down to the same horizon forcibly, followed by the right which forms a larger semicircle and ceases before the left as previously. (*Kaiowa* I; *Comanche* III; *Apache* II; *Wichita* II.)

Darkness. (Compare **Night.**)

Make the sign for **Sun,** then extend the hands horizontally forward, backs upward, and pass one over the other two or three times touching it. (*Long.*)

Both hands spread out flat and cross vertically past each other before the face and neck. (*Cheyenne* I.)

Same as the sign for **Night** or **Clouds.** (*Dakota* I.)

Deaf-mute natural sign.—Point at your eyes partly shut and then turn your face to the sky with your hand waving over your head. (*Cross.*)

Daughter. See **Relationship, daughter.**

Day. (Compare **Sun** and **East**; and **Long,** in **lapse of time.**)

Place both hands at some distance in front of the breast, apart and back downward, elevate the index finger and move it forward to indicate one, twice for two, etc. (*Wied.*) This is still in use. The holding up of the index-fingers is not essential. When the hands are held as described they are first placed near to one another and then moved apart, as you suggest. Since my memory is assisted I remember this sign well. (*Matthew.*) There is no observable difference in either execution or conception between the sign (*Oto and Missouri* I) and *Wied's.* (*Boteler.*)

Make a circle with the thumb and forefinger of both hands in sign of the sun. (*Burton.*)

Bring both hands simultaneously from a position in front of the body, fingers extended and joined, palms down one above the other, forearms horizontal, in a circularly separating manner, to their respective sides, palms up and forearms horizontal; *i. e.*, "everything is open." This sign is the reverse of that for **Night.** (*Arapaho* I.)

Another sign may be indicated by making the sign for **Sleep**, and one finger touched or held up above, being the equivalent of "one sleep," one day. (*Arapaho* I.)

Forefinger of right hand crooked and held toward the east to represent the sun, hand elevated, finger uppermost and passed in a semicircle down toward the west. Both hands slightly spread out and elevated to a point in front and considerably above the head, then brought down in semicircle to level below shoulders ending with outspread palms upward. (*Cheyenne* I.)

When speaking of a day, they pass the finger slowly along the entire vault of heaven, commencing at the east and terminating in the west. This is the sign for "one day." (*Ojibwa* I.)

Both hands loosely extended, palms down, the right lying over the left; then pass them outward toward their respective sides turning the palms up in so doing. (*Absaroka* I; *Shoshoni and Banak* I; *Wyandot* I.)

Both hands (**W**) are raised above the head the extended fingers horizontal, pointing toward each other (meeting), palms down, arms necessarily somewhat bowed. Open up the hands so that the fingers point upright and at once carry the arms out to their full extent to the sides on the level of the shoulders, bringing the palms up (**X**). (*Dakota* I.) "The opening of the day from above." "The dispersion of darkness."

From positions a foot or eighteen inches in front of the lower part of the chest, the open hands pointing forward, near together, palms upward, are to be separated by carrying them out a foot or eighteen inches. (*Dakota* IV.) "All open."

Another: From positions a foot or eighteen inches in front of the lower part of the chest, pointing obliquely forward and inward, palms downward, the right two or three inches above the left, separate the hands about two feet, carrying each one outward through an arc, gradually turning the palms until they are upward. Uncovering the sun. (*Dakota* IV.)

A day is indicated by making the sign for **Sun**, and moving the hand thus formed from the left horizon to the right, forming a half circle, and indicating the course of the sun through the sky. (*Dakota* V.)

Right hand closed, forming a circle with the index and thumb, move the hand from east to west, following the course of the sun. (*Dakota* VII.)

The head is turned toward the orient and eyes wonderingly upwards. The right arm is then elevated semi-flexed to level of left shoulder. The hand in position (**J** 1) modified by index being a little more opened (and horizontal palm inwards). The hand thus pointing toward eastern horizon, is made to traverse the arc of the vault of the heavens, followed by the eyes until it stops on level with right shoulder, arm extended, crooked index pointing west. (*Oto and Missouri* I.) "The time between the rise and encompassed by the course and setting of the sun."

Both flat hands with palms down are held horizontally before the breast, the right over the left, then throw them outward toward their respective sides, turning the palms up in doing so. (*Kaiowa* I; *Comanche* III; *Apache* II; *Wichita* II.)

Close the right hand leaving the index bent in the form of a half circle, the index also extended, then pass the hand from east to west. (*Pai-Ute* I.)

Sign made for **Sun**, and pass across the zenith from east to west. (*Apache* I.)

Thumb and index circled; sweep hand from east to west across the sky. In rapid use thumb and index are often parted, and the hand is swept through a very short arc, not above shoulder. (*Apache* III.)

The French deaf-mutes fold the hands upon each other and the breast, then raise them, palms inward, to beyond each side of the head.

——— To-day. (Compare **Now.**)

Touch the nose with the index tip, and motion with the fist toward the ground. (*Burton.*)

(1) Both hands extended, palms outward; (2) swept slowly forward and to each side, to convey the idea of openness. (*Cheyenne* I.) This may combine the idea of *now* with *openness*, the first part of it resembling the general deaf-mute sign for **Here** or **Now.**

Designate the hour simply. See **Hour.** (*Arapaho* I.)

Point with the hand to the east, and carry it slowly overhead to the west. (*Iroquois* I.)

First make the sign for **Now,** and then the sign for **Day.** (*Dakota* I.) "Now with openness."

Make the sign for **Day,** to signify the period of time distinct from night. This being completed, the right hand was brought in modified position (**I**), from the stop at right to the center of forehead and made to describe a semicircle forward from the body toward the ground. The finger remains hooked and palm outward and downward. (*Oto and Missouri* I.) "The day that is now before me, or present time."

——— Evening.

Forefinger of right hand crooked as in sign for **Morning,** and lowered toward the west, followed by the sign for **Night.** (*Cheyenne* I.)

Make the sign for **Sun,** and hold it toward the western horizon. (*Absaroka* I; *Shoshoni and Banak* I.)

The right hand and arm, the former in modified position (**I**), index is more opened; hand is horizontal and palm inward; execute the sign for **Day.** At the completion of this sign the hand is quickly everted and assumes type position (**M**); index is moved to the west. In this sign, as well as that for *noon*, *morning*, *day*, *to-morrow*, *yesterday*, &c., the subject must be with his back to the north and right hand west. (*Oto and Missouri* I.) "When the sun goes down or at the conclusion of the day."

Point the extended index to the western horizon, or a little lower, by curving the index in that direction, the palm being below and still horizontal with the shoulder. (*Pai-Ute* I.)

Make the sign for **Sun,** passing the hand slowly from the western horizon to a short distance below it, holding it there a moment. (*Kaiowa* I; *Comanche* III; *Apache* II; *Wichita* II.)

——— Hour, time of day.

Join the tips of the thumb and forefinger of the same hand the interior outline approximating a circle, and let the hand pause at the proper altitude east or west of the assumed meridian. (*Arapaho* I.)

Is indicated by making the sign for **Sun,** and holding the hand in that portion of the course followed by the sun, to indicate the time to be expressed. (*Absaroka* I; *Shoshoni and Banak* I.)

Can only be approximately told by placing the sign for **Sun** in the position in the heavens corresponding with the hour, dividing the time between sunrise and sunset into as many equal spaces as there are hours. (*Dakota* I.)

——— Morning.

Make the sign for **Sun,** and hold it toward the eastern horizon. (*Absaroka* I; *Shoshoni and Banak* I.)

First make the sign for **Night** and then the sign for **Day**. Morning can likewise be made by simply using the sign for day. (*Dakota* I.) "Darkness has gone—daylight has come."

Make the sign for **Sun,** and hold the hand below the eastern horizon, moving it slowly to or a little above it. (*Dakota* VII.)

The arm and hand (right) are elevated to left shoulder and the hand in modified position (**I**) is made to describe the sign for **Day.** The right arm is then brought extended across the upper part of chest, with hand in position (**J** 1), horizontal; both arm and index fully extended toward the east, followed by the eyes. *This morning* is described in same manner after making the sign for *To day.* (*Oto and Missouri* I.) "The beginning of a day or to day."

Make the sign for **Sun,** and hold the hand toward the eastern horizon a short time, or bring it above the horizon from a short distance below it, slowly. (*Kaiowa* I; *Comanche* III; *Apache* II; *Wichita* II.)

Point with the extended index to the eastern horizon. (*Pai-Ute* I.)

Deaf-mute natural signs.—Point to the eastern horizon and move the forefinger a little way upward. (*Ballard.*)

Open your eyes, and, from the open hand, raise your head to its erect position, as if you have just now arisen from bed. (*Cross.*)

——— Noon.

Make the sign for **Sun,** and hold it toward the zenith, so that the eye can see through the circle formed by the thumb and index (*Absaroka* I; *Shoshoni and Banak* I.)

Make the sign for **Sun,** holding the hand overhead, the outer edge uppermost. (*Dakota* IV.)

The hand and arm are elevated to left shoulder and, with fingers in (**I**) position, modified by index being more open, horizontal, and palm to the breast, begin the sign for **Day.** This sign is then half executed, and the hand stopped in a line with the middle of the forehead. The index is then made to point to the zenith. The hand, when arrested, is in type position (**I**); it is quickly everted and assumes position (**J**) to complete the sign. (*Oto and Missouri* I.) "The half course of the sun or middle of the day."

——— To-morrow.

Describe the motion of the sun from east to west. Any number of days may be counted upon the fingers. (*Burton.*)

Sign for **Night** followed by sign for **Sunrise.** (*Cheyenne* I.)

Join the tips of the thumb and forefinger of the same hand, the interior outline approximating a circle, and describe therewith a space from left to right, corresponding to the supposed course of the sun during twenty-four hours. To distinguish an hour, let the hand pause at the proper altitude, east or west of the assumed meridian. See **Hour.** (*Arapaho* I.)

First make the sign for **Sleep;** then follow by counting one (see **Counting**), and finish with the sign for **Sunrise.** "One night's sleep."

Bring the hand to side of head, and with head reposing in right palm, eyes closed, head and hand inclining to the right, which is the sign for **Sleep.** This completed, the right arm and hand are brought across the top of the chest, and describe the sign for **Day** or **To-day.** This sign, as is evident, is a compound sign, comprising that for *Sleep*, *Day*, or *To-day*. (*Oto and Missouri.*) "The day after we sleep."

Make the sign for **Day** once, then reverse it, and stop at the point in the heavens indicating the intended time of the day. If noon, point directly upwards. To express two or more days, make the sign for **Day** and hold up the proper number of fingers. (*Iroquois* I.)

Deaf-mute natural sign—Place the hand on the cheek, incline the head, and shut the eyes, to denote **Sleep,** and then raise the head and open the eyes to signify **Awake,** and hold up the forefinger to denote **One;** that is to say, in one day. (*Ballard.*)

——— Yesterday.

Make with the left hand the circle which the sun describes from sunrise to sunset, or invert the direction from sunset to sunrise with the right hand. (*Burton.*)

Sign for **Night** and **Sunset.** (*Cheyenne* I.)

The sign for **Day, To-morrow,** the motion reversed. (*Arapaho* I.)

Omit the sign for **Sunrise;** otherwise as **Day, To-morrow.** (*Dakota* I.) "Have slept one night."

The right hand and arm are elevated, and, with hand in type position (**I**), modified by index being more opened, horizontal, and palm to the breast, made to execute the sign for **Day.** At the conclusion of this sign the fingers are all collected droopingly extended, touch at points slightly curved. In this position the hand makes a sudden forward movement to the ground. (*Oto and Missouri* I.) "The day or sun that has gone down."

Make the sign for **Night,** followed by that for **Before, in time.** (*Kaiowa* I; *Comanche* III; *Apache* II; *Wichita* II.)

Dead, death.

Throw the forefinger from the perpendicular into a horizontal position towards the earth, with the back downward. (*Long.*)

Hold the left hand flat over the face, back outward, and pass with the similarly held right hand below the former, gently striking or touching it. (*Wied.*) The sign given (*Oto and Missouri* I) has no similarity in execution or conception with *Wied's*. (*Boteler.*) This sign may convey the idea of "under" or "burial," quite differently executed from most others reported. Dr. *McChesney* conjectures this sign to be that of wonder or surprise at hearing of a death, but not a distinct sign for the latter.

The finger of the right hand passed to the left hand and then cast down. (*Macgowan.*)

Place the palm of the hand at a short distance from the side of the head, then withdraw it gently in an oblique downward direction and incline the head and upper part of the body in the same direction. (*Ojibwa* II.) This authority notes that there is an apparent connection between this conception and execution and the etymology of the corresponding terms in Ojibwa: "he dies," is *nibo;* "he sleeps," is *niba.* The common idea expressed by the gesture is a sinking to rest. The original significance of the root *nib* seems to be "leaning;" *anibeia,* "it is leaning; *anibekweni,* "he inclines the head sideward." The word *niba* or *nibe* (only in compounds) conveys the idea of "night," perhaps as the falling over, the going to rest, or the death, of the day. The term for "leaf" (of a tree or plant), which is *anibish,* may spring from the same root, leaves being the leaning or downhanging parts of the plant. With this may be compared the Chahta term for "leaves," literally translated "tree hair."

Hold both hands open, with palms over ears, extend fingers back on brain, close eyes, and incline body a little forward and to right or left very low, and remain motionless a short time, pronouncing the word *Ke-nee-boo* slowly. (*Ojibwa* IV.)

Left hand flattened and held back upward, thumb inward in front of and a few inches from the breast. Right hand slightly clasped, forefinger more extended than the others, and passed suddenly under the left hand, the latter being at the same time gently moved towards the breast. (*Cheyenne* I.) "Gone under."

The left hand is held slightly arched, palm down, nearly at arm's length before the breast; the right extended, flat, palm down, and pointing forward is pushed from the top of the breast, straight forward, underneath, and beyond the left. (*Shoshoni and Banak* I.)

Both hands horizontal in front of body, backs outward, index of each hand alone extended, the right index is passed under the left with a downward, outward and then upward and inward curved motion at the same time that the left is moved inward toward the body two or three inches, the movements being ended on the same level as begun. "Upset, keeled over." Many deaths, repeat the sign many times. The sign furnished you before, I have since ascertained is not used in the sense of dead, death. The sign credited to *Titchkemátski* (*Cheyenne* I) expresses "gone under," but is not used in the sense of death, dead, but going under a cover, as entering a lodge, under a table, etc. (*Dakota* I.)

Make the sign for **Alive,** then the sign for **No.** (*Dakota* IV.)

Hold the left hand, palm downward and backward, about a foot in front of the lower part of the chest, and pass the right hand from behind forward underneath it. Or from an upright position in front of the face, back forward, index extended and other fingers closed, carry the right hand downward and forward underneath the left and about four inches beyond it, gradually turning the right hand until its back is upward and its index points toward the left. (*Dakota* IV.) "Gone under or buried."

Hold the left hand slightly bent with the palm down, before the breast, then pass the extended right hand, pointing toward the left, forward under and beyond the left. (*Dakota* VI, VII.)

Hold the right hand flat, palm downward before the body, then throw it over on its back to the right, making a curve of about fifteen inches. (*Dakota* VI; *Hidatsa* I; *Arikara* I.)

Extend right hand, palm down, hand curved. Turn the palm up in moving the hand down towards the earth. (*Omaha* I.)

The countenance is brought to a sleeping composure with the eyes closed. This countenance being gradually assumed, the head next falls toward either shoulder. The arms, having been closed and crossed upon the chest with the hands in type positions (**B B**) are relaxed and drop simultaneously toward the ground, with the fall of the head. This attitude is maintained some seconds. (*Oto and Missouri* I.) "The bodily appearance at death."

Place the open hand, back upward, fingers a little drawn together, at the height of the breast, pointing forward; then move it slowly forward and downward turning it over at the same time. (*Iroquois* I.) "To express 'gone into the earth, face upward.'"

The flat right hand is waved outward and downward toward the same side, the head being inclined in the same direction at the time, with eyes closed. (*Wyandot* I.)

Hold the left hand loosely extended about fifteen inches in front of the breast, palm down, then pass the index, pointing to the left, in a short curve downward, forward and upward beneath the left palm. (*Kaiowa* I; *Comanche* III; *Apache* II; *Wichita* II.)

Bring the left hand to the left breast, hand half clinched (**H**), then bring the right hand to the left with the thumb and forefinger in such a position as if you were going to take a bit of string from the fingers of the left hand, and pull the right hand off as if you were stretching a string out, extend the hand to the full length of the arm from you and let the index finger point outward at the conclusion of the sign. (*Comanche* I.) "Soul going to happy hunting-grounds."

Close both eyes, and after a moment throw the palm of the right hand from the face downward and outward toward the right side, the head being dropped in the same direction. (*Ute* I.)

Touch the breast with the extended and joined fingers of the right hand, then throw the hand, palm to the left, outward toward the right, leaning the head in that direction at the same time. (*Apache* I.)

Palm of hand upward, then a wave-like motion towards the ground. (*Zuñi* I.)

Deaf-mute natural signs.—Place the hand upon the cheek, and shut the eyes, and move the hand downward. (*Ballard.*)

Let your head lie on the open hand with eyes shut. (*Cross.*)

Use the right shut hand as if to draw a screw down to fasten the lid to the coffin and to keep the eyes upon the hand. (*Hasenstab.*)

Move the head toward the shoulder and then close the eyes. (*Larson.*)

The French deaf-mute conception is that of gently falling or sinking, the right index falling from the height of the right shoulder upon the left forefinger, toward which the head is inclined.

——— Die, To.

Right hand, forefinger extended, side up, forming with the thumb an (**U**); the other fingers slightly curved, touching each other, the little finger having its side towards the ground. Move the hand right and left, then forward, several times; then turn it over suddenly, letting it fall towards the earth. (*Omaha* I.) "An animal wounded, but staggering a little before it falls and dies."

——— Dying.

Hold the left hand as in dead, pass the index in the same manner underneath the left, but in a slow, gentle, interrupted movement. (*Kaiowa* I; *Comanche* III; *Apache* II; *Wichita* II.) "Step by step; inch by inch."

——— nearly, but recovers.

Hold the left hand as in *dead;* pass the index with a slow, easy, inter rupted movement downward, under the left palm, as in dying, but before passing from under the palm on the opposite return the index in the same manner to point of starting, then elevate it. (*Kaiowa* I; *Comanche* III; *Apache* II; *Wichita* II.)

Deaf.

The tip of the right index is inserted in the right ear, withdrawn and rotated around the organ. (*Dakota* I.) "This would seem to indicate 'a noise in the ear preventing hearing.'"

Deep.

Right hand with fingers extended, joined, back outward, ends of fingers pointing straight down, is carried downward in front of the right side of the body to near the ground, the body being inclined forward at the same time, touching the ground, indicating that although deep the bottom had been reached. (*Dakota* I.) "Finding the depth of water, etc."

Place the flat hand, palm down, several feet from the earth, or at such elevation to show depth, and pass it slowly to one side. (*Dakota* VI; *Hidatsa* I; *Arikara* I.)

Deer.

The right hand extended upwards by the right ear with a quick puff from the mouth. (*Dunbar.*)

The forefinger of the right hand is extended vertically, with the back toward the breast; it is then turned from side to side, to imitate the motion of the animal when he walks at his leisure. (*Long.*)

Pass the uplifted hand to and fro several times in front of the face. (*Wied.*) I have given you much the same sign for "white-tailed deer," but I have said that all fingers, except the index, were flexed. This may not be absolutely essential in making the sign, which is simply designed to imitate the peculiar motion of the tail when the animal stands observing. (*Matthews.*) The right arm is elevated to the front of the body and the right index-finger, in position (**M**), projected forward several times. The hands, with extended and divergent fingers, as in position (**P**), are now placed aside of the head to represent the animal's horns. The right index-finger is now extended full length and wagged behind. The above *Oto* sign differs from that of *Wied's* in execution, and in the latter the conception is wanting, which is the animal with branching horns that runs in jumps. (*Boteler.*)

Extend the thumbs and the two forefingers of each hand on each side of the head. (*Burton.*)

Both hands, fingers irregularly outspread and elevated to sides of head to represent outspread horns of deer. (*Cheyenne* I.) This sign is made by our deaf-mutes.

With the right hand in front of the body on a level with the breast and about eighteen inches from it, back of hand to the right (**S** 1), make quick sidewise motions with the hand in imitation of the motion of the deer's tail when running. The wrist is fixed in making this sign. Corrected from sign before given. (*Dakota* I.) "Movements of the deer's tail when running."

Hands applied to each temple, fingers spread and pointing upward. (*Dakota* II.) "Horns."

Bring right hand to a level with the shoulder, thumb and forefinger curved, three fingers curved and nearly closed (×). Move thumb and forefinger forward, wrist motion alone, imitating the movements of the animal. (*Omaha* I.) "The deer runs."

Similar to the preceding, but with wrist and arm motion, with hand raised high above the head. (*Omaha* I.) "The deer bounds away."

Hold the right hand down, extend index-finger, thumb tip touching tip of middle finger; shake index-finger rapidly. (*Omaha* I.) "The deer goes along rapidly, making its tail shake."

Hold right hand next to left shoulder, thumb crossing middle finger at first joint; move index-finger back and forth to and from left shoulder. (*Omaha* I.) "The deer's tail shows bright or red in the distance as it leaps away."

Similar to the preceding, but the hand is held in front on a level with the face and moved right and left. (*Omaha* I.) "The deer's tail shows white suddenly."

The right arm is elevated and the right index extended is thrown forward several times. The hands with extended and divergent fingers as in position (**I**), are placed aside the head to resemble the branching horns. Finally the extended index-finger is wagged from the seat of body. (*Oto and Missouri* I.) "The branching horns, short tail and leaping motion of the animal."

Imitate the motion of a deer running by closing the hand, palm downward, except the two forefingers, which are only a little bent downward. Then move the hand forward from you with a rather slow up-and-down motion, slightly moving the two forefingers—the whole designed to mimic the long jumps with which a deer starts off. If necessary to explain this further, place one hand on each side of the head to represent horns. (*Iroquois* I.)

Place both hands, flat, with fingers and thumbs spread, on either side of the head and a short distance from it. (*Ute* I.)

Place the hands with fingers fully extended and spread about twelve inches from either side of the head and slightly above it. (*Apache* I.)

——— Black-tailed [*Cariacus macrotis* (Say), Gray].

First make the gesture for **Deer** then indicate a tail. (*Wied.*) When he says "indicate a tail," I have little doubt that he refers to the sign I have already given you for *deer, blacktailed.* (*Mandan and Hidatsa* I.) I do not think it is either essential or common to "make the preceding gesture" either "first" or last. (*Matthews.*)

Extend both hands, fingers close together, place them with palms to front on each side of the forehead, fingers upward, and then make short motions back and forth in imitation of the movements of the large ears of the animal. (*Dakota* II.)

With the right index, its palm inward, on the right side, at the height of the hip, pass the left index, back forward, from its middle forward to its end. (*Dakota* IV.) "That much black."

The left hand is held pendent a short distance in front of the chest, thumb inward, finger ends approximated to each other as much as possible (*i. e.*, with the 1st and 4th drawn together under the 2d and 3d). The right hand is then closed around the left (palm to back and covering the bases of the left-hand fingers) and drawn downward, still closed, until it is entirely drawn away. This sign seems to represent the act of smoothing down the fusiform tuft at the end of the animal's tail. (*Mandan and Hidatsa* I.)

——— White-tailed [*Cariacus virginianus macrurus* (Raf.) Coues].

Move the right hand, its palm obliquely forward and downward, from side to side two or three times about a foot, through an arc of a circle, at the height of the hip, on the right side (the tail of the deer); then with the palm inward and the fingers pointing forward, cast the hand forward several times through an arc of about a foot to imitate the jumping of a deer. (*Dakota* IV.)

The right hand is held upright before the chest, all fingers but the index being bent, the palm being turned as much to the front as possible. The hand is then wagged from side to side a few times rather slowly. The arm is moved scarcely or not at all. This sign represents the motion of the deer's tail. (*Mandan and Hidatsa* I.)

Elevate the forearm to the height of the elbow, pointing forward, extend the flattened hand, pointing upward, with the palm forward; then throw the hand right and left several times, the motion being rotation of the forearm. (*Arikara* I.) "From the motion and white appearance of the deer's tail in running."

Defiance. I defy you.

Point to the person you defy to do the act with the right index (others closed), and then turn the hand, extend the fingers so that they will appear as figured in (**V**), when the hand is drawn in to the body with considerable force. This would indicate *Come and do it*, but the emphasis of the motion and accompanying facial expression indicate something stronger than invitation. (*Dakota* I.)

The right hand closed with the index only extended and slightly crooked, palm facing front; hold about twenty inches in front of the chest and wave the finger from side to side, movement being made at the wrist. (*Kaiowa* I; *Comanche* III; *Apache* II; *Wichita* II.)

Italian sign.—Every tyro in Latin knows that extending the middle or little finger, gestures still made every day at Rome, was a token of scorn or defiance.

Bite the forefinger, commonly with the joint nearest the end bent—a gesture which throws light on "the biting the thumb at me," with which Romeo and Juliet commences.

When one would kill the hopes of a mendicant and say he will give nothing with emphasis, he blows on his hand and shows it wide open to the petitioner or elevates one forefinger, shaking it gently to and fro. (*Butler.*)

Deity; God; Great Spirit; Great Father; Master of Life.

Blow upon the open hand, point upward with the extended index-finger whilst turning the closed hand hither and thither, then sweep it above the earth and allow it to drop. (*Wied.*) I have never seen this sign. I once extracted a bullet from the leg of a deaf and dumb Indian of Sioux descent who had lived long among the Arikaras. When the operation was completed he made some preliminary sign (for thanks) which I did not observe well, and then pointed to me and upward to the sky. One of the best interpreters in the country, who was a bystander, told me that the Indian thanked me and the Great Spirit. I say "me" first as he first pointed in my direction. (*Matthews.*) There is no similarity in the sign (*Oto* I) and *Wied's*. (*Boteler.*)

When speaking of the Great Spirit they usually make a reverential or timid glance upwards, or point the forefinger perpendicularly but gently to the sky. (*Ojibwa* I.)

First make the sign for **Sacred** and then make the sign for **Big.** (*Dakota* I.) "The great Sacred Being."

First make the sign for **Medicine-man,** and then after placing the opened relaxed hands, palms inward, abont six inches apart, upright, just above the head, move them apart to arm's-length. (*Dakota* IV.) "A thunder cloud coming up and spreading."

Point toward the zenith, allowing the eyes to follow the same direction. (*Dakota* VII.)

The arms are flexed and both hands elevated open as in position represented (**W**). When hands are elevated on sides of head on a level with the eyes a uniform swaying "to and fro" movement is performed, followed by an upward movement of right hand as in hand position (**J**). (*Oto* I.) "Him above who is an angel 'on the wing.'"

Close the right hand, leaving the index straight and extended (or slightly curved); hold it before the face, move the hand quickly forward and downward for a distance of about six inches, then pass the index vertically upward before the face about as high as the top of the head. (*Ute* I.)

Elevate hand toward sky, deliberately; looking upward. (*Apache* III.)

Extend the right hand with the index pointing upward. the eyes also being turned upward. (*Wichita* I.)

Deaf-mute natural signs.—First close the hand except its forefinger, and then move it up slowly and also turn the eyes toward the clouds with a solemn expression. (*Hasenstab.*)

To look up to heaven, at the same time to point with the forefinger as if to point to heaven. (*Zeigler.*)

Depart. See **Go.**

Destroyed; all gone; no more.

The hands held horizontally, and the palms rubbed together, two or three times round, the right hand is then carried off from the other, in a short, horizontal curve. (*Long.*) Rubbed out. This resembles the Edinburg and our deaf-mute sign for "forgive" or "clemency," the rubbing out of offense.

Left hand held in front, outspread, palm upward, right-hand fingers extended, palm down, swept rapidly across palm of left. Right hand sometimes held out after passing over the left with fingers wide spread and shaking as if expressing *bad* in the sense of *no good.* (*Cheyenne* I.)

Move both hands as if in act of rending asunder or tearing in pieces and throwing aside with violence and sudden jerks of hands and arms. (*Ojibwa* IV.)

——— Exhausted, consumed, completed.

The left hand, extended, palm upward, pointing diagonally forward and to the right, is placed before the chest. Then the right hand, palm downward, is laid transversely on the left, and, while the left remains stationary, the right is carried forward a foot or more with a rapid sweep. To show a gradual diminution and then exhaustion, the right palm revolves on the left once or oftener with a gradual motion, as if some plastic substance were made spherical between the palms. (*Mandan and Hidatsa* I.)

Place the open left hand in front of the navel, palm backward, and move the opened right hand, palm downward, in a horizontal circle above it. (*Oto* I.) "All caught, killed, or destroyed. That's the end of it."

This sign resembles that for **Go.** The right arm, flexed, and the hand, in position (**B** 1), modified by being more horizontal, is brought to the epigastrium. The arm is then suddenly extended, hand likewise, with fingers extended approximated, palms downward. The left hand in the same position approaches the side of the right, both now being in type-position (**W**), diverge and sweep backward. (*Oto and Missouri* I.) "All gone or swept away."

——— By force.

Imitate the breaking of a stick in the two hands and throwing the pieces away, then lightly strike the palms and open fingers of the hands together as if brushing dust off them. The amount of force used and the completeness of the destruction is shown by greater or less vigor of action and facial expression. (*Dodge.*)

——— Anything of little importance, and by accident or design.

Indicate the object, then slightly strike the palms and open fingers of the hands together, as if brushing dust off from them. (*Dodge.*)

Rotate the right palm upon the left as if rubbing something into smaller fragments. (*Kaiowa* I; *Comanche* III; *Apache* II; *Wichita* II.) "Rubbing out; grinding to atoms."

Raise both hands to position on right of face, fingers extended, separate, and pointing upward, palms facing each other (**R** 1, right and left), then with an energetic movement throw both to left side, as though throwing something violently to the ground; then place hands near together and make sign for **All gone.** (*Sahaptin* I.) "Broken into pieces; nothing of its former self remaining."

Deaf-mute natural sign.—Imitate the act of breaking and move the hands in a curve in opposite directions. (*Ballard.*)

——— Ruined.

An article might be ***destroyed***, ***ruined***, by breaking, when the sign for ***break*** might be used in connection with the sign for the particular article destroyed; same by ***fire:*** but the idea of ***rubbed out***, as contained in ***gone***, appears to be the prevalent one. (*Dakota* I.)

——— Spent.

Bring both hands together in front of the breast, the left hand under the right, horizontal, flat, palm upward, fingers extended obliquely toward the right (**X**), right hand with fingers extended obliquely toward the left, flat, palm downward (**W**) on the palm of the left, slide the two palmar surfaces over each other, moving the hands slightly inward and outward. (*Dakota* I.) "Rubbed out."

Dialogue. See **Speaking.**

Different, contrasted.

First and second fingers of right hand extended, separated (others closed), is passed from the right breast outward, with back of hand toward the right, forefinger pointing obliquely upward, and the second finger pointing straight outward or forward. (*Dakota* I.) "The idea of this is contained in contrasting the appearance of the two fingers; one of the fingers is so and the other is not so—*i. e.*, not the same, different."

Dirty.

Point to the ground with the right index, and then carry the hand in front of the face, with fingers separated, hand upright, back outward (**R**), move up and down and around, as though covering the face and breasts. (*Dakota* I.) "Dirt from the ground covering the face," etc.

Discharge of a gun. See **Gun.**

——— of an Arrow. See **Arrow.**

Disgust. (Compare **Contempt.**)

Extend both hands quickly outward from near the face, palms out; turn away the face slightly from the object of disgust; extend the hands and arms but partly. (*Ojibwa* IV.)

Tap the left breast (heart) with the right hand, fingers extended, back outward, then the hand is carried forward outward in front of the right breast, so that the palm is up, fingers extended, pointing outward, hand horizontal (**X**), where it is slightly rotated or curved a few times, and then carried directly outward toward the right, back outward. (*Dakota* I.) "I am not pleased or satisfied."

Sign as for **Surprise, Wonder,** then turn the head over the left shoulder, retaining the hand over the mouth. (*Omaha* I.) "As at the sight of a dead body."

Shake the head slowly from side to side, at the same time throwing the open right hand, palm down, outward toward the right side. (*Wyandot* I.)

Avert the head and make the sign of **Negation.** (*Apache* III.) N. B. In narrative, simply *disapproval.*

Dissatisfaction, or Discontent.

The extended finger placed transversely before the situation of the heart, rotate the wrist two or three times gently, forming a quarter of a circle each time. (*Long.*)

The index right hand held transversely before the heart and rotated from the wrist several times. (*Dakota* I) "Heart ill at ease; disturbance of the organ." Our aborigines, like modern Europeans, poetically regard the heart as the seat of the affections and emotions, not selecting the liver or stomach as other peoples have done with greater physiological reason.

Distance, Long; Far.

Place the hands close together and then move them slowly asunder, so slowly that they seem as if they would never complete the gesture. A *Cheyenne* sign. This sign is also made to indicate great antiquity in time. (Report of Lieut. *J. W. Abert* of his examination of New Mexico in the years 1846–'47. Ex. Doc. No. 41, 30th Congress, 1st session, p. 426.)

(1). Head drawn back or elevated, eyebrows contracted as if looking to great distance, right hand raised to level of chin, palm upward: (2) pushed forward with a curved motion in the direction in which the speaker is looking. (*Cheyenne* I.)

A slowly ascending movement of the extended hand, fingers joined, from the body and in the direction desired to be indicated. (*Arapaho* I.)

Raise the right hand to a natural position, thumb below the two first fingers, then the arm with rising motion as high as top of head, stretching it out as far as possible; then bend the hand downward, the arm falling slowly, advance the body slightly without moving feet. (*Ojibwa* IV.)

Right hand, forefinger of which is extended and points forward (other fingers closed) (**M**), palm toward the left, is extended in front of the breast as far as the arm will reach, the body being inclined forward at the same time, the extended position maintained a moment and then the body and hand are brought back with a moderately quick movement. To a limited and very indefinite extent distance is sought to be expressed by the distance the arm is extended and the amount of leaning forward. (*Dakota* I.) "Distance."

From an upright position just in front of the right shoulder and a little above it, palm forward, fingers relaxed and thumb against the index, move the right hand forward and upward through an arc to arm's length and to the height of the head, gradually turning the palm downward.

Or with the fingers at right angles with the palm, pointing toward the left and their backs forward, thumb in palm, move the right hand from the right shoulder forward and upward to arm's length. The left hand, its back forward, to be held in front of the right breast. (*Dakota* IV.)

Elevate the right hand to a position in front of the chest, drop the index-finger toward the ground, then move it forward and upward, resting it on a line slightly above the horizon, the eyes following the direction indicated. The idea of much greater distance, or to intensify the extreme distance intended, is done by making the gesture a little quicker, turning the finger higher at the end of the sign, and throwing back the head slightly. (*Dakota* V.)

Place the flat hand in front of the chest, pendent, then gently indicate a course from before the body to arm's length, fingers pointing above the horizon. (*Dakota* VII.)

Describe the curve by raising the hand above and in front of the head (**J**), index extended more to the right or left according to the direction intended and the hand that is used. (*Omaha* I.) "Go around in that way."

Another: Throw the right hand backward over the shoulder, index extended, then upward and forward. (*Omaha* I.)

Another: Raise the arm above and in front of the head, then pointing forward with index, shoot the hand forward to arm's length horizontally. (*Omaha* I.)

The arms are folded and the hands, in type position (**C**), are approximated before the chest. The arms and hands then widely diverge from the body to signify intended space existing between two objects or persons, as the case may be. (*Oto and Missouri* I.) "Wide extent of space between."

Point with the extended index at arm's length a little above the horizon, the eyes following the same direction. (*Pai-Ute* I.)

Push the hand forward and a little downward (**T** on edge, palm in); repeat with hand a little higher, again and again, each time higher and farther forward. (*Apache* III.) "Over several mountains."

Deaf-mute natural signs.—Extend the forefinger forward, and look into the distance. (*Ballard.*)

Bending slowly your body forward, move your outstretched hand, with your eyes looking over a great space, in the direction the hand moves. (*Cross.*)

Move the open hand up in a horizontal line from back to front and, at the same time, blow lightly from the mouth. (*Hasenstab.*)

Separate the two fists from each other. (*Larson.*)

——— Half way.

Make sign of for **Far away** and then bring the hand half way back to the shoulder. (*Dakota* IV.)

——— Short.

Same motion of hand as **great distance,** only project arm forward a little, do not raise so high and drop more quickly, wrist and hand bent down more, no movement of body. (*Ojibwa* IV.)

The forefinger left hand extended straight, upright (**J** except palm outward, edge of fingers sidewise) is held on the level of the eyes 18 inches in front, and then the right hand in the same position (**J**, etc.) is carried upward close to the body as high as the right eye, and then directly forward to *near* the left hand (which is stationary), a little to the right side and behind it, so that the extended forefingers are *nearly* on a line and with their palmar surfaces outward. (*Dakota* I.) "Approaching, coming near any person or object."

Hold the right hand as for **Far away,** and place it in front of the right breast and close to it. (*Dakota* IV.)

Do, To. I have done it.

Throw the opened right hand, palm inward, from an upright position in front of the right shoulder forward and downward until it is horizontal and eighteen inches in front of the right breast. (*Dakota* IV.)

——— Do it again. (Compare **Repeat.**)

Pass the opened right hand, palm backward, straight across from right to left eighteen inches in front of the chest, beginning a little outside of the line of the right side and stopping in front of the right breast. (*Dakota* IV.)

Doctor, Physician.

Make motions and movements of head and body as if hunting and examining herbs and roots, also by signs of smelling and tasting, and; as if holding the thing gathered, point with the right hand, motion as if to drink or swallow. (*Ojibwa* IV.)

Right hand closed, leaving the first two fingers extended and slightly separated, elevate to before the forehead and move the fingers circularly, passing the hand slightly upward at the same time. (*Absaroka* I.) "Superior knowledge."

Make the signs for **White man**, and **Shaman.** (*Dakota* VI, VII; *Hidatsa* I; *Arikara* I.)

The left hand is extended as in (**W**) and the back of it rubbed by index of right hand extended as in (**J**). Both hands are then brought tremblingly to sides of chest as in type (**Q.**) The hands are then carried to the sides of head and extended indices press the temples. The right hand is then swept vertically edgewise up before the face and retained thus several seconds; left falls to side. (*Oto* I.) "One distinguished, who rubs together or writes that which removes inward distress."

First make the sign for **White man,** which must be quickly followed by placing the closed right hand before the face, leaving the index and second fingers extended and separated, then rotate the hand in passing it upward and forward to the height of the top of the head. (*Kaiowa* I; *Comanche* III; *Apache* II; *Wichita* II.) "Superior knowledge."

Right hand closed with the index only extended, elevate to before the forehead, and move circularly, passing the hand slightly upward at the same time. (*Shoshoni and Banak* I.) "Superior knowledge."

——— Indian. See **Medicine Man.**

Dog.

Pass the flat hand from above downward, stopping at the height of a dog's back. (*Wied.*) In the *Oto* sign the hand is opened, palm downward, the whole then held about the height of the animal from the ground in passing from side to side before the body. It is evident at a glance that this sign and *Wied's* are similar in conception and execution. The slight difference may be attributed to the contributor's misconstruction. (*Boteler.*)

Is shown by drawing the two forefingers slightly opened horizontally across the breast from right to left. (*Burton.*) This sign would not be intelligible without knowledge of the fact that before the introduction of the horse, and even yet, the dog has been used to draw the tent-poles in moving camp, and the sign represents the trail. Indians less nomadic, who built more substantial lodges, and to whom the material for poles was less precious than on the plains, would not, perhaps, have comprehended this sign, and the more general one is the palm lowered as if to stroke gently in a line conforming to the animal's head and neck. It is abbreviated by simply lowering the hand to the usual height of the wolfish aboriginal breed, and suggests *the* animal *par excellence* domesticated by the Indians and made a companion.

Right hand lowered, palm downward, as if to stroke a dog's head and back, and moved along from before backward horizontally, conforming to the head, neck, and back of a dog, elevated or depressed to express difference of size. (*Cheyenne* I.)

Extend and spread the right, fore, and middle fingers, and move the hand about 18 inches from left to right across the front of the body at the height of the navel, palm downward, fingers pointing toward the left and a little downward, little and ring fingers to be loosely closed, the thumb against the ring-finger. (*Dakota* IV.) "Represents the lodgepoles and 'travois' which were formerly dragged by the dogs."

Fore and second fingers of right hand (others closed) extended, separated V-shaped, carried with a downward winding motion from about the left shoulder in front of the body to the right, the hand stopping right side of the body well to the front at about the height of a good-sized dog. (*Dakota* I.) "From the use of the dog in carrying the lodgepoles."

Cross the thumb over the middle finger, three fingers being closed, back of hand down. The forefinger curved represents the tail. (*Omaha* I.) "The tail moving up and down as he walks."

Another: Hold right hand in front of you, thumb over first joint of middle finger (not crossing it), forefinger straight and pointing up; shake it right and left, moving it about 6 inches. (*Omaha* I.) "The tail, elevated, shakes in the air, as when he scents any game."

Imitate the quick, running movement of a dog, by moving the hand from the breast forward, palm downward, and at the same time partly closing and opening all the fingers together quite rapidly. (*Iroquois* I.)

Close the right hand, leaving the index and second fingers only extended and joined, hold it forward from and lower than the hip, and draw it backward, the course following the outline of a dog's form from head to tail. (*Kaiowa* I; *Comanche* III; *Apache* II; *Wichita* II.)

The French and American deaf-mutes specifically express the dog by snapping the fingers and then patting the thigh, or by patting the knee and imitating barking with the lips.

——— Or wolf.

Bring both arms together at wrists and hands together in position (**W**). Sometimes the sign is made further plain by the Indian making an accompanying bow-wow with mouth. (*Oto* I.) "Height of animal and size."

Done, finished.

The hands placed edge up and down, parallel to each other, the right hand without, which latter is drawn back as if cutting something. (*Dunbar.*) "An *end* left after cutting is suggested; perhaps our colloquial 'cut short.'"

A motion of cutting with the right hand. (*Macgowan.*)

Both fists clinched, placed before the chest, palms facing, then drawn apart and outward toward their respective sides. (*Absaroka* I; *Shoshoni and Banak* I.)

The hands placed in front of body horizontal, with fingers extended, arched, meeting near the tips, thumbs resting on tips of forefingers, back of hands outward, separate the two hands by carrying to the right and left slowly. (*Dakota* I.) "Drawn apart, an end left."

From positions about 4 inches apart and a foot in front of the upper part of the chest, the upright fists, palms facing, are to be separated about 3 feet, each one being made to describe an arc downward. Or, after placing the half-closed hands near together and opposite each other, obliquely upward and inward, about a foot in front of the upper part of the chest, quickly separate them about 3 feet. (*Dakota* IV.) "We will part."

The right arm is flexed a little over a right angle and brought closer to the front of chest. The hand in position (**S** 1), modified by being horizontal with palm toward the breast and tips of index and ring fingers resting on biceps flexor muscle of opposite arm. Sometimes the arm is held in same posture away from the body. The sign is completed by a cutting stroke with hand, edges up and down, from left to right. (*Oto* I.) "We cut it short; are done."

Hold the left fist horizontally in front of the body, then pass the flat and extended right hand, edgewise, quickly downward in front of the left. Sometimes the right is passed down in front of and by the knuckles of the left. (*Kaiowa* I; *Comanche* III; *Apache* II; *Wichita* II.) "Cut off."

Deaf-mute natural signs.—Hold both hands slightly extended, with the palms downward, and then turn the hands over, at the same time giving a side movement to right and left with each arm. (*Hasenstab.*)

Similar to the sign for **None,** meaning "nothing more." (*Ballard.*)

The French and our deaf-mutes give a cutting motion downward with the right hand at a right angle to the left.

Door, entrance, &c.

The arms are elevated and semi-flexed before body. The hands are then collected in type-position (**U**) and tips of index fingers made to touch. The arms, with finger-tips in contact, then approach and diverge several times. (*Oto* I.) "The triangular entrance to the wigwam opens and shuts."

Doubt. See **Indecision.**

Drawn out.

Both hands extended in front of body, the left on the outside and at a lower elevation than the right, both hands fists, the left (**B**), the right with back looking toward the right; draw both arms in toward the body, and then carry out again, repeating the movements several times as though drawing out for some distance. (*Dakota* I.) "From drawing out of the water or hole a person or thing."

Dream.

First make the sign for **Sleep,** and then the hand is carried downward from the head and curved upward and inward to the right breast, and then thrown out from the body (turned over) with a downward curved movement, *i. e.*, hand turned partially over so that the extended fingers point toward the left, palm of hand nearly flat, thumb outside, pointing obliquely downward. (*Dakota* I.)

Drink, drinking.

The hand is partially clenched, so as to have something of a cup shape and the opening between the thumb and finger is raised to the mouth as in the act of drinking. (*Long.*)

Scoop up with the hand imaginary water into the mouth. (*Burton.*)

Forefinger of right hand crooked, thumb side of hand inward and brought to mouth in upward curve, then suddenly curved outward and downward several inches. (*Cheyenne* I.)

Do the hand in the shape of a cup. (*Macgowan.*)

With the right held in front of the body, make with the thumb and fingers a circle resembling a cup, thumb and forefinger forming the top or rim, and then carry to the mouth, hand horizontal, back outward. (*Dakota* I.) "Drinking from a cup or glass."

The right arm is flexed and the hand, in type-position (**Y** 1), modified by collecting the fingers and letting the thumb rest against palms of the index and middle fingers, is then brought to and from the mouth several times successively. The right or left hand may be used. The motion resembles much an upward rotary movement of a spoon from a dish, the hand being quite hollowed in the center to resemble a cup-shaped vessel in which water must be taken. (*Oto and Missouri* I.) "To take up in a hollow vessel to the mouth."

Collect the fingers of the right hand to a point, and bring it to the mouth, palm first. (*Wyandot* I.)

The right hand with tips of fingers and thumb brought nearly to a point is brought to the mouth once or twice, as if drinking from a cup. (*Apache* I.)

Hand half closed, supinated, and an up-and-down motion in front of the mouth. (*Zuñi* I.)

Italian sign.—Is imitated with the fist. (*Butler.*)

Drowned.

First make the sign for **River,** then make the sign for **Gone under;** should the person not be rescued make the sign for **Dead, death;** should he be rescued, however, make the sign for **Drawing out,** and if, after getting the body out, resuscitation should occur, the sign for **Life, living,** should be made to indicate that he has been raised (up) to life again. (*Dakota* I.) "This is a graphic picture."

Drum.

Make the sign for **Kettle,** with hands further apart, and omit the part indicating placing on the fire; then hold the left hand stationary and raise the right hand upward (Fist **B**, except back upward and inward); strike down with it to near the left hand, and repeat several times as though striking the head of a drum. (*Dakota* I.) "From beating the drum."

Duck.

The sign for **Turkey,** then the sign of **Water,** and lastly the sign of **Swimming.** (*Dunbar.*)

Earth. Ground; Land.

The two hands, open and extended, brought horizontally near each other opposite to either knee, then carried to the opposite side and raised in a curve movement until brought round and opposite to the face. (*Dunbar.*)

Right hand elevated to level of face, flattened, palm upward, thumb pointing forward, little finger pointing to left at right angle to thumb; hand moved horizontally forward and outward to represent extended surface. Sometimes both hands employed, left hand flat, palm upward, fingers pointing to the right. (*Cheyenne* I.)

First point toward the ground with the right index, and then bring both hands together in front of the chin, fingers extended, pointing toward the front, palms down, flat horizontal (**W**), and carry to the right and left with a curved motion, arms nearly extended. (*Dakota* I.) "The entire earth, without end."

Eyes cast vaguely about, the right arm is flexed to the front of the body, the hand in type-position (**P** 1) modified by being inverted horizontal with palm downward. The hand is not held rigidly; it is then moved forward and downward, and the palm point of the right middle finger is made to touch the ground. The whole arm is then raised and the hand in type-position (**W**) describes a circle before the body. (*Oto and Missouri* I.) "The spherical object touched; beneath me."

Deaf-mute natural sign.—Point at it and then move the open hand horizontally, meaning how extensive the land is. (*Cross.*)

——— As soil.

Right index points toward the ground in front of the right side of the body. (*Dakota* I.) "Designating the ground."

East. (Compare **Day, morning.**)

Point to the east; *i. e.*, point of sunrise. (*Arapaho* I.)

Forefinger of right hand crooked to represent half of the sun's disk and pointed or extended to the left, then slightly elevated. (*Cheyenne* I.)

Simply point toward the east with the extended right index. (*Dakota* I.) "Direction."

Eat, eating; I have eaten.

The fingers and thumb are brought together in opposition to each other, into something of a wedge shape, and passed to and from the mouth four or five times, within the distance of three or four inches of it, to imitate the action of food passing to the mouth. (*Long.*)

Imitate the action of conveying food with the fingers to the mouth. (*Burton.*)

Bringing the fist to the mouth. (*Macgowan.*)

Join the tips of the fingers and thumbs, and move them back and forth towards the mouth. *Cheyenne* sign. (Report of Lieut. *J. W. Abert*, loc. cit., p. 431.)

Fingers and thumb of right hand placed together as if grasping a morsel, brought suddenly upward to level of mouth and moved toward it and downward in the direction of the throat to suggest the act of cramming. (*Cheyenne* I.)

Right hand scoop-fashion, back of hand outward at the front of body, height of abdomen, is carried to the mouth as though conveying food, and repeated several times. (*Dakota* I.) "From the movements of the hands in eating."

——— I have eaten.

After making the above sign the extended thumb and forefinger of the right hand (other fingers closed) is passed, back of hand outward, horizontal, &c., from the stomach upward in front of body and mouth and above the latter. (*Dakota* I.) "I am full."

Close the hand, allowing the forefinger extended, then move it up and down before the face several times toward and from the mouth, as if ramming food into the mouth with the finger. (*Dakota* V.)

Bring the thumb, index, and second finger to a point, and make repeated motions downward before the face toward the mouth. (*Dakota* VII.)

The hands, with arms flexed at elbow, and fingers, as in type (**E** 1), modified by hand being held horizontal, palm up, are brought alternately to the mouth and back, as it were, to the table or dish, as motion of Chinese eating rice. (*Oto* I.) "To fill up in parts."

Collect the thumb, index and second fingers to a point, hold them above and in front of the mouth, and make a repeated dotting motion toward the mouth. (*Kaiowa* I; *Comanche* III; *Apache* II; *Wichita* II.)

Place the thumb across the palmar surface of the partly extended fingers of the right hand, then bring the tips of the fingers to the mouth quickly several times. (*Pai-Ute* I.)

Extend the index (or index and second finger) of the right hand, hold it in front of and a little higher than the mouth, palm towards the neck, then make repeated thrusts toward the mouth with the finger. (*Ute* I.) "Indicates the direction in which food goes."

Close the right hand, leaving the index extended but slightly curved; place the hand in front of and a little above the mouth, making a quick motion of pointing to and back from the mouth several times, as if ramming down anything. (*Apache* I.)

Italian sign.—Is imitated with the open hand (*Butler.*)

——— Something to eat.

Join the ends of the fingers and thumb of the right hand, place them upright six or eight inches in front of the mouth, backs forward; bend the hand at the wrist and turn the ends of the fingers and move them toward the mouth and then downward to the upper part of the sternum (breastbone). (*Dakota* IV.) "Putting food into the mouth and swallowing it."

Egg.

The right hand held up with the fingers and thumb extended and approaching each other as if holding an egg within. (*Dunbar.*)

With the finger and thumb of the right hand suggest the outline of an egg. The fingers and thumb of the left hand are also sometimes shaped in the same manner, and placed over the points of the right as they assume the position described. (*Cheyenne* I.)

Elk (*Cervus canadensis*).

Stretch the arms high and alongside of the head. (*Wied.*) This sign is still in use. (*Matthews.*) In the *Oto* sign both arms are elevated and the hands opened, but fingers approximated, are then placed aside the head. The hands are in type position (**T**), palms outward; the ball of the thumb rests against parietal ridge. Though there is an evident incompleteness in *Wied's* description, a marked identity in position of the arms and the probable conception is observable. The sign is completed by the right index marking on the extended left index the animals short tail. (*Boteler.*)

Is signified by simultaneously raising both hands with the fingers extended on both sides of the head to imitate palmated horns. (*Burton.*)

All or most of the fingers of both hands held together and brought to the sides of the head to represent the palmated horns of an elk. (*Cheyenne* I.)

The same as **Deer,** except that after the first position both hands with fingers spread are carried upwards and outwards to imitate the branching horns of the animal. (*Dakota* II.)

Extend and widely separate the fingers and thumbs of both hands, place them upright, palms inward, just above and in front of the ears, and shake them back and forth three or four times. (*Dakota* IV.) "The elk's antlers.

The arms are elevated and the hands placed aside of the head with the fingers extended but approximated as in position (**W** 1)—more erect. The extended right index-finger marks off on the extended left index the length of the tail. (*Oto and Missouri* I.) "The short tail and broad horns of the animal."

Place both flat hands with fingers and thumbs spread upward and outward from either side of the head at arm's length. (*Ute* I.)

Embroidery.

Same as that part of the sign for **Beads** where the hands are in front of the body, left stationary, and the right holding the sinew moved to and over it. (*Dakota* I.) "From embroidering with beads."

End. See **Done.**

Enemy.

With the right fore and middle fingers spread, pointing toward the left, and backs forward, throw the hand about a foot forward and outward two or three times from near the face ("I don't want to see him"); touch the chest over the heart and afterward make the sign for **Bad** ("my heart is bad"); then, from just in front of the right eye, its palm forward, push the upright fist forward six inches and a little toward the left, at the same time turning the palm backward ("I am angry"). *Dakota* IV.)

The Italian sign for enmity.—Opposition in the ends of the middle fingers touching each other, and all the rest of the fingers clinched. (*Butler.*)

Enough, a belly full. (Compare **Glad and Full.**)

The sign for **Eating** is first made, then the thumb and forefinger are opposed to each other so as to form a semicircular curve, which is elevated along the body from the belly to the neck, in order to indicate that the interior is filled with food up to that part. (*Burton.*)

First make sign for **Eating,** then stretch the forefingers and thumbs apart, as if to span something; then place the hand near the stomach, and move it up along the body until the muscle connecting the thumb and forefinger rests in the mouth. *Cheyenne* sign. (Report of *J. W. Abert*, loc. cit., p. 431.)

Right hand brought to front of body, forefinger pointing to and resting against stomach and gently drawn along upward to the throat and continued upward and outward. (*Cheyenne* I.)

The Sioux Indians express **Enough** by **End, Done.** (*Dakota* I.) "I am done; have had enough."

Make the sign **Desire,** then the sign for **No.** (*Dakota* IV.)

Deaf-mute natural signs.—Move the hand (palm upward) in a gentle curve downward, with a suitable expression of countenance. (*Ballard.*)

Bend your head a little forward and move the hand (held horizontally) upward on the throat. (*Cross.*)

Move the forefinger across the front of the neck from side to side, so as to indicate that the throat is full. (*Hasenstab.*)

Move to and fro the outstretched hand over the other hand. (*Larson.*)

Entering a house or lodge. See **Lodge.**

Equal. See **Same.**

Exchange. See **Trade.**

Excited; excitement.

Same sign as for **Coward, Fear.** The heart being the primary seat of the emotions from the standpoint of the Indian, he acknowledges no such thing as excitement of mind, therefore this word is included in *fear*, for every Indian who allows his heart to *flutter* is considered by his people a coward. (*Dakota* I.) "From fear, coward."

Eye.

Simply touch the eye with the right index. (*Dakota* I.)

Face.

The hand is passed downward in front of the face, once only, from forehead to chin or a little below the chin. (*Mandan and Hidatsa* I.)

Fail, to.

Left hand stationary, horizontal, fingers nearly closed, back of hand outward about eighteen inches in front of the breast, and the right index (others closed) is brought upward close in front of the body to the breast and then carried out to the left hand with slight up and down jerking motions, finger upright, back of hand inward, and then draw the right hand back directly to the breast, and make the sign for **No.** (*Dakota* I.) "He did it not. He did not come; did not keep his appointment."

Failure.

Hold the left hand edgewise before the breast, pointing forward, then bring the extended index, pointing, toward the left palm; touch it, and throw the index in a short curve over and downward on its back, resembling the sign for *Dead.* (*Kaiowa* I; *Comanche* III; *Apache* II *Wichita* II.) "Interrupted in progress; defeated efforts."

Fall, to.

Left hand extended to the front, palm down, then bring tips of all the fingers together, open and shut, as if letting something fall. (*Omaha* I.)

Fall, first (of the leaves).

Raise the left hand above the head, forefinger extended, move right and left with a waning and trembling motion. (*Omaha* I.)

False. See **Lie.**

Far, a long way off. See **Distance, far.**

Fasting.

Blacken face, neck, and hands. In camp sit cross-legged, eyes bent on the earth; fold hands in front, palms up, remaining motionless, frequently uttering in a low voice, progress (?). When standing perfectly erect and motionless. (*Ojibwa* IV.)

Fat.

Raise the left arm with fist closed, back outward, grasp the arm with the right hand and rub downward thereon. (*Wied.*)

(1) Both hands, loosely closed, brought near to and on a level with the shoulders; (2) pushed a few inches straight forward and slightly upward. (*Cheyenne* I.) "Probably lumps of fat taken from a large or broad animal."

Both fists clinched, placed before the breast, thumbs touching and palms downward; then draw them outward and downward, forming the upper half of a circle. (*Absaroki* I; *Shoshoni and Banak* I.)

Sign for **Man,** and then the sign for **Big** made in front of the abdomen. (*Dakota* I.) "Big in body."

Hold the opened left hand obliquely upward toward the right, a foot in front of the breast, palm upward, backward, and to the right; grasp the ulnar side of the hand between the fingers and thumb of the right hand, the thumb on the palm, and rub it lightly from the base of the fingers to the wrist several times. (*Dakota* IV.) "Thick, and therefore fat."

Both arms are flexed inward and both hands brought before the body, divergent and extended fingers drooped. The hands are much in the position (**Q**) on type plates. In this position the hands describe a semicircle over the abdomen, and are carried over the limbs severally. If the sign is applied to any other object than man, the sign corresponding to said object is first made. (*Oto and Missouri* I.) "Of increased dimensions or that which increases one's size."

Father. See **Relationship.**

Fear, cowardice; coward.

The two hands with the fingers turned inward opposite to the lower ribs, then brought upward with a tremulous movement as if to represent the common idea of the heart rising up to the throat. (*Dunbar.*)

The head stooped down and the arm thrown up to protect it; a quick motion. (*Long.*)

Point forward with the index, followed by the remaining fingers; each time that is done draw back the index. (*Wied.*) Impossible to keep the coward to the front.

(1) Fingers and thumb of right hand, which droops downward, closed to a point to represent a heart; (2) violently and repeatedly beaten against the left breast just over the heart to imitate palpitation. (*Cheyenne* I.)

May be signified by making the sign for a **Squaw,** if the one in fear be a man or boy. (*Arapaho* I.)

Crook the index, close the other fingers, and, with its back upward, draw the right hand backward about a foot, from eighteen inches in front of the right breast. (*Dakota* IV.) "Drawing back."

Make the sign for **Brave,** then throw the right hand, open and flat, outward toward the right. (*Absaroka* I; *Shoshoni and Banak* I.) "Not brave."

Right hand (**Q**) in front of left breast, back outward, and carried forward for about six inches with a tremulous motion of the fingers. Many of the Sioux, however, do not move the hand from the breast. (*Dakota* I.) "Excitement; fluttering of the heart."

Cross the arms over the breast, fists closed; bow the head over the crossed arms, but turn it a little to the left. (*Omaha* I.)

Strike the right side of the breast gently with the palmar side of the right fist; then throw the hand downward and outward toward the right, suddenly snapping the fingers from the ball of the thumb, where they had been resting, as if sprinkling water. (*Wyandot* I.)

Deaf-mute natural signs.—Represented by shuddering with the shoulders and moving the body slightly backward. (*Ballard.*)

Run backward. (*Larson.*)

Place the forefinger between the upper and lower teeth. (*Zeigler.*)

The French deaf-mutes, besides beating the heart, add a nervous backward shrinking with both hands. Our deaf-mutes omit the beating of the heart, except for excessive terror.

Female. (Compare **Woman.**)

Bring the two hands open toward the breast, the fingers approaching, and then move them outward. (*Dunbar.*)

Elevate the open right hand, pointing forward, to the level of and to the right of the shoulder; draw the fingers back, keeping them together until the tip of the forefinger rests against the tip of the thumb, forming an almond-shaped opening between the thumb and forefinger. (*Cheyenne* I.) "Probably from its resemblance to the appearance of the external female genitals."

——— Applied to animals.

Same as the sign for **Woman.** (*Dakota* I.) "Designating sex. From the flowing hair of woman."

——— Generic.

Fingers and thumbs of both hands separated and curved; place the hands over the breasts and draw them forward a short distance. (*Ute* I.)

Fence.

Both hands extended, fingers spread; place those of the right into the spaces between those of the left; then indicate a zigzag course forward with the extended index. (*Wyandot* I.) "Position of rails in a fence, and the zigzag course."

Fight, fighting. (Compare **Battle.**)

Make a motion with both fists to and fro, like a pugilist of the eighteenth century, who preferred a high guard. (*Burton.*)

Joining hands rapidly. (*Macgowan.*)

Same sign as for **Battle.** (*Dakota* I.)

Both hands clinched, holding them palm to palm at a distance of about four inches from one another; form short vertical circles, as if "sawing" a hard-mouthed horse. (*Dakota* VI, VII.)

Both hands closed, forefingers elevated and extended, facing one another; move palms toward and from each other. (*Omaha* I.)

Close both hands, except the index-fingers. Hold them before the breast, the index-fingers upright, and move these from side to side, opposite each other, like two antagonists facing one another and avoiding each other's blows. This movement, followed by that for **Dead,** would express "They fought, and one was killed." If more than one, hold up two or more fingers. (*Iroquois* I.)

Both hands brought up nearly as high as the face, about twelve inches apart, fingers pointing toward those of the opposite hand; the fingers are then moved carelessly as the hands are brought toward and from one another. (*Wyandot* I.)

Deaf-mute natural signs.—Act as if you strike your left open hand with your right clinched hand, and do the act of striking several times successively. (*Hasenstab.*)

Raise the fists as fighters begin to fight. (*Larson.*)

——— Indian.

Extend both arms, hands clinched (**J**); place the tips of the index-fingers together, and push them first one way and then the other, still keeping the tips of the forefingers together. Facial expressions add greatly to the intensity of all signs. (*Comanche* I.) "Wrestling."

Finished. See **Done.**

Fire, burning.

The two hands brought near the breast, touching or approaching each other, and half shut; then moved outward moderately quick, the fingers being extended and the hands a little separated at the same time, as if to imitate the appearance of flame. (*Dunbar.*)

The act of striking fire with the flint and steel is represented; after which the ascent of the smoke is indicated by closing the fingers and thumb of the right hand, holding them in a vertical position, with the hand as low as convenient; the hand is then gradually elevated, and the fingers and thumb a little expanded to show the ascent and expansion of the volume of smoke. (*Long.*)

Hold the fingers of the right hand slightly opened and upward and elevate the hand several times. (*Wied.*) The body is bent somewhat forward—the right index, middle finger, and thumb are then approximated at their points and hand is held, droopingly, near the ground. The hand thus cup-shaped is made to open and close successively and rise by jerks, like the jumping and cracking of a flame. Although the position of the hand is reversed in *Wied's* sign, there is a marked similarity of execution and conception in the two. (*Boteler.*) The sign may portray the rising forked tongues of the flame.

Blow it and warm the hands before it. To express the boiling of a kettle, the sign of **Fire** is made low down and an imaginary pot is eaten from. (*Burton.*)

Scratching the breast. (*Macgowan.*)

Raise and lower the hand alternately, palm up, the fingers extended upward and moving in imitation of tongues of flame. (*Arapaho* I.)

Right forearm in vertical position, and hand slowly elevated, the fingers and thumb pointing upward, being meanwhile opened and closed two or three times. (*Cheyenne* I.)

Right hand (**P**) extended in front to the ground, fingers pointing upward, raise the hand slowly, with a tremulous motion, not more than a foot from the ground to denote a small fire, such as Indians build in their lodges. For a larger fire raise the hand higher with the tremulous motion. Of course the body will have to be inclined forward in making this sign. (*Dakota* I.) "From the flame and smoke of a fire."

Raise the right hand several times from near the ground, its back forward, fingers pointing upward and a little bent and separated. (*Dakota* IV.) "Flames shooting upward."

The hand is brought near the ground in hand type (**G**), the body at the same time inclined forward. The fingers and thumb are then successively opened and closed as the hand is elevated by jerks, as it were. (*Oto* I.) "Jumping and crackling of a flame."

Right hand, palm toward and a short distance before the waist, fingers extended upward and separated, moved upward about eight or ten inches several times. (*Ute* I.) "Tongues of fire, flame."

Deaf-mute natural signs.—Blow through the mouth, and move the hand up and down rapidly. (*Ballard*,)

Open both hands freely, move them up fast and down slowly several times, and at the same time keep the mouth, half open, blowing. (*Hasenstab.*)

Move the fingers of both hands together upward with a little blow sent from the mouth. (*Larson.*)

——— To light a.

Hold the left hand before the body, palm down and arched, collect the fingers and thumb of the right hand to a point and pass them quickly along the thumb of the left from the basal joint toward the tip; then place the right hand quickly under the arched left for a moment; then suddenly closing both hands, side by side, move them upward a few inches, quickly extending the fingers and thumbs in so doing; palms forward. (*Ute* I.) "Striking a match, kindling the brush, and smoke."

Hold both hands before the body, straight, fingers curved sufficient for the tips to be directed toward their respective hands; strike from above downward with the right, so that the finger-nails strike those of the left in passing. (*Apache* I.) "From the old method of obtaining fire with flint."

——— To make a.

First make the sign for **Fire**; then hold the arched left hand close to the ground with the palm downward, placing the right, fingers and thumb directed to a point, underneath the left. (*Apache* I.) "Kindling grass or other combustibles."

Fish.

Hold the upper edge of the hand horizontally, and agitate it in the manner of a fan but more rapidly, in imitation of the motion of the tail of the fish. (*Long.*)

Make the sign for **River, Lake,** or **Pond,** and then with the right index in front of the body move in imitation of a swimming fish. (*Dakota* I.) "From the manner of swimming."

The extended right hand, thumb upward, fingers pointing forward, is held near the body, in front and to the right of the median line; it is then moved rather gently forward with a laterally waving motion, so as to represent the movements of a fish. (*Mandan and Hidatsa* I.)

Flame. See **Fire.**

Flat. (Compare **Big** in the sense of **Flat.**)

Bring both hands together in front of the breast, fingers extended and pointing outward, forward, palms upward (**X**), flat, and carry the arms out to the sides of the body as far as they can be extended and as nearly on the same level as possible. (*Dakota* I.) "A level or flat piece of ground."

Deaf-mute natural sign.—Move one hand horizontally over the other. (*Ballard.*)

Flour.

Pantomimic.—Simulate kneading dough. (*Arapaho* I.)

With the right hand (**Q**) in front of body as though holding flour or any pulverized substance and sprinkling or sifting it through the thumb and forefinger. (*Dakota* I.) "A fine substance."

The arms are extended in front of the body, parallel with the hands in positions (**S, S**), modified by being held horizontal. The palms are then approximated as in slapping together, and opened several times successively. Finally the hands are made to turn over, the palms loosely in contact, in a tumbling manner, the right and left alternately on top. (*Oto and Missouri* I.) "The sign represents the Indian's mode of flattening the dough for cakes."

Fly, To (as a bird).

Imitate with crooked elbows the motion of wing during flight. (*Arapaho* I.)

Bring the hands slightly in front of their respective sides of the head (hands as in **T**), and, by bending the wrists, make the forward and backward movements as nearly as possible in imitation of the movements of a bird's wings in flying. (*Dakota* I.)

Fool, Foolish.

The finger is pointed to the forehead, and the hand is then held vertically above the head and rotated on the wrist two or three times. (*Long.*) Rattle-brained.

Place the hand in front of the head, back outward, then turn it round in a circle several times. (*Wied.*) Still used. Also for **Crazy,** I think. (*Matthews.*)

Sign for **Man;** right hand extended downward, palm outward, fingers unclosed and shaken. When referring to a particular person the finger of the right hand is pointed at him, eyes resting on him critically, brows raised and contracted as in pity and aversion. (*Cheyenne* I.) The shaking of the opened fingers gives the idea of "looseness" without reference to the head.

Bring the right hand to the medial line of the forehead (**R**), fingers but slightly separated, where it is rotated several times two or three inches in front. (*Dakota* I.) "Rattle-brained."

Move the opened right hand through a small circle two or three times in front of the forehead, the palm toward the left, fingers separated a little and pointing upward. (*Dakota* IV.) "Head turned."

Rotate the extended and separated index and second fingers of the hand upward and toward the left before the forehead. (*Dakota* VI; *Hidatsa* I; *Arikara* I.)

French deaf-mutes shake the hands above the head after touching it with the index.

——— He is the greatest fool of all.

Sway the hand (**W** 1), palm downward "over all," then point to the person (fig. 1), then place end of fingers on forehead (**H** 1), and then swing hand around in circle in front of forehead, hand and fingers upright, joined, and palm oblique to face (**T** 1, palm oblique), and lastly make the sign **Bad.** (*Sahaptin* I.) "Of all, his brain whirls worst."

——— Your words are foolish.

After pointing to person addressed and making sign for words (as in **G**), the hand is moved to a point in front, but little to right, of forehead, fingers all naturally relaxed, pointing upward, palm quartering to face (**Y** 1, changed to vertical position), then swung around in small circle several times, then dropped to a point in front of body and thrown vigorously downward to side and rear. At beginning of this last motion the fingers are hooked, second resting against thumb, palm downward (**G** 1), but during backward movement the fingers are gradually thrown open, palm outward (**R** 1, fingers pointing obliquely downward). Sign **Bad** (**C**). (*Sahaptin* I.) "Words from whirling brain—bad."

Forest.

Slightly spread and raise the ten fingers, bringing the hands together in front of the face; then separate them. (*Wied.*) The numerous trees and their branches may be indicated, for a time obscuring the vision.

Spread the fingers of the right hand slightly; raise the hand on a level with the face, and while moving it from side it is gently thrust up and down. (*Dakota* V.) "The fingers represent the appearance of the trunks of the trees visible along the edge of the forest."

Raise the hand vertically, palm up, fingers partially closed and extended upward, in a manner indicative of the growth of trees. (*Arapaho* I.)

Make the sign for **Tree,** and then the sign for **Many.** (*Dakota* I.) 'Many trees."

Make the signs for **Ax** and **Trees.** (*Dakota* VI; *Hidatsa* I; *Arikara* I.) "Timber that is fit for cutting with an ax."

Make the sign for **Tree** several times, then throw the back of the upright and flat right hand toward the right, front, and left. (*Kaiowa* I; *Comanche* III; *Apache* II; *Wichita* II.) "Trees, trunks close" (represented by joined fingers), "all around."

Forget, forgotten.

Hands outspread, opened, palms downward; crossed as in sign for **Night or darkness** at a level of and angle with the elbows, left hand being over the right. (*Cheyenne* I.) Darkness in the memory.

Deaf-mute natural signs.—Put the hand on the forehead and draw it away. (*Ballard.*)

Have the head up suddenly, and open the mouth a little at the same time, and then nod, and the upper teeth rest on the lower lip. (*Hasenstab.*)

Place the forefinger on the forehead and then strike the lap with the shut hand. (*Zeigler.*)

Fort.

On level of the breasts in front of body, both hands with fingers turned inward, straight, backs joined, backs of hands outward, horizontal, turn outward the hands until the fingers are free, curve them, and bring the wrists together so as to describe a circle with a space left between the ends of the curved fingers. (*Dakota* I.) "A circularly fortified place."

Found, discovered. (Compare **See.**)

First make the sign for **See, to,** and then carry the right hand (**Q**) in front of body toward the ground and back to body, as though having picked something up. (*Dakota* I.) "Seeing and picking up anything."

Bring the left hand opposite the breast a foot or so away, fingers closed and slightly bent, palm downward as if it were concealing something; bring the right hand over it, hand in the position of being just ready to pick something up with the fingers; then pass the right hand over the left, the latter remaining still, and bring the fingers of the right hand together as if you had picked up something. (*Comanche* I.)

Deaf-mute natural signs.—Touch the eyelid, then bend your body, and, having pointed at the ground with your hand, clinch it and bring it up and disclose it to the eye. (*Cross.*)

Lower the open hand toward the ground, and then raise it shut up as if to indicate that something is picked up. (*Hasenstab.*)

Fowl. See **Bird.**

Fraction.

Indicate with the forefinger of the right hand the equivalent length of the left forefinger. (*Arapaho* I.)

Freezing. (Compare **Frost.**)

Make a closing movement, as if of the darkness, by bringing together both hands with the dorsa upward and the fingers to the fore; the motion is from right to left, and at the end the two indices are alongside and close to each other. (*Burton.*)

No sign separate from **Cold, It is; Ice;** or **Frost,** which denote different degrees of cold. (*Dakota* I.)

Deaf-mute natural signs.—Shudder from head to foot; then set in motion your feet after the manner of skating, and then move the hand about horizontally. (*Cross.*)

Raise the arms toward the breast and shake the fists, and then move the outstretched hands in a horizontal line. (*Larson.*)

Friend. (Compare **Salutation.**)

(1) Tips of the two first fingers of the right hand placed against or at right angles to the mouth; (2) suddenly elevated upward and outward to imitate smoke expelled. (*Cheyenne* I.) "We two smoke together."

Hold the extended left hand before the body, and grasp it with the right. (*Sac, Fox, and Kickapoo* I.)

Point forward and a little upward with the joined and extended fore and middle fingers of the right hand, which is to be placed a foot or so in front of the right breast, the little and ring fingers closed, thumb on middle joint of ring finger; move the hand upward to the right side of the face, then straight forward about eight inches, and then a little upward. Or hook the bent right index, palm downward, over the bent left index, palm upward, the hands to be about a foot in front of the body. This last they call a Mexican sign. (*Dakota* IV.)

Extend the right hand as if reaching to shake hands. (*Hidatsa* I; *Arikara* I.)

The left and right hands are brought to the center of chest open, then extended, and the left hand, with palm up, is grasped crosswise by right hand with palm down, and held thus several seconds. The hands are then in double position (**Y** 1), right inverted. Hands are now unclasped, and right fist is held in left axilla, by which it is firmly grasped. (*Oto* I.) "One whom I will not let go."

Bring both hands together in a full clasp of all the fingers, after the ordinary manner of handshaking. (*Comanche* I.)

——— Extraordinary.

Bring the two hands near each other in front, and clasp the two index-fingers tightly, so that the tips of the finger and thumb of each touch. (*Comanche* I.)

Clasp the two hands after the manner of our congratulations. (*Wichita* I.)

Friendly; friendship.

Raise both hands, grasped, as if in the act of shaking hands, or lock the two forefingers together while the hands are raised. This sign given by parties meeting one another to ascertain intentions. For more general idea of friendship clasp the left with the right. (*Burton.*)

Hands clasped in front of body, palm of left up, palm of right resting in that of left, hands shaken up or down one or more times. (*Cheyenne* I.)

Pantomimic grasping or shaking of the hand, or a pantomimic embrace. (*Arapaho* I.)

The left hand held horizontal, palm inward, fingers and thumb extended and pointing toward the right about a foot and a half from and in front of breast, is clasped by the right, carried up in front of body and out from breast, thumb and fingers pointing downward and drawn directly into the body. (*Dakota* I.) "Grasping the hand of a friend."

Our deaf-mutes interlock the forefingers for "friendship." clasp the hands, right uppermost, for "marriage," and make the last sign, repeated with the left hand uppermost, for "peace." The idea of union or linking is obvious.

Frost. (Compare **Freezing** and **Snow.**)

Begin with the sign of **Water,** then with the sign of **Night** or **Darkness,** then the sign of **Cold,** then the sign of **White,** and, lastly, the **Earth.** (*Dunbar.*)

First make the sign for **Grass**, and then the right hand (**W**) is moved as though waving it over the grass. (*Dakota* I.) "A covering of the grass."

Fruitless; in vain.

The left arm is brought forward, hand as in (**L** 1), modified by thumb being closed. The right hand then takes position (**S** 1), modified by being held horizontal. Now the left index, extended as above, punches the right palm, and is then swept backward and downward by left side. (*Oto* I.)

Full, as a box or sack.

Right and left hands (**W**, with fingers slightly bent) are brought together in front of body, ends of fingers pointing outward, then carry the right hand quickly over the back of the left, and back as though brushing off the surplus. (*Dakota* I.) "It is full; brushing off the surplus."

Deaf-mute natural signs.—Place the hand down, and raising it, and moving it right and left as if at the top of the sack or box. (*Ballard.*)

Clinch your hand in the form of the letter **C**, and over the supposed convex surface above it pass the other hand somewhat clinched also. (*Cross.*)

The same as **Enough.** (*Larson.*)

——— Appetite satisfied. (Compare **Enough.**)

Finger and thumb rising from the mouth. (*Macgowan.*)

Make the sign for **Eat**, and when completed and hand brought before abdomen, as in type (**T** 1) modified by being held horizontal and arched with back outward, it then describes an arc over abdomen as to indicate fullness. (*Oto* I.) "Filled up; distended."

Make the sign of **Eat**, then close the right hand, spreading the index and thumb wide apart, palm toward the body, then pass it from the breast upward to before the mouth. (*Dakota* VII; *Kaiowa* I; *Comanche* III; *Apache* II; *Wichita* II.)

Future, to come (in time).

Right index upright (**J**, except back of hand toward the body) is pushed straight forward, outward, from the shoulder and drawn back three or four times, arm extended to its full capacity. Seems to be connected with **Far.** (*Dakota* I.) "Far in time."

Gap; cañon.

Indicate the walls thereof with the hands, in front of the body, palms toward each other, fingers extended and pointed downward. (*Arapaho* I.)

Both hands eighteen inches in front of the breast, separated about six inches, fingers and thumbs pointing upright; with the palms facing each other (**S,** with edge of hands outward) thus (| |); draw the right hand inward about a foot, turn it so that the palm is downward, flat, fingers joined, pointing straight outward, and then push the hand forward so that it would go through the middle of the space formed by the hands in the first position (|). Often made with the fingers curved. (*Dakota* I.) "Sides of a cañon or ravine; passing through a cañon."

Gelt.

Bring the fingers and thumb of the left hand together as if something was held by them, and then approach the right hand and make the motion of cutting across what is supposed to be held in the left hand, and then draw off the right hand as if pulling away what has been cut. (*Dunbar.*)

Generous.

Hold both hands open, the palms above and held in front of breasts or body, then present toward the other party an open smiling countenance. (*Ojibwa* IV.)

The sign for **Good Heart** or **Big Heart** is made. (*Dakota* I.)

Ghost.

Sign for **Dead, Death,** and then the sign for **Man.** (*Dakota* I.) "Dead man from the spirit land."

Gimlet.

Index pointing to the center of the left hand forefinger (which indicates the handle or boring part). Then the screw motion with the right hand conveys the idea clearly. (*Zuñi* I.)

Give me, or **Bring to me.**

The hand half shut with the thumb pressing against the forefinger, being first moderately extended either to the right or left, is brought with a moderate jerk to the opposite side, as if something was pulled along by the hand. Consequently the sign of water preceding this sign would convey the expression, "Give me water." (*Dunbar.*)

The hand extended in a pointing position toward the object in request, then brought toward the body with the fingers raised vertically, and laid against the breast. (*Long.*)

Object wished for pointed to, the right hand being held as in sign for **Bring**; brought with two or three jerky motions toward the face or breast of the speaker. (*Cheyenne* I.)

Place the right hand nearly at arm's length before the breast, palm up, and make a short oscillating motion to and fro. (*Absaroka* I; *Shoshoni and Banak* I.)

After placing the right hand about eighteen inches in front of the neck, and turning it so that the palm will be outward and the little finger toward the neck, fingers overlapping and upright, thumb in the palm, move the hand toward the neck. (*Dakota* IV.)

——— Bring, to.

Left hand extended ten or twelve inches in front of the left breast (**P** 1, palm upward); right hand (**P**), extended in front of right breast, arms full extent, is moved over the left as though carrying something to the left hand with the right. (*Dakota* I.)

——— Bring to me.

Continue the sign for **To Bring** by bringing the hands close to the left breast. (*Dakota* I.) "Bring and give to me."

Point to or otherwise indicate the person directed and the object or article desired, and imply approach by beckoning. (*Arapaho* I.)

Deaf-mute natural signs.—Extend the open hand and draw it back as if conveying something to one's self. (*Ballard.*)

Stretch out the forefinger and then move it toward the breast. (*Larson.*)

Close the hand and move it to one's self. (*Ballard.*)

——— Give to me or us.

The right hand extended in front of body (**Q**) as though taking hold of anything, and then brought back to body with fingers pointing upward as though holding the article in it (**P**). (*Dakota* I.) "Taking the article from the donor."

Deaf-mute natural sign.—Close the hand except its forefinger, with it point to something, and then move it toward the breast. (*Hasenstab.*)

——— Give to him or another.

Reverse **Give to me**—handing the article to the person. (*Dakota* I.) "I give it to you."

The right arm is semi-extended, also the hand (as in type-position **Y** 1, modified by middle, ring, and index fingers being more collected or closed); arm and hand thus point to object. Both are then brought toward the body, the index becoming more hooked as hand approaches, and finally its end concludes the arc of approach by touching center of breast. (*Oto* I.) "Evident in sign."

——— I will give.

First make the sign for **I**, personal pronoun, and then the sign for **Give,** as contained in **Give,** as **Give to him.** (*Dakota* I.) "I hand it to you; I give it to you."

Hand held in position for "give me," near to the chin or breast, extended quickly toward the person addressed. (*Cheyenne* I.)

From an upright position in front of the chin, palm turned toward the right hand; throw the right hand forward eighteen inches, or until the ends of the fingers point obliquely forward and upward. (*Dakota* IV.)

Deaf-mute natural signs.—Point to the bosom, meaning the speaker, stretch the closed hand and then open it. (*Ballard.*)

First point to your breast with your finger, then move forward the hand clinched, and set free the fist. (*Cross.*)

Italian sign.—The motion that one is willing to give something, and which may be called the bribing gesture, is to put one hand into the money pocket. (*Butler.*)

Girl. See **Child.** (Compare **Woman.**)

Glad; content; pleased; satisfied. (Compare **Good.**)

With the raised right hand pass with a serpentine movement upward from the breast and face above the head. (*Wied.*) Heart beats high. Bosom's lord sits lightly on its throne.

Wave the open hand outward from the breast to express "good heart." (*Burton.*)

The sign for **Pretty,** not made in immediate juxtaposition to the face or any part of the body, is significant generally of *content, satisfaction, complaisance,* etc., expressed by the Indian phrase *Good,* or *It is good.* (*Arapaho* I.)

Strike the chest over the heart lightly two or three times with the palm of the right hand; then make the sign for good. (*Dakota* IV.) "Heart good."

The right hand, extended horizontally, palm downward, is held in front of and near or touching the throat, and is then moved forward a few inches. This denotes a comfortable feeling of fullness or satisfaction; but to indicate the more intense feelings of being cloyed or glutted the hand may be held at the chin or at the mouth, the sign being otherwise unchanged. These signs may be used to denote *satiety* from other causes besides eating and drinking. (*Mandan and Hidatsa* I.)

With the right hand (**S**) tap the left breast several times, and then carry the hand forward and toward the right, with palm downward (**W**). Content, glad, good, happy, satisfied, are all expressed by this sign. (*Dakota* I.) "The heart feels good."

Make an inclination of the body forward, moving at the same time both hands forward from the breast, open, with the palm upward, and gradually lowering them. (*Iroquois* I.) "I give you thanks."

Extend both hands outward, palms turned down, and make a sign exactly similar to the way ladies smooth a bed in making it. (*Wichita* I.) Smooth and easy.

Glass.

Left hand arched, the thumb and forefinger meeting at their tips, forming a circle, is held in front of the left breast, horizontal back of hand upward, thumb and forefinger toward the body; then the right index is brought up close to the body in front and passed forward from the breast, describing a series of circles to the center of the circle formed by the left hand, but not placed within it. Daylight is included in this. (*Dakota* I.) "From the same rays of light passing through a pane of glass into the house."

Glutton; Parasite.

Italian sign.—Pinch the cheek with the finger. (*Butler.*)

Go; go away; depart; leave here.

The back of the hand stretched out and upward. (*Dunbar.*)

Like **Come** (*Wied*), but begin near the face and extend the hands with a number of gentle jerks. (*Wied.*) The same remarks apply to this sign as to that for **Come.** (*Matthews.*) The right arm is bent and the hand in position (**B** 1) horizontal is brought to the epigastrium and suddenly arm and hand are extended. The identity of the conception of this *Oto* sign and *Wied's* is evident. The movement of extension in the latter description "by jerks' has little relevancy, and may be reasonably explained by the caprice of the subject. (*Boteler.*)

Move both hands edgeways (the palms fronting the breast) toward the left, with a rocking-horse motion. (*Burton.*)

Right hand held toward left shoulder forearm across the breast, fingers and thumb extended, palm upward and inward; brought with elbow for a pivot suddenly to the right. (*Cheyenne* I.)

The right hand is carried to the left side of the body, level of the breast, horizontal, palm outward, thumb below (**W** 1), extended fingers pointing toward the left; carry the arm out to full extent in front of body and to the right. This is emphatic. "Dismissing the person." Many Indians make the sign for **Go, Go away,** by using the index held upward, the rest of the fingers being closed and carrying from left to right or directly out in front of body, but this is not near so emphatic as the above. **Go,** in the sense of **Gone, Departed,** would be denoted preferably by using the latter sign and by stopping the motion of the hand several times in carrying it out to its final extended position, in resemblance of the walking away of the person. (*Dakota* I.) "He has walked away."

The opened right hand being advanced about a foot and at the height of the navel, palm toward the left fingers separated a little, hand bent at the wrist and pointing downward and forward, move it straight ahead about a foot, and at the same time raise the ends of the fingers until the hand is horizontal. (*Dakota* IV.)

This is indicated by a motion the reverse of **Come,** and when the sign is completed the hand stands as in the beginning of the former gesture. (*Mandan and Hidatsa* I.)

Place the closed hand, knuckles upward, before the breast; elevate the index and pass the hand slowly, in a jerking or interrupted movement toward the front and left, palm facing the front. (*Kaiowa* I; *Comanche* III; *Apache* II; *Wichita* II.)

With the index only extended, point to the earth and trace a course along the earth toward and above the horizon. (*Apache* I.)

——— Go away.

Place the open left hand twelve inches or so in front of the lower part of the chest, pointing forward toward the right, palm looking obliquely upward and backward; then quickly pass the palm of the right hand forward and upward across that of the left and beyond it about a foot. (*Dakota* IV.)

The hand, with the palm facing downward and backward, is held close to the body and about on a level with the stomach; it is moved upward to a level with the top of the head, a foot or so in front of it, describing an arc whose convexity is forward. (*Mandan aud Hidatsa* I.)

The right hand is closed as in type-position (**B**), and arm semiflexed, bringing hand to center of body. The arm is now suddenly extended to full length and hand expanded in the movement. (*Oto* I.) "To remove from."

The right arm is flexed, and the hand, in position (**B** 1), more horizontal, is brought to the epigastrium. The arm is then suddenly extended, hand likewise, with the index finger pointing directly from the body. Accompanying this sign there is generally a repulsive or forbidding frown assumed by the countenance when one is ordered to depart from displeasure. *To go on a message*, the countenance assumes a smile, but sign is identical and is combined with sign for **Speak.** (*Oto and Missouri* I.) "To withdraw or move from."

Place the right hand at the height of and in front of the abdomen, pointing upward, palm outward, fingers slightly separated and bent; then move the hand off toward the left. (*Pai-Ute* I.)

Throw the right hand over the right shoulder so that the index points backward. (*Wichita* I.)

Deaf-mute natural signs.—A slight movement of the feet on the floor and moving the hand forward. (*Ballard.*)

Open the hand and move up, and at the same time point it forward. (*Hasenstab.*)

Close the hand, except its forefinger, and move it forward, pointing in the direction you wish the person to go. (*Hasenstab.*)

Move the hand forward and forward. (*Larson.*)

——— To a place, to go.

Make the sign for **Go,** and when the hand is near at arms' length extend the left and place it horizontally before the moving right so that they come together audibly. (*Kaiowa* I; *Comanche* III; *Apache* II; *Wichita* II.)

Goat.

Pass both hands, with the tips of the finger and thumbs brought to a point, in a curve backward and downward from the ears in the direction of the horns, then place both hands at the lower part of the abdomen, palms about 3 inches apart and facing, with fingers separated and curved so that the tips touch. (*Apache* I.) "Curved horns and large testicles."

Going, traveling, journey.

To describe a journey on horseback the first two fingers of the right hand are placed astride of the forefinger of the left hand, and both represent the galloping movement of a horse. If it is a foot journey wave the two fingers several times through the air. (*Ojibwa* I.)

The kind of locomotion may be indicated, as on horseback, &c., after that pantomimic, the arms and hands being made to represent the legs and feet. (*Arapaho* I.)

Touch the nose with the right index for I, then make the sign for **Go.** (*Dakota* IV.)

Touch the heart with the right index, and then with a wave of the hand outward, point in the direction to which you intend to go. (*Iroquois* I.)

Strike the palms of both hands together obliquely and gliding past one another. (*Zuñi* I.)

Place the hands thumb to thumb, palms forward, fingers extended, separated, and pointing upward, then move them from above the right hip forward and toward the left. (*Pai-Ute* I.)

Deaf-mute natural sign.—No general sign, but for riding, whether on horseback or on wheels, the sign was made by bending the arms and raising and depressing the elbows in imitation of a man riding on horseback, and moving the hand horizontally forward to signify *away.* (*Ballard.*)

——— Running.

Hold the hand in the same position as **Walking,** and, with the forefinger extended forward, advance the hand, keeping the closed portion underneath, and turning it slightly and rapidly to right and left by a quick motion of the wrist. (*Iroquois* I.)

——— Traveling; marching.

The hand held vertically as high as the neck, with slightly divided fingers, and rocked edgewise forward and backward upon the wrist, extending the arm a little forward. This sign resembles that for **Question,** but differs in the direction of the motion of the hand. (*Long.*)

Hands placed in the position for making the sign for **Battle,** except that they are further separated, and then describe a series of half circles or forward arch like movements with both hands. (*Dakota* I.) "Person walking."

Hand on edge, extended, tips forward, palm in (**T** on edge, forward); wave it vertically and forward. Both hands used, indicate two parties, *especially* the double column in which troops always march. (*Apache* III.) "Ambulant, forward motion."

——— Walking.

Close the hand except the index finger, and with that extended, at the height of the breast, move the hand forward, bringing it down a little, at regular intervals, to imitate the steady movement of a walker. (*Iroquois* I.)

Gone under, disappeared from view.

Left hand flattened and held back upward, thumb inward, in front of and a few inches from the breast, right hand slightly clasped, forefinger more extended than the others, and passed suddenly under the left hand, the latter being at the same time gently moved toward the breast. The idea here is that the person has gone under a covering of some kind, as a table, tent, blanket, &c., or gone under the water, *i. e.*, disappearance from view, not used in the sense of dead, death, or permanent disappearance. (*Dakota* I.)

——— **All; no more.** See **Destroyed.**

Gonorrhœa.

Close the right hand, allowing the index to remain half closed, pointing downward; hold the hand at the height of the hip, either at the side or in front, making quick downward movements extending an inch or two. (*Absaroka* I; *Shoshoni and Banak* I.) "Flaccid glans penis and dropping of gonorrhœal discharge."

Good. (Compare **Glad** and **Yes.**)

The hand held horizontally, back upward, describes with the arm a horizontal curve outward. (*Long.*) This is like our motion of benediction, but may more suggestively be compared with several of the signs for **Yes,** and in opposition to several of those for **Bad** and **No,** showing the idea of acceptance or selection of objects presented, instead of their rejection.

Place the right hand horizontally in front of the breast and move it forward. (*Wied.*) This description is essentially the same as the one I furnished. (*Mandan and Hidatsa* I.) I stated, however that the hand was moved outward (*i. e.*, to the right). I do not remember seeing it moved directly forward. In making the motion as I have described it the hand would have to go both outward and forward. (*Matthews.*) The left arm is elevated and the hand held in position (**W**). The arm and hand are thus extended from the body on a level with the chest; the elbow being slightly bent, the arm resembles a bent bow. The right arm is bent and the right hand in position (**W**), sweeps smoothly over the left arm from the biceps muscle over the ends of the fingers. This sign and *Wied's* are noticeably similar. The difference is, the *Oto* sign uses the left arm in conjunction and both, *more to the left.* The conception is of something that easily passes; smoothness, evenness, etc., in both. (*Boteler.*)

Wave the hand from the mouth, extending the thumb from the index and closing the other three fingers. This sign also means **I know.** (*Burton.*)

(1) Right-hand fingers pointing to the left placed on a level with mouth, thumb inward; (2) suddenly moved with curve outward so as to present palm to person addressed. (*Cheyenne* I.)

Pass the opened right hand, palm downward, through an arc of about 90° from the heart, 24 inches horizontally forward and to the right. (*Dakota* IV.) "Heart easy or smooth."

Place the flat right hand, palm down, thumb touching the breast, then move it forward and slightly upward and to the right. (*Dakota* VI, VII; *Kaiowa* I; *Comanche* III; *Apache* II; *Wichita* II.)

Pass the flat hand, palm down, from the breast forward and in a slight curve to the right. (*Dakota* VI; *Hidatsa* I; *Arikara* I.)

The extended right hand, palm downward, thumb backward, fingers pointing to the left, is held nearly or quite in contact with the body about on a level with the stomach; it is then carried outward to the right a foot or two with a rapid sweep, in which the forearm is moved but not necessarily the humerus. (*Mandan and Hidatsa* I.)

Move right hand, palm down, over the blanket, right and left, several times. (*Omaha* I.)

Another: Hit the blanket, first on the right, then on the left, palm down, several times. (*Omaha* I.)

Another: Point at the object with the right forefinger, shaking it a little up and down, the other fingers being closed. (*Omaha* I.)

Another: Same as preceding, but with the hand open, the thumb crooked under and touching the forefinger; hand held at an angle of 45° while shaking a little back and forth. (*Omaha* I.)

Another: Hold the closed hands together, thumbs up; separate by turning the wrists down, and move the fists a little apart; then reverse movements till back to first position. (*Omaha* I.)

Another: Hold the left hand with back toward the ground, fingers and thumb apart, and curved. Hold the right hand opposite it, palm down; hands about six inches apart. Shake the hands held thus, up and down, keeping them the same distance apart. (*Omaha* I.)

Another: Hold the hands with the palms in, thumbs up, move hands right and left, keeping them about six inches apart. (*Omaha* I.)

Another: Look at the right hand, first on the back, then on the palm, then on the back again. (*Omaha* I.)

The above eight signs were all taken from one Omaha, who stated that they all gave the idea of the word uda^n, **Good.**

The flat right hand, palm down, is moved forward and upward, starting at a point about twelve inches before the breast. (*Wyandot* I.)

Throw right hand rom front to side, fingers extended and palm down, forearm horizontal. (*Sahaptin* I.)

Same sign as for **Glad, Pleased.** (*Iroquois* I.)

Bring both hands to the front, arms extended, palms outward; elevate them upward and slightly forward; the face meanwhile expressive of wonder. (*Comanche* I.)

Another: Bring the hand opposite the breast, a little below, hand extended, palm downward (**W**), and let it move off in a horizontal direction. If it be very good, this may be repeated. If comparatively good, repeat it more violently. (*Comanche* I.)

Deaf-mute natural signs.—Smack the lips. (*Ballard.*)

Close the hand, while the thumb is up, and nod the head and smile as if to approve of something good. (*Hasenstab.*)

Use the sign for **Handsome,** at the same time nod the head as if to say "yes." (*Zeigler.*)

Italian signs.—The fingers gathered on the mouth, kissed and stretched out and spread, intimate a dainty morsel. (*Butler.*)

The open hand stretched out horizontally, and gently shaken, intimates that a thing is so so, not good and not bad. (*Butler.*)

Some of the signs appear to be connected with a pleasant taste in the mouth, as is the sign of the French and our deaf-mutes, waving thence the hand, back upward, with fingers straight and joined, in a forward and down ward curve. The same gesture with hand sidewise is theirs and ours for general assent: "Very well!"

——— Good, Heart is.

Strike with right hand on the heart and make the sign for **Good,** from the heart outward. (*Cheyenne* I.)

Touch the left breast two or three times with the ends of the fingers of the right hand, then make the sign for **Good.** (*Dakota* IV.)

Place the fingers of the flat right hand over the breast, then make the sign for **Good.** (*Dakota* VII.)

Move hand to position in front of breast, fingers extended, palm downward (**W** 1), then with quick movement throw hand forward and to the side to a point 12 or 15 inches from body, hand same as in first position. (*Sahaptin* I.) "Cut it off."

——— Good, Very.

Place left hand in position in front of body with all fingers closed except first, thumb lying on second (**M** 1 changed to left), then with forefinger of right hand extended in same way (**M** 2) point to end of forefinger of left hand, move it up the arm till near the body and then to a point in front of breast to make the sign **Good.** (*Sahaptin* I.)

Grandmother. See **Relationship.**

Grass. (Compare **Forest.**)

Point to the ground with the index, and then turn the fingers upward to denote growth. If the grass be long, raise the hand high; and if yellow, point out that color. (*Burton.*)

Sign for **Forest,** but distinguished therefrom by relative height to which the hand is raised. (*Arapaho.*)

Extend the right hand in front of body and near the ground in the same position as the first part of the sign for fire (hand **P**); raise the hand but slightly, however, turn it over flat (**W**), and carry it with a sort of waving motion to the right and left as in imitation of the waving of a large field of grass. (*Dakota* I.) "The springing up of the grass."

Hold the right hand near the ground or higher, according to the height of the grass, its back forward, fingers pointing upward, and a little bent or separated. For grass growing, raise the hand a foot or so three or four inches at a time. (*Dakota* IV, V, VI, VII; *Hidatsa* I; *Arikara* I.)

Hold the left hand extended and flat, palm down before the breast, the right underneath it at a distance to show height of grass; move the right hand forward and backward from the tip to the wrist of the left (maintaining previous distance) to show the spreading of the growth over the surface. (*Ute* I.) "This sign is used only in conversation, when it is known that grass is meant by previous reference; otherwise the following sign is used."

Another: Is also represented by holding the flat right hand, palm downward, a few inches from the ground, different elevations representing varying heights of growth. (*Ute* I.)

Hold the right hand at the height of the knee, back down, fingers pointing upward and diverging; then indicate the height by placing the flat and extended palm downward at the required height. (*Apache* I.)

Deaf-mute natural sign.—Imitate the manner of mowing with a scythe, and, having touched the lips while the jaws are in motion, hold the hands apart, one over the other, and then move about the hand horizontally. (*Cross.*)

——— Fine or nice.

Make the sign for **Grass** and **Good.** Literally, good grass. (*Dakota* VI.)

Grazing. See **Feeding.**

Grease.

Left hand held carelessly to level of and in front of left breast; two first fingers of the right hand slightly extended and rubbed against the palm of the hand at the base of the thumb. (*Cheyenne* I.) Probably a reference to the manner in which marrow is used in painting.

Right hand with fingers bent in resemblance to the bowl of a spoon, and then in this position brought over the left, as though emptying contents of spoon into it, which is held level of stomach horizontal, palm upward, fingers extended, curved upward, pointing outward, right hand then joining the left and in same position, both hands are pushed out in front of body as though holding a pan in them, and at the same time incline the head forward and blow on the hands. (*Dakota* I.) "From cooling hot grease, or food cooked in it."

Great. (Compare **Big,** in the sense of great.)

The two hands open placed wide apart on each side the body and moved forward. (*Dunbar.*)

The sign for **Big,** in the sense of *large around*, used to represent the quality of greatness. (*Cheyenne* I.)

Deaf-mute natural sign.—Separate the hands widely apart. (*Ballard.*)

Great Father: President; also Secretary of Interior.

Make the signs for **White man, Chief,** and **Father.** (*Kaiowa* I; *Comanche* III; *Apache* II; *Wichita* II.)

Green. See **Color.**

Grief.

Bring the right hand up to the left breast with fingers and thumb together (**U**, fingers downward), forefinger against breast, make a downward movement, the hand turning as it goes down till the palm is upward. (*Cheyenne* II.) Down-hearted; lost heart.

Place the ends of the fingers over the eyes with both hands at the same time, and gently let the hands drop, imitating dropping of tears by repeating the motion two or three times, the hands falling about to the breast, the head bowed forward. (*Ojibwa* IV.)

The sign for *Weeping, Crying* is a part of the sign for grief, which would hardly be complete, however, without reference to the heart, and the following is the sign for grief or sorrow of the heart, or, as the Sioux say, "*Heart is down, upset.*" The palmar surface of the right hand horizontal, with fingers extended, pats the left breast several times, and is then turned over three or four inches in front of the left breast so as to bring the palm upward, thumb outward. (*Dakòta* I.) "Upset heart, weeping."

Place the palmar side of the extended fingers of the right hand to the front side of the head, close the eyes, and drop the head forward upon the breast. (*Wyandot* I.)

Grieved ; Wounded feelings.

With the index only extended and held horizontally in front of the breast, pointing toward the left, pretend to puncture the heart at repeated movements. (*Kaiowa* I; *Comanche* III; *Apache* II; *Wichita* II.)

Ground. See **Earth.**

Grow or Vegetate, To.

Turn hand and forefinger up from the ground, indicating the growing or coming from the ground or out of the earth, close the balance of fingers over the upturned palm of the right hand, raise hand by a quick motion a few inches from the earth. (*Ojibwa* IV.)

Commence with the sign for **Life, living,** &c., but stop the upward motion of the hand at the proper time and turn it over, and hold as seen in (**W**). (*Dakota* I.) "Designating the amount of growth by height from the ground."

The hand is collected somewhat cup-shaped, as in type position (**O**), the fingers more closed and hand horizontal, and made to approach the mouth, as in the act of eating. The extended right and left indices are then crossed and brought near the ground. From this position the hands *in situ* are uniformly and successively raised in jerks. (*Oto and Missouri* I.) "The coming up of something to eat from the earth."

Gun.

Hold out the left hand, as in the act of supporting the gun when directed horizontally, and with the right appear to cock it. (*Long.*)

Close the fingers against the thumb, elevate the hand, and open the fingers with a quick snap. (*Wied.*) It seems to me that here he only describes in slightly varied language the motion already referred to in **Arrow, To shoot with**, and **Gun, Discharge of,** the firing of a gun and not the gun itself. The only sign I ever remember to have seen for **Gun** was made by the arms being held in the position of shooting a gun; but whether this sign was conventional or improvised I do not remember. (*Matthews.*)

The dexter thumb and fingers are flashed or scattered, *i. e.*, thrown outward or upward, to denote fire. (*Burton.*)

Right-hand fingers pointing upward, partially closed against the thumb, held to level of shoulder, suddenly elevated above the head, and the fingers and thumb at the same time snapped open to imitate the quick discharge of smoke from a gun. (*Cheyenne* I.)

Both index-fingers extended (others closed), thumbs resting on second fingers, the right brought to the right eye, with back of hand toward the right, horizontal finger pointing straight outward, and the left arm extended to its full capacity in front of the eye, so that the extended forefinger of that hand is on a line with the extended forefinger of the right, back of left hand toward the left, hand horizontal, finger pointing forward. Edge of the fingers of both hands downward. (*Dakota* I.) " From the act of aiming with the gun."

Semi-flex the fingers and thumb of the left hand, and place the hand in front of the chest, with its palm inclining downward, backward, and toward the right at an angle of 45°; pass the right fist upward along the palm of the left until the right wrist lies between the left thumb and index, then suddenly open all the fingers. (*Dakota* IV.)

The right hand brought to the right side of the chest's top in position (**E** 1), horizontal; the fingers suddenly opened, as in grasping a trigger; the left arm and index-finger extended, and the right hand in above position brought to the middle of the left arm; the right hand suddenly expanded upward, as the smoke from the explosion of the cap when the gun is discharged. (*Oto and Missouri* I.) " That which is discharged by a trigger from the shoulder."

——— Cap, percussion.

Left hand held in front of the breast, half closed, fingers of the right hand clasped as though grasping a small object and brought down to left hand; motion made with the right hand as though cocking a gun; right hand slapped into the palm of the left hand, making a sharp clap. (*Cheyenne* I.)

——— Discharge of a.

Place both hands as in **Arrow, to shoot an**; extend the left arm, contract the right before the face, then snap the ends of the fingers forward. (*Wied.*) The description is so ambiguous that I can not satisfy myself as to his meaning. Perhaps he is trying to describe a form of the sign for **Volley,** which I gave you. (*Matthews*) There is indisputable similarity in the (*Oto and Missouri* I) sign and *Wied's* as to conception and execution, the slight difference being in executing it. (*Boteler.*)

Left hand with all the fingers extended, horizontal, back outward, edge of fingers downward, is held about 18 inches in front of the breast, and the back of the right with fingers extended, upright, joined, back outward (**S**), is carried out from the breast, struck quickly against the palm of the left, *i. e.*, clapping the hands. This is for a single discharge. For a continuous discharge of musketry, clap the hands many times in rapid succession. (*Dakota* I.) " From the noise of the discharge of fire-arms."

Gun-flint.

With the index-finger of the right cut off a piece of the extended thumb, so that the finger is laid across the thumb-nail. (*Weid.*) I have seen this sign made. (*Matthews.*)

Gunpowder.

Appear to take up a pinch of the powder and to rub it between the finger and thumb, then turning the hand spring the fingers from the thumb upward, so as to represent the exploding of the powder. (*Long.*)

Rub the thumb and index finger together repeatedly. (*Wied.*) I remember having seen this sign made; but I think that when the fingers were rubbed together they were held points downward, an inch or so above the open palm of the left hand. (*Matthews.*)

The left hand horizontal, palm upward, fingers to the right (**X**), is held in front of the breast; right hand (**U**), with fingers pointing downward, back outward, is held over the palm of the left, with a slight motion of the thumb and forefinger, as though feeling the grains of powder with them. (*Dakota* I.) "Examining the grains of powder."

Rub the thumb and index of the right hand together for several seconds. (*Dakota* IV.)

Gun-screw.

Elevate the hand to indicate the gun and twist the fingers spirally around the thumb. (*Wied.*)

Gun-shot.

First make the sign for **Explosion** with the right hand, which is made by the fist (**B** 2), excepting that the thumb is under the fingers *i. e.*, resting in the palm of the hand, the fingers are snapped forward and upward from the level of the breast to that of the head, where all the fingers are closed but the index and second, which are separated and point upward (**N**). (*Dakota* I.) "Indicates the explosion or discharge of both barrels of a double-barreled shot-gun."

——— To hit with a.

Place the tips of the fingers downward upon the thumb, then snap them forward and strike the hands together. (*Wied.*) Probably when he says, "strike the hands together," he wishes to describe my sign for **Shot.** When the person whom the prince saw making this sign, raised the finger, he may have done so to indicate a **Man,** or **One** shot. I do not think that the raising of the finger is an integral part of the sign. (*Matthews.*)

Same as **Bow,** but in making the first sign have both hands opened and the fingers pointing toward the left. (*Dakota* IV.)

Place the left hand at arm's length in front of the breast, the right at the same elevation but in front of the right shoulder, then snap the fingers from the thumb simultaneously. When the sign is used in connection with other gestures in a sentence it is thus abbreviated; when used alone, the backs of the fingers of the right hand are struck flat against the palm of the left. (*Pai-Ute* I.)

Extend the left hand, closed, nearly at arm's length. Place the closed right hand before the right shoulder, first two fingers resting on the ball of the thumb; then simultaneously snap forward the index and second fingers of the right hand and the forefinger of the left; finally, throw the open right hand over and downward toward the right. (*Ute* I; *Absaroka* I; *Shoshoni and Banak* I.) "To hit and kill."

——— To hit the target.

With the hands in the position stated at the completion of the sign for **Gun, discharge of a,** draw the right hand back from the left, that is, in toward the body; close all the fingers except the index, which is extended, horizontal, back toward the right, pointing straight outward, is pushed forward against the center of the stationary left hand with a quick motion. Should the target not be hit it can be indicated by pushing the index above, below, or to the side of the left hand, as the case may require. (*Dakota* I.) "From the striking of the target. Bullet comes to a stop."

Hail.

Begin with the sign of **Water,** then the sign of **Cold,** next the sign of a **Stone,** then that for **Same** or **Similar,** then the sign of **White,** and lastly conclude with the sign of an **Egg;** all of which combined gives the idea of hail. (*Dunbar.*)

With the forefinger of right hand resting against the ball of the thumb, with the second and third phalanges crooked in such a manner as to form a small round opening resembling a hail-stone, other fingers extended, make in front of face and above and around it the upward and downward movements, describing the falling of hail-stones. (*Dakota* I.) "From the falling of hail-stones."

Deaf-mute natural signs.—Represented by showing the ball of the thumb to indicate the size, pointing to the shirt-bosom to signify the color, and moving the extended fingers down repeatedly to denote the fall. (*Ballard.*)

Move the hands outstretched upward and downward, and then strike the fist on the head. (*Larson.*)

Hair.

The movement of combing. (*Dunbar.*)

(1) Left hand naturally closed, elevated to the front of the person, right hand held carelessly against base of thumb of left hand; (2) sign for **Grass** made. This sign is sometimes made over other portions of the body, the idea being skin grass. (*Cheyenne* I.)

Touch the hair of the person or animal with the right index. (*Dakota* I.) "Designating the hair."

Halt! Stop!

Raise the hand, with the palm in front, and push it backward and forward several times—a gesture well known in the East. (*Burton.*)

Palm of right hand down, move sideways two or three times, during the extension of arm slowly extending to natural length (not stretched), then stopped and held still a moment, gently lower a little, with a careful movement to the right of said shoulder. (*Ojibwa* IV.)

Right hand brought in front of the right breast a few inches, hand and fingers upright, joined, palm outward (**T**), carry it in this position directly outward from the body with a quick movement, and when the arm is extended about two-thirds of its capacity, come to a sudden stop and hold there a moment. (*Dakota* I.) "From the act of stopping suddenly."

Both hands clinched as in (**A**), held in front of the body, the right hand above the left. Bring the right hand down quickly on top of the left. (*Dakota* III.)

Shake the upright opened right hand four to eight inches from side to side a few times, from twelve to eighteen inches in front of the right shoulder, the palm forward, fingers relaxed and separated a little; then close the hand and lower the fist about eighteen inches, back outward. (*Dakota* IV.)

Another: Incline the body forward, place the right fist at arm's length forward, a little higher than the navel, bent upward at the wrist, back outward, then move the fist downward about eighteen inches. (*Dakota* IV.)

Extend the left hand, palms turned inward, fingers closed. Extend the right in the same way and bring the extended palm of the right smartly across the tips of the fingers of the left, just missing them. While traveling, if they want to stop and camp, this sign is used with the following addition: Clinch the right fist (**F**) and bring it smartly down. This indicates they will *sit down* there. (*Comanche* I.) "Cut it off."

Close the right, leaving the index fully extended; place the tip to the mouth, then direct it firmly forward and downward toward the ground. (*Pai-Ute* I.)

Deaf-mute natural sign.—Lift both hands up. (*Zeigler.*)

Italian sign.—The open hand stretched out with the fingers up. (*Butler.*)

Halt. A stopping-place.

Must be indicated by the proper name of the place (as Bad River, Standing Rock, Big Woods, &c.) at which the halt (which is also the sleeping-place for the night) is made. Indians have no "halts" in the English sense of the word, but would say instead, "I slept at such a river or rock." In only one way can Indians be said to have halting-places, *i. e.*, as on a long journey over a well-known country the same camps are made time after time as long as timber and water last, and distance is roughly calculated or estimated by so many camps or days' journey. (*Dakota* I.)

Hammer.

Same as the sign for **Stone, Hard,** &c. Sometimes a distinction is sought to be made in the case of *Hammer* by pounding in the palm of the left hand with the lower part of the right fist (**A**). (*Dakota* I.) "From the use of the hammer."

Handsome. (Compare **Good** and **Pretty.**)

Right hand touching the left side of forehead, hand horizontal, palm inward, fingers joined, and the thumb pointing obliquely upward (**W** 2), bring the hand down over the face to the left breast, there turn the hand so that it will be as in (**W**) with tips of extended fingers pointing toward the left, and carry it outward and toward the right. (*Dakota* I.) "Pretty face, good heart."

Deaf-mute natural sign.—Draw the outstretched palm of the right hand down the right cheek. (*Zeigler.*)

Harangue. See **Speaking**.

Hard.

Open the left hand and strike against it several times with the right (with the backs of the fingers). (*Wied.*) Still used for hard in certain senses. (*Matthews.*)

Same as the sign for **Stone.** (*Dakota* I.) "Would seem to refer to the time when the stone hammer was the hardest pounding instrument these Indians knew."

Rotate the right fist a very little from side to side while raising it about a foot from just in front of the right breast, its palm inward. The fist represents the heart which is hard, and the motion its beating. (*Dakota* IV.)

Strike the palm of the left hand with the front of the right fist several times. (*Kaiowa* I; *Comanche* III; *Apache* II; *Wichita* II.)

Push the tip of the right index against the tightly-clinched left fist (**A**) at several points. (*Apache* III.) "Impenetrability."

Deaf-mute natural signs.—Strike the fists together. (*Larson.*)

Hold the forefinger and thumb as if pressing something between, at the same time smack the lips as if there was a pasty substance between. (*Zeigler.*)

——— Excessively hard; harder than anything.

Make the sign for **Hard,** then place the left index-finger upon the right shoulder, at the same time extend and raise the right arm high, extending the index-finger upward perpendicularly. (*Wied.*) This was said by an Ogalala to mean "a hardy man." (*Corbusier.*)

Strike the palm of the left hand with the back of the right fist (to denote the resistance); then make the sign for **Hard.** (*Dakota* IV.)

Hare.

With the fore and second fingers (**N,** with the two fingers considerably separated) of each hand on their respective sides of the head, and with the extended fingers pointing upward, backs of hands right to right and left to left, wag them forward and backward, and then with both hands open, fingers straight, slightly separated, pointing upright, backs of hands in the same relative position as above, in front and at the level of the face, and with the left hand seven or eight inches forward of the right, make the forward movements with both at the same time in imitation of the running jumps of the rabbit by moving them forward on a short curve, and then at the level of the breast lay the extended straight and joined first and second fingers of the right hand across the backs of of the similarly extended fore and second fingers of the left at right angles. (*Dakota* I.) "From the ears, manner of running, shadow, &c., of the rabbit."

Hat.

Pass the parted thumb and index-finger about both sides of the head where the hat rests upon it. (*Wied.*) This sign is still in use among the Indians. (*Matthews.*) The latter movement of the hand in *Wied's* sign is probably the same as the first motion of the (*Oto and Missouri* I) sign. (*Boteler.*)

With the right index extended and pointing toward the left, and the thumb extended and pointing backward, the other fingers closed, move the hand from the top of the head forward and downward to the eyebrows. (*Dakota* IV.) "Covering the head."

Place the extended index-finger and thumb on the forehead and right side of the head, then draw them together on a horizontal line where the hat rests upon the head. (*Dakota* V.)

The arms and hands are raised to the forehead, the thumbs are erect, the extended indices meet at the middle of the forehead—the hands are in position (**L**)—and diverge, drawing a line to the ears. The fingers are now closed and seemingly pull something from the top of the head downward. (*Oto and Missouri* I.) "Something that is pulled down on and marks the forehead."

Hatchet. See **Tomahawk.**

Hate, I. (Compare **Contempt.**)

Shake the head, make outward quick motion of both hands as if pushing back with the right hand, palms out, extend thumb and fingers partly, and sometimes shut the hands or fists closely, indicating more intense feeling. (*Ojibwa* IV.)

The sign is nearly the same as for **Contempt, Disdain,** but the thumb is held under the hand, its tip projecting between the second and third fingers, and the hand is thrown out in front of the body and fingers separated, back upward, with great force. This expresses genuine hate, hatred, and the case is always a serious one where this sign is used by an Indian, who also expresses his hatred by appropriate facial emotions. (*Dakota* I.)

Have; I Have. See **Possession.**

He, or **another.**

The forefinger extended and hands shut, and fingers brought over one another, or nearly touching, and then separated moderately quick. (*Dunbar.*)

Indicate one's self by touching or otherwise, followed by sign of **Negative;** *i. e.*, "not myself, another." (*Arapaho* I.)

Indicate the particular person or persons to which reference is made by pointing toward him or them with the right index (**M**). (*Dakota* I.) "Designating a particular person."

(1) Only by pointing with right index (2) and extending left index. (*Apache* III.) (2) The person (1) over there.

Deaf-mute sign.—Point the thumb over the right shoulder.

——— In the phrase of, *Another speaks.*

Precede the gesture for **Speak,** by placing the hand not near the mouth, but beginning farther away, drawing it nearer and nearer. (*Wied.*)

Heap, pile, mound.

The hands are brought from their natural positions by a curved motion with the palms downward, fingers extended separated, nearly together in front of the breasts, 18 inches, fingers pointing outward, hands horizontal, flat, &c., when the curved motions cease, and the hands (**W**) are carried straight downward six or seven inches and brought to a stop suddenly as though patting down the top of a pile. (*Dakota* I.) "From the shape of a mound, heap, or pile."

Hear, to; hearing; heard; listen. (Compare **Understand.**)

Place the open thumb and index-finger over the right ear, and move them hither and thither. (*Wied.*) I have given you this sign with a similar meaning, but, as in signs for *Coat* and *Leggings*, I say the finger tips are in contact, which he says they are not. The remarks I make about this difference in the former signs may apply as well to this. (*Matthews.*)

Tap the right ear with the index tip. (*Burton.*)

A twirling at the ear. (*Macgowan.*)

First and middle finger of right hand pointing upward held close to right ear; moved gently downward and forward, eyes looking askance, eyebrows contracted as if in the act of listening. (*Cheyenne* I.)

Place the hand partially closed, palm to the front, behind the ear. (*Arapaho* I.)

First make the sign for **Attention,** and then carry the hand back in the same position to near the right breast, and at the same time incline the head toward the sound or the direction in which it comes. (*Dakota* I.) "Same as with us; first attracting attention as an indication to keep quiet and then listening."

Close the right hand, leaving the index and thumb fully extended and separated; place the hand upon the right ear with the index above it and the thumb below; then pass forward and slightly downward by the cheek. (*Absaroka* I; *Hidatsa* I; *Arikara* I; *Shoshoni and Banak* I.)

Forefinger right hand extended, curved, (others closed); thumb resting on second finger is carried directly to the right ear. (*Dakota* I.) "From the act of hearing."

Holding the fingers as for **I know,** place the right index, back outward, in front of the right ear, pointing upward and a little forward. (*Dakota* IV.)

The tips of the forefinger and thumb being opposed, the hand is held a few inches from the ear and then caused to approach the latter. The hand may then be restored to its original position and motion repeated. (*Mandan and Hidatsa* I.)

Spread the thumb and index widely apart, remaining fingers closed; place the hand, palm forward, to the side of the head, the thumb below the ear, and the index above it; then move the hand forward and downward. (*Kaiowa* I; *Comanche* III; *Apache* II; *Wichita* II.)

Place the hand behind the ear in the most natural manner for a listener. (*Comanche* I.) "I want to hear."

Fingers and thumb of the right hand closed, index crooked, placed opposite and pointing to the ear, palm toward the shoulder, and moved toward the ear several times. (*Apache* I.)

——— To listen.

Hold the right index in front of the ear as for **I Hear,** and then turn the hand a little from side to side two or three times. (*Dakota* IV.)

Bring the hand (**R** with last three fingers shut) near the ear, thumb and index raised, other fingers closed; turn the head to bring the ear toward source of sound; facial expression inquiring. (*Apache* III.) "Sound coming to ear."

Deaf-mute natural signs.—Place the forefinger on the ear; at the same time incline the head as if to listen to something. (*Zeigler.*)

Move the forefinger nearly to the ear. (*Ballard.*)

To point the finger to the ear. (*Larson.*)

——— Do not. I do not understand.

Make the sign for **I Understand, I Hear,** and then the sign for **No,** or throw the hand outward from the ear. (*Dakota* IV.)

Point the forefingers of both hands to the external meatus of the ears. (*Zuñi* I.)

——— With one's own ears, To.

Make the sign for **Hear,** then pass the index of the right hand from the left ear outward toward the left. (*Arikara* I.)

Heard, I have.

Open wide the thumb and index-finger of the right hand, place them over the ear, and in this position move them quickly past the chin and nose. (*Wied.*)

Bring the extended palm (**W**) to the ear a time or two, as if fanning the ear. (*Comanche* I.)

Heart.

Same as the sign for **I,** personal pronoun. (*Dakota* I.) The heart is selected as the seat of all the emotions.

——— Bad. See **Bad heart.**

——— Good. See **Good heart.**

Heat.

The two hands raised as high as the head and bending forward horizontally, with the points of the fingers curving a little downward. (*Dunbar.*)

(1) Both hands, palm downward, elevated to the level of the eyes and extended outward; (2) brought downward with a rapid motion, half opened while descending. Reference to the sun's rays, for more particular idea "it feels hot," right hand held with little finger against the part affected and sign for **Fire** made. (*Cheyenne* I.)

Place both hands over the head, palms down, fingers pendent and hanging downward at a short distance from the forehead. (*Kaiowa* I; *Comanche* III; *Apache* II; *Wichita* II.) This sign if made a little lower and in front of the face, the hands also being lowered and raised again several times quickly, signifies **Rain.** "Rays of light and heat."

Heavy.

Place both flat and extended hands before the chest, pointing forward with the palm up, about four inches apart, as if supporting a large body; then move them simultaneously upward and downward about two or three inches, the upward motion being made more rapidly than the downward. (*Kaiowa* I; *Comanche* III; *Apache* II; *Wichita* II.)

Point at an imaginary object; seize with both hands and lift with great effort; also, take up an imaginary object in one hand and lift slowly two or three feet (**Y**). (*Apache* III.)

With both hands clinched (**B**), arms more than half extended, draw them upward in front of the body from the level of the abdomen to that of the face, with a slow and more or less interrupted movement, as though the hands seized and conveyed upward a heavy body; at the same time the effort is denoted by contraction of the facial muscles (*Dakota* I.) "From the act of lifting a heavy body."

Help, To assist.

Bring right hand up to right breast, forefinger straight (**J** in upright position, palm out); the left to the left breast, same position, hands about six inches apart. Move both forward. (*Oto and Missouri* I.)

Only by others coming to join in doing some special work. (*Apache* III.)

First commence with the open right hand (**Y**) back outward, pointing obliquely upward, about two feet from the right side of the front of the body and draw inward nearly to the body, describing a series of circular movements, then without stopping the movement of the hand, carry it with moderate force, back upward, horizontal, fingers extended, straight, joined, etc., against the palm of the left hand, (inner side of the right index striking it), which is held horizontal, about a foot in front of the body, with fingers extended, straight, joined, etc., edge downward. back toward the right, and then the right hand is carried outward from the body by a series of circular movements the reverse of the first movements given above. (*Dakota* I.) "Come and help me."

Here.

Right hand closed, fist, back outward, upright, is moved upward and downward in front of right side of body from the level of the breast to that of the top of the head. (*Dakota* I.) "It is right here; at the place."

The right arm is flexed toward the body, and the hand in type-position (**A**) describes a circle before the breast. The sign continuing from the completion of the circle, the hand is fully opened as in type-position (**S 1**), horizontal, and edges being held upward and downward, and made to strike the ground forcibly. (*Oto and Missouri* I.) "On this very spot of earth."

Only by pointing, as to ground or into my tent as I sat at its entrance. (*Apache* III.)

——— Where we stand.

Same position of hand as **Great distance,** and point down directly in front to the feet or between them with sudden dropping of head or quick bow, the eyes following the direction of the hand in every instance. (*Ojibwa* IV.)

Hide, To; Conceal. (Compare **Steal.**)

Place the hand inside the clothing of the left breast. This means also to put away or to keep secret. (*Burton.*)

(1) Sign for **Steal;** (2) left hand flat, palm downward, placed near some part of the body; (3) right-hand fingers hooked; covered by left. (*Cheyenne* I.)

The left hand stationary about eight inches in front of left breast, horizontal, back outward, edge downward, fingers closed, and then pass the right hand, with fingers hooked, back of hand outward, edge of fingers downward, horizontal, quickly between the left hand and the body to the left side, as though passing it under a blanket or the coat. Although not identical with the sign for *Stealing* this sign resembles it very much, and it is used to denote concealment of any article from view, much the same as we would put out of sight any article we did not wish seen for any reason other than that the article was stolen. (*Dakota* I.) "Placing it out of sight."

Hold the opened left hand, palm downward, fingers pointing toward the right a foot or eighteen inches in front of the lower part of the chest, and pass the opened right hand, palm downward, over it, and along the forearm to the elbow; then close both hands and carry the right fist under the left arm, as if hiding it. (*Dakota* IV.)

Grasp the forefinger of the right with the palm of the left. Sometimes, when desiring to express *Theft* they go through the motion of concealing something under their blanket. (*Comanche* I.)

Deaf-mute natural signs.—First hold the open left hand in front of the body, next pass slowly the open right hand beside the left, and at the same time incline the head, with the mouth closely shut and the eyes half opened, toward the left, and then point to the hiding-place. (*Hasenstab.*)

To put the outstretched hands together toward the head. (*Larson.*)

To incline the head and face as if seeking some place of refuge. (*Zeigler.*)

——— Secret, secreted.

Deaf-mute natural signs.—Having touched the lower lip, the mouth opening and shutting alternately, shake your head. (*Cross.*)

Place the forefinger on the mouth, at the same time moving the lips as if speaking, and then shake the head as if to say "no." (*Zeigler.*)

Hide, skin.

Sign for **Animal**; both hands closed, palms facing but not touching each other; quickly but slightly drawn apart. (*Cheyenne* I.)

High. See **Big** in the sense of **High.**

Hill, bluff, mountain.

A clinched hand held up on the side of the head, at the distance of a foot or more from it. To signify a range of mountains, hold up the fingers of the left hand a little diverging from each other. (*Long.*)

Close the finger tips over the head; if a mountain is to be expressed, raise them high. To denote an ascent on rising ground, pass the right palm over the left hand, half doubling up the latter, so that it looks like a ridge. (*Burton.*)

Both hands outspread near each other, palms downward, and elevated to the level of the face; brought downward to represent the slope of a hill, the motion at the base being somewhat more rapid than at the first part. (*Cheyenne* I.)

Hold the left hand clinched at some distance before the face, the knuckles pointing upward, representing the elevation or hill. (*Dakota* VII.)

——— Going over a.

First make the sign for **Hill** as contained in **Mountain**, with both hands or with the left hand, in front of breast, and then the right hand is drawn back toward the body (left, representing the hill, stationary), and then carried outward with short, jerking motions over the back of the left, as though a man was riding or walking over a hill. (*Dakota* I.) "Going over a hill and passing out of view."

Hold the left hand about a foot in front of the upper part of the chest, back outward and forward, and pass the slightly-flexed right hand forward over it, about twelve inches, through an arc beginning two or three inches behind it, back upward, the fingers at first pointing a little upward, then forward, and toward the last a little downward. (*Dakota* IV.)

——— Peak.

Place the left fist, with the knuckles pointing upward, at some distance before the face. (*Ute* I.)

Close the left hand loosely, the thumb resting upon the middle joint of the forefinger, palm toward the face, and hold it as high as the shoulder. (*Apache* I.)

Hoe; Hoeing.

Pantomime of handle by extended left arm, blade by adjusted right hand, and the action of using a hoe. (*Apache* III.)

Hog.

Right-hand fist (**B**, turned downward) is moved around in various directions below the level of the body, pointing downward, with upward movements in imitation of the hog's manner of rooting in the ground with its snout. (*Dakota* I.) "From the hog's manner of eating."

Both hands are brought to the sides of the head in type-position (**W**), and made to vibrate to and from, the thick of hand being stationary. The right hand then approaches the mouth, and is made to scoop successively forward from the mouth, in imitation of the animal plowing the ground. (*Oto and Missouri* I.) "A being with large ears that plows up earth with its nose."

(1) Hand on edge, forward (**T** on edge), waved vertically and forward, short, quick motion; (2) hand nearly extended, fingers gathered together (**U**, more loosely and bent a little), and placed in front of the mouth, tips forward. (*Apache* III.) "Gait of animal, and snout. The first part of the sign is an ideally perfect reproduction of the trot of the half-grown porkers scavenging the agency rubbish."

Honest, honesty. (Compare **Truth.**)

Right hand held with thumb inward against the heart; forefinger extended, knuckle placed against the mouth; thrust straight forward and outward in a slight downward curve to express "straight from the heart." (*Cheyenne* I.)

——— An honest man.

First make the sign for **Man** and then the sign for **Truth.** This relates to *True, truth*, in speaking. *He tells no lies.* It is such a rare occurrence for one Indian to steal from another of the same tribe, that the Sioux have no separate sign for *An honest man*, as implying the opposite of *Thief.* (*Dakota* I.)

Indicate the person with the index, then place both hands, flat and extended, about eighteen inches apart, with palms facing, as high as the head, and move them eastward to arm's length. (*Wyandot* I.) "Uniform from head to foot, or, literally, the same from end to end."

Indicate object. (1) Grasp toward it (**P**, closing more and prone); (2) sign of negation; (3) wave the hands off forward and down (**Q**). (*Apache* III.) "(1) Taking; (2) no; (3) leaves it where it is."

First point to the person, then make sign for **Good**; then place fist of left hand at a point in front of body (**A** 1, changed to left), and make a pass under it with right hand, as though grasping something and pulling it away from where it belongs, fingers and thumb naturally relaxed at first (**Y** 1, palm down), but before the return movement is made the fingers and thumb are closed (**A** 2, palm inward), as though laying hold of something; then the final motion is making the sign of **Negation, Not.** (*Sahaptin* I.) "He is a good man; will steal nothing at all."

Horror.

The palm of the right hand (**W** 2) laid over the mouth, and at the same time the sign for **Surprise** is made by drawing the head and body backward. I have seen a few Indians use both hands in making this sign, laying one over the other crosswise, thus +, covering the mouth. One of these signs would be used as above if an Indian walking along should unexpectedly see the body of a dead person lying on the ground, when the sign for *Surprise* would be made simultaneously as expressing his emotions. One of the signs would also be used by a person on hearing of an unexpected death; and I believe it was some occurrence of this kind that misled the Prince of Wied-Neuwied and caused him to give substantially the second sign above as the one for *Dead*, *Death*. (*Dakota* I.)

Horse.

The right hand with the edge downward, the fingers joined, the thumb recumbent, extended forward. (*Dunbar.*)

Place the index and third finger of the right hand astraddle the index finger of the left. (*Wied.*) By the "third" he means the "middle" finger, as appears in another connection. He counts the thumb as the first.—*Ed.* I have described this sign in words to the same effect. (*Matthews.*) The right arm is raised, and the hand, opened edgewise, with fingers parallel and approximated, is drawn from left to right before the body at the supposed height of the animal. There is no conceivable identity in the execution of this sign and *Wied's*, but the sign for **Horse** by the *Prince of Wied* is nearly identical with the sign for **Ride a Horse** among the Otos. (*Boteler.*)

Left-hand thumb and forefinger straightened out, held to the level of and in front of the breast; right-hand forefinger separated from the middle finger and thrown across the left hand to imitate the act of bestriding. They appear to have no other conception of a horse, and have thus indicated that they have known it only as an animal to be ridden. (*Cheyenne* I.)

A hand passed across the forehead. (*Macgowan.*)

Draw the right hand from left to right across the body about the heart, the fingers all closed except the index. (*Dodge.*) This probably refers to the girth. It has a resemblance to *Burton's* sign for **Dog**, and is easily confounded with his sign for **Think, Guess.**

Place the first two fingers of the right hand, thumb extended (**N** 1), downward, astraddle the first two joined and straight fingers of the left hand (**T** 1), sidewise to the right. Many Sioux Indians use only the forefinger straightened. (*Dakota* I.) "Horse mounted."

The first and second fingers extended and separated, remaining fingers and thumb closed; left forefinger extended, horizontal, remaining fingers and thumb closed; place the right-hand fingers astride of the forefinger of the left, and both hands jerked together, up and down, to represent the motion of a horse. (*Dakota* III.)

The two hands being clinched and near together, palms downward, thumbs against the forefingers, throw them, each alternately, forward and backward about a foot, through an ellipsis two or three times, from about six inches in front of the chest, to imitate the galloping of a horse, or the hands may be held forward and not moved. (*Dakota* IV.)

Place the extended and separated index and second fingers of the right hand astraddle of the extended forefinger of the left. (*Dakota* VI, VII; *Hidatsa* I; *Arikara* I.)

The left hand is placed before the chest, back upward, in the position of an index-hand pointing forward; then the first and second fingers of the right hand (only) being extended, separated, and pointing downward, are set one on each side of the left forefinger, the inter-digital space resting on the forefinger. The palm faces downward and backward. This represents a rider astride of a horse. (*Mandan and Hidatsa* I.)

Close hands, except forefingers, which are curved downward; move them forward in rotation, imitating the fore feet of the horse, and make puffing sound of "Uh, uh!" (*Omaha* I.) "This sign represents the horse racing off to a safe distance, then puffs as he tosses his head."

The arm is flexed and with the hand extended is brought on a level with the mouth. The hand then assumes the position (**W** 1), modified by being held edges up and down, palm toward the chest, instead of flat. The arm and hand being held thus about the usual height of a horse are made to pass in an undulating manner across the face or body about one foot distant from contact. The latter movements are to resemble the animal's gait. (*Oto* I.) "Height of animal and movement of same."

The index and second fingers of the right hand are placed astraddle the extended forefinger of the left. (*Wyandot* I.)

Hold the right hand flat, extended, with fingers joined, lay the thumb inward against the palm, then pass the hand at arm's length before the face from left to right. (*Kaiowa* I; *Comanche* III; *Apache* II; *Wichita* II.)

Another: Place the extended and separated index and second fingers astraddle the extended and horizontal forefinger of the left hand. (*Kaiowa* I; *Comanche* III; *Apache* II; *Wichita* II.) "This sign is only used communicating with uninstructed white men, or with other Indians when whose sign for *Horse* is specifically distinct."

Place the right hand, palm down, before the right side of the chest; place the tips of the second and third fingers against the ball of the thumb, allowing the index and little fingers to project to represent the ears. Frequently the middle fingers extend equally with and against the thumb, forming the head of the animal, the ears always being represented by the two outer fingers, viz., the index and little finger. (*Ute* I.)

Elevate the right hand, extended, with fingers joined, outer edge toward the ground, in front of the body or right shoulder, and pointing forward, resting the curved thumb against the palmar side of the index. (*Apache* I.) "This sign appears also to signify *Animal* generically, being frequently employed as a preliminary sign when denoting other species."

Deaf-mute natural signs.—Imitate the motion of the elbows of a man on horseback. (*Ballard.*)

Act in the manner of a driver, holding the lines in his hands and shouting to the horse. (*Cross.*)

Move the hands several times as if to hold the reins. (*Larson.*)

Our instructed deaf-mutes indicate the ears, followed by straddling the left hand by the fore and middle fingers of the right. The French deaf-mutes add to the straddling of the index the motion of a trot.

——— A man on a.

Same sign as for **Horse,** with the addition of erecting the thumb while making the gesture. (*Dodge.*)

——— Bay.

Make the sign for **Horse,** and then rub the lower part of the cheek back and forth. (*Dakota* IV.)

——— Black.

Make the sign for **Horse,** and then point to a black object or rub the back of the left hand with the palm of the fingers of the right. (*Dakota* IV.)

——— Bronco. An untamed horse.

Make the sign **To ride,** then with both hands retained in their relative positions, move them forward in high arches to show the bucking of the animal. (*Ute* I.)

——— Grazing of a.

Make the sign for **Horse,** then lower the hand and pass it from side to side as if dipping it upon the surface. (*Ute* I.)

——— Packing a.

Hold the left hand, pointing forward, palm inward, a foot in front of the chest and lay the opened right hand, pointing forward, first obliquely along the right side of the upper edge of the left hand, then on top, and then obliquely along the left side. (*Dakota* IV.)

——— Racer, fast horse, etc.

The right arm is elevated and bent at right angle before the face; the hand, in position (**S** 1) modified by being horizontal, palm to the face, is drawn across edgewise in front of the face. The hand is then closed and in position (**B**) approaches the mouth from which it is opened and closed successively forward several times, finally it is suddenly thrust out in position (**W** 1) back concave. (*Oto and Missouri.*) "Is expressed in the first sign for **Horse,** then the motion for quick running."

——— Racing.

Extend the two forefingers and after placing them parallel near together in front of the chest, backs upward, push them rapidly forward about a foot. (*Dakota* IV.)

Place both hands, with the forefingers only extended and pointing forward side by side with the palms down, before the body; then push them alternately backward and forward, in imitation of the movement of horses who are running "neck and neck." (*Ute* I; *Apache* I, II.)

——— Saddling a.

Hold the left hand as in the sign for **Horse, Packing a,** and lay the semi-flexed right hand across its upper edge two or three times, the ends of the right fingers toward the left. (*Dakota* IV.)

——— Spotted; pied.

Make the sign for **Horse,** then the sign for **Spotted.** (*Dakota* IV.)

Horseback, To ride.

Make the sign for **Horse,** with the difference that hand extends farther and the gesture is made quickly. (*Wied.*)

Separate the fore and middle fingers of the right hand, over the fingers of the left extended and joined, both palms toward the body, the forefinger of the right along the back of the left hand. (*Arapaho* I.)

Place the fore and middle fingers across the forefinger of left hand, both advanced in front of breast, both hands advancing motions as if riding, by up and down motions on finger and left hand. (*Ojibwa* IV.)

Place the first two fingers of the right hand (**N** with thumb resting on third finger) astraddle the two joined (many Sioux use only the forefinger straightened) and straight first finger of the left (**T** 1), then make several short arched movements forward with hands so joined. (*Dakota* I.) "The horse mounted and in motion."

Double the fists and make a succession of plunging motions, alternately with either hand, forward and downward in imitation of the motion of a horse's forefeet in trotting or galloping. The sign of straddling the fingers for *riding* is also in use among the Sioux, but is not so common as the above. (*Dakota* II.)

Extend and spread the right fore and middle fingers and place them, their ends pointing directly downward, astride the fingers of the left hand; the little and ring fingers of the right hand to be semi-flexed, thumb against index; the radial side of the left hand to be upward, fingers extended and joined, pointing forward, thumb in palm, then raise and lower the end of the left hand several times. This sign is also used for *Horse* when the hands are kept still. (*Dakota* IV.)

Extend the first two fingers of the left hand before the body, then straddle the fore and second fingers of the right hand across those of the left; in this position make a series of short jumps or jerks from left to right, imitating the gallop of a horse. (*Dakota* V.)

Make the sign for **Horse,** and as the hands are retained in this position, move them forward in short curves to represent motion of riding. (*Dakota* VI; *Hidatsa* I; *Arikara* I.)

The hands are arranged as in the sign for **Horse,** and then moved forward. (*Hidatsa* I.) This indicates in a general way a journey on horseback; but different modes of riding may be shown by appropriate modifications of this sign, thus: a slow journey is shown by moving the hands slowly forward, a race by moving them rapidly, a gallop by moving them in a series of small arcs whose convexity is upward, a jog-trot by moving them in a series of small angles with a slight arrest of motion between each angle, etc.

Left hand represents the horse, forefinger held up a little from the other fingers. Right forefinger and middle finger astride left forefinger; right thumb curved upward on left of left forefinger. (*Omaha* I.)

Place the first two fingers of the right hand astride the left hand, and move both forward. (*Iroquois* I.)

Place the hands as in the sign for **Horse,** and move them forward in short interrupted arched curves. (*Wyandot* I.)

Throw the index and middle fingers of the right hand astraddle the forefinger of the left. (*Sahaptin* I.)

Place the extended and separated index and second fingers of the right hand across the extended forefinger of the left, back of the hand forward. (*Pai-Ute* I.)

The index and second finger of the right hand alone extended and separated, placed astraddle the extended forefinger of the left, the palm of the right hand facing the back of the left hand. (*Apache* I.)

Hot, Hot weather. (Compare **Heat.**)

Hands at the height of the head or sometimes over it, horizontal, flat, with fingers and thumbs extended, separated, pointing toward the front, palm of hands down (**W**), make a slight tremulous motion with the fingers, without moving the hand. Sign is also often made with the extended fingers pointing toward each other. (*Dakota* I.) "Glimmer during hot weather. Reflection and refraction through the atmosphere often seen during hot weather."

Hour, Time of day. (Compare **Day.**)

To indicate any particular time of day, the hand with the sign of the sun is stretched out toward the eastern horizon, and then gradually elevated until it arrives in the proper direction to indicate the part of the heavens in which the sun will be at the given time. (*Long.*)

Forefinger of right hand, crooked as in sign for **Morning,** made to describe an arc over the head from east to west, being stopped at any point in the arc according to the time of morning or evening. (*Cheyenne* I.)

Indicate the spot at which the sun stood when the event to which they are alluding occurred. Point fixedly to that point and hold the arm in that position for several moments. (*Ojibwa* I.)

Curve the index of the closed right hand in the form of a half circle; move it from the eastern horizon, following the course of the sun, and allowing it to rest at the position occupied by that body at the time to be indicated. (*Dakota* V.)

The sign for **Sun** being made, the hand is held in the direction of the place which the sun would occupy at the time to be indicated, or the hand is made to describe an arc corresponding to the course of the sun during the lapse of time referred to. Thus the forenoon is shown by stretching the hand (in position of sign for *sun*) toward the horizon, and then slowly sweeping it up toward the zenith; the afternoon is shown by a reverse motion; noon, by holding the hand toward the zenith. (*Hidatsa* I.)

Deaf-mute natural signs.—Indicated by striking the air with the forefinger, signifying the stroke of the clock. (*Ballard.*)

Move the forefinger in a circle, indicating the motion of the minute-hand, and then indicate the number of hours. (*Hasenstab.*)

House. (Compare **Lodge.**)

The hand half open and the forefinger extended and separated; then raise the hand upward and give it a half turn, as if screwing something. (*Dunbar.*)

Partly fold the hands, the fingers extended in imitation of the corner of an ordinary log-house. (*Arapaho* I.)

Both hands outspread near each other, elevated to front of face; suddenly separated, turned at right angles, palms facing; brought down at right angles, suddenly stopped. Representing square form of a house. (*Cheyenne* I.)

The fingers of both hands extended and slightly separated, then those of the right are placed into the several spaces between those of the left, the tips extending to about the first joints. (*Absaroka* I.) "From the arrangement of the logs in a log building."

Cross the ends of the extended fingers of the two hands, the hands to be nearly at right angle, radial side up, palms inward, thumbs in palms. (*Dakota* IV.) "Represents the logs at the end of a log-house."

Both hands extended, fingers spread, place those of the right into the spaces between those of the left, then move the hands in this position a short distance upward. (*Wyandot* I.) "Arrangement of logs and elevation."

Both hands are held edgewise before the body, palms facing, spread the fingers, and place those of one hand into the spaces between those of the left, so that the tips of each protrude about an inch beyond. (*Hidatsa* I; *Kaiowa* I; *Arikara* I; *Comanche* III; *Apache* II; *Wichita* II.) "The arrangement of logs in a frontier house." In ordinary conversation the sign for *white man's house* is often dropped, using instead the generic term employed for *lodge*, and this in turn is often abbreviated, as by the Kaiowas, Comanches, Wichitas, and others, by merely placing the tips of the extended forefingers together, leaving the other fingers and thumbs closed, with the wrists about three or four inches apart.

Both hands held pointing forward, edges down, fingers extended, and slightly separated, then place the fingers of one hand into the spaces between the fingers of the other, allowing the tips of the fingers of either hand to protrude as far as the first joint, or near it. (*Shoshoni and Banak* I.) "From the appearance of a corner of a log-house—protruding and alternate layers of logs."

Deaf-mute natural signs.—Draw the outlines of a house in the air. (*Ballard.*)

Put the open hands together toward the face, forming a right angle with the arms. (*Larson.*)

——— Going into a.

Hold the open left hand a foot or eighteen inches in front of the breast, palm downward or backward, fingers pointing toward the right, and pass the right hand, palm upward, fingers bent sidewise and pointing backward, from before backward underneath it, through a curve until near the mouth. Some at the same time move the left hand a little forward. (*Dakota* IV.)

——— Going out of a.

Hold the open left hand a foot or eighteen inches in front of the breast, palm downward or backward, fingers pointing toward the right, and pass the right with index extended, or all of the fingers extended, and pointing forward, about eighteen inches forward underneath the left through an arc from near the mouth. Some at the same time move the left hand toward the breast. (*Dakota* IV.)

——— Stone, Fort.

Strike the back of the right fist against the palm of the left hand, the left palm backward, the fist upright (idea of resistance or strength); then with both hands opened, relaxed, horizontal, and palms backward, place the ends of the right fingers behind and against the ends of the left; then separate them, and moving them backward, each through a semicircle, bring their bases together. (*Dakota* IV.) "An inclosure."

Humble or **meek.**

Express by bent body, the right hand holding the mouth, or over it, the hands also sometimes blackened. (*Ojibwa* IV.)

First make the sign for **Poor, in property,** and then the extended forefinger of both hands (others closed), pointing upright (**J**), with backs inward, are carried straight outward from about a foot in front of their respective eyes as far as the arms can be extended. (*Dakota* I.) "Closely related to *poor in property;* and possibly means not seeing anything belonging to the person."

Hungry.

A sawing of the breast. (*Macgowan.*)

Touch the epigastrium with the forefinger of the right hand, and then opening the mouth point down the throat with the same finger. (*Dakota* II.) "The first motion indicates the emptiness of the stomach and the second the mode of remedying it."

The arm is flexed at the elbow, the hand collected into shape of a spoon, fingers and thumbs approximated and forming a hollow in the hand, not closed at points. With the palm up, the hand is then drawn edgewise across the epigastrium twice. (*Oto* I.) "Am empty or exhausted."

Another: The arm is flexed and the hand brought to the pit of stomach, as in position (**S** 1), modified by being horizontal and the back concave outward. The hand is then passed semicircularly downward and outward from the abdomen. (*Oto* I.) "Hollowed out or empty."

Pass the outer edge of the flat right hand across the epigastrium with a sawing motion. (*Kaiowa* I; *Comanche* III; *Apache* II; *Wichita* II.) "The craving of an empty stomach."

Both hands placed near together in front of stomach, fingers pointing toward the body, then each hand quickly jerked aside as though tearing something apart. (*Sahaptin* I.)

Place the flat right hand transversely to the pit of the stomach accompanied by an expression of weariness. (*Apache* I.)

Italian sign.—Tap the side with the open hand. (*Butler.*)

Hunting or **searching for.**

The forefinger is brought near the eye and placed in the attitude of pointing; it is then wagged from side to side, the eye following its devious motion, and seeming to look in the direction indicated. Sometimes the hand is extended far before the eye, and the same motion is given to the finger. (*Long.*)

Right forefinger extended (others closed) (**M**), is carried outward from the right eye, with considerable up-and-down and right-and-left movements, as though searching for something lost, the eyes following the course of the finger. (*Dakota* I.) "From the act of hunting or searching for anything."

With the index (or index and second fingers separated) only extended, place the hand nearly at arm's length before the face, the finger pointing slightly above the horizon; move it from side to side, with the eyes intently following the movement. (*Dakota* VII; *Ute* I.)

The hand is held as in the sign for **See,** and is then moved forward with a laterally zigzag motion. (*Hidatsa* I.)

With the right hand extended at arm's length, palm down, fingers pointing to the front and slightly above the horizon, move it horizontally from side to side, allowing the eyes to follow the motion, with an expression of inquiry. (*Apache* I.)

Deaf-mute natural signs.—Knit the eyebrows and move the head in different directions, bending the eye upon vacancy. (*Ballard.*)

Bring your head forward a little and change your look, showing that you are looking around for something not yet found. (*Cross.*)

Another: Having touched the eyelid, move horizontally the finger, with an expression of hunting for something. (*Cross.*)

Place the forefinger on the eye; at the same time incline the head as if hunting for something. (*Zeigler.*)

——— For game.

Same as the sign **Hunting for.** This is a general sign, and if hunting for a particular kind of game it must be specified by its proper sign, as *deer*, *antelope*, *buffalo*, etc. (*Dakota* I.) "From the act of seeking, searching."

Hurry.

Close the right-hand, index extended and elevated, pointing upward, back of hand forward, and beckon by drawing the hand toward the body several times excitedly. (*Omaha* I.)

Place the hands, palms up, near the stomach and in front of it, then make an up-and-down motion as if tossing a large light body a short distance. (*Kaiowa* I; *Comanche* III; *Apache* II; *Wichita* II.) "Evidently from the movement of the stomach sometimes experienced when running."

Husband. (Compare **Companion; Same; Married.**)

The two forefingers are extended and placed together with their backs upward. This sign is also used for *Companion.* (*Long.*)

Make the sign in front of the privates for **Man,** and then move the right fist, back outward, forward a foot or eighteen inches from six inches in front of the navel. (*Dakota* IV.) "Man I have."

——— And Wife.

The same sign frequently used for both: Lay the two forefingers together, side by side, straight and pointing forward, the other fingers loosely closed. (*Dakota* IV.) "Two joined as one."

——— Or Wife.

Extend the forefingers of each hand and bring them together side by side in front of the breast and a foot therefrom. (*Comanche* I.)

I, me, myself.

The fingers of the right hand laid against the breast. (*Dunbar.*)

The clinched hand struck gently, and with a quick motion, two or three times upon the breast. Or, the fingers brought together are placed perpendicularly upon the breast. (*Long.*)

Touch the nose-tip, or otherwise indicate self with the index. (*Burton.*)

Touch or otherwise indicate one's self. (*Arapaho* I.)

Right-hand fingers drooping, forefinger separated from the others, gently touched once or twice to the right breast. (*Cheyenne* I.)

The fingers of the right hand are collected to a point, the thumb lying against the palms of the fingers, then bring the hand, pointing upward, slowly toward the breast. Also used to express *to me.* (*Absaroka* I; *Shoshoni and Banak* I.)

Right hand (**S** 1) thumb and fingers extended horizontal, back outward, tapping the left breast. When the gesturer desires to be very emphatic, the clinched right hand is struck repeatedly against the right breast. (*Dakota* I.) (Compare **Heart,** *Dakota* **I.**)

Touch the end of the nose with the radial side of the right forefinger, the forefinger pointing upward. (*Dakota* IV.)

Place the extended index against the middle or upper portion of the breast. (*Dakota* VII.)

Touch the middle of the breast with the index. (*Hidatsa* I; *Arikara* I.)

Strike the left breast with index-finger of right hand, the other fingers being closed. (*Omaha* I.)

With the right hand arched, so that the thumb rests along the side of the index, place the inner side of the hand against the breast, with the fingers pointing downward. (*Kaiowa* I; *Comanche* III; *Apache* II; *Wichita* II.)

Another: Place the index or the ends of the extended fingers against the breast. (*Kaiowa* I; *Comanche* III; *Apache* II; *Wichita* II.)

Place the index upon the breast rather quickly. If this gesture is made slowly, and when in connection with other signs of a narrative, or preceding the expression of want, or desire, it expresses *to me.* (*Pai-Ute* I.)

Sweep the hand up the body and raise the right forefinger up with vigor just before face. (*Apache* III.)

Deaf-mute natural signs.—Indicated by pointing to one's self—to the person speaking. (*Ballard.*)

Rest the tip of the forefinger upon the breastbone, and at the same time nod the head. (*Hasenstab.*)

Put the right forefinger on the breast. (*Zeigler.*)

Some deaf-mutes push the forefinger against the pit of the stomach, others against the breast, and others point it to the neck for this personality.

——— Objective.

With the fingers placed closely together at the tips, the thumb resting alongside of the index, bring the hand, pointing upward, slowly to and against the middle of the breast. (*Kaiowa* I; *Comanche* III; *Apache* II; *Wichita* II.)

Ice.

Begin with the sign of **Water,** then of **Cold,** then of **Earth,** and lastly a **Stone,** with the sign of **Sameness** or **Similarity.** (*Dunbar.*)

Same as the sign for **Stone.** (*Dakota* I.) "A hard substance."

Ignorant.

Tap the forehead with the slightly-curved index, followed by the sign for **Lie.** (*Apache* I.)

Ill. See **Sick.**

Imprecation.

Italian sign.—The forefinger turned down is a motion of a girl at Thrasymene, who was refused alms, as she cried *va a l' inferno.* (*Butler.*)

Imprudent—Rash.

Shade the eyes with left hand, relaxed (**Y** palm inward), right hand in front of breast, forefinger straight upright (**J** palm outward); move forward, making three or four short stops in the movement to represent the motion of a person walking. This is the general description; if rereferring to *rash charge in battle*, the sign with left hand is first made, then sign for *charge* is made with both. (*Cheyenne* II.) "Going blindly, without looking."

Place the tips of the extended forefingers to the temples, then throw the hands outward and downward. (*Wyandot* I). "No judgment—*literally.*"

In, Within.

Forefinger and thumb of the left hand are held in the form of a semicircle, opening toward and near the breast, and the right forefinger, representing the prisoner, is placed upright within the curve, and passed from one side to another, in order to show that it is not permitted to pass out. This is the sign for **Prisoner,** as given in "Introduction to the study of sign-language," etc. This sign is the one made by the Sioux for *In, Within,* and also to indicate *Prisoner,* but when so used the semicircle would be continued to a circle after passing the finger within it. This sign, however, is not limited to *Prisoner.* (*Dakota* I.)

The left-hand fingers extended side by side, the thumb facing but an inch or two from them, pointing toward the right (forming a U held sidewise); the bent index is then pushed partly into the space between the thumb and fingers of the left. (*Apache* I.)

Indecision, Doubt. (Compare **Question,** and **Know, I don't.**)

The index and middle finger extended diverged, place them transversely before the situation of the heart, and rotate the wrist two or three times gently, forming each time a quarter of a circle. (*Long.*) "More than one heart for a purpose."

Combine the signs of **Affirmative** and **Negative,** *i. e., Yes—No.* (*Arapaho* I.)

The right hand brought to the region over the heart, with the first two fingers extended, pointing obliquely downward toward the left, thumb resting on third finger, which with the fourth is closed, back of hand outward, make several quick tremulous motions, then extend the thumb and fingers, and carry the hand out in front of the right side of body, turning the hand so that it is brought, horizontal, flat, palm upward (**X**), extending the forearm from the elbow only. (*Dakota* I.)

The right index extended, back upward, pointing toward the left, in front of the left breast, the other fingers half closed, thumb on middle finger; move the hand through an arc forward and toward the right until it is in front of the right breast. (*Dakota* IV.) "Going around, therefore not certain."

Make the sign for **Have,** and then, with its back upward, fingers separated a little, slightly flexed, and pointing forward, rotate the right hand to the right and left, describing an arc upward (to imply doubt.) (*Dakota* IV.) "Perhaps I will get it, or have it."

Index and second fingers straight and separated, remaining fingers and thumb closed, place the tips near the region of the heart, pointing to the left; move to and from the heart repeatedly as if puncturing it, at each thrust rotating the hand slightly so that the position of the fingers will be similar at each alternate movement. (*Kaiowa* I; *Comanche* III; *Apache* II; *Wichita* II.)

——— Perplexity.

Italian sign.—Open hand shaken before the forehead as if an agitation of the brain. (*Butler.*)

——— Doubt.

Italian sign.—Both hands with fingers apart and palms forward, raised by the head. (*Butler.*)

Indian Agency.

First make the sign for **White man,** and then the sign **Give,** then designate the particular agency (see local names in "PROPER NAMES, PHRASES, ETC.") by its proper sign, if desiring to be specific. (*Dakota* I.) "The place where government provisions are issued, and the white man who issues them."

Indifference. None of my business.

Italian signs.—Both hands held down by the thighs. (*Butler.*)

The hand waved under the chin. (*Butler.*)

Indigent. See **Poor.**

Infant. See **Child.**

Inquiry. See **Question.**

It is so. See **Yes.**

Journey. See **Going.**

Kettle.

Same sign as for **Village,** but is made closer to the earth. (*Wied.*) The configuration of a common kettle (the utensil obtained from the whites in trade being, of course, the one referred to) is the same as that of the stockaded villages of the Mandans and Hidatsa, the intervals left between the hands representing in this case the interruption in the circle made by the handles. The differentiation is effected by the position closer to the earth.

First make the sign for **Fire,** and then place the fingers and thumbs of both hands together in front of the breast so as to describe a circle looking downward, and then move the hands still held in this position as though putting a kettle over the fire. (*Dakota* I.) "From one of the uses to which they put a kettle."

Make the sign for **Eating,** and then make a circle by holding the ends of the partly bent forefingers and thumbs near together, the palms of the hands inward (*Dakota* IV.)

Kill, Killing. (Compare **Knife, to kill with.**)

The hands are held with the edge upward, and the right hand strikes the other transversely, as in the act of chopping. This sign seems to be more particularly applicable to convey the idea of death produced by a blow of the tomahawk or war-club. (*Long.*)

Clinch the hand and strike from above downward. (*Wied.*) I do not remember this. I have given you the sign for killing with a stroke. (*Matthews.*) There is an evident similarity in conception and execution between the (*Oto and Missouri* I) sign and *Wied's.* (*Boteler.*) This motion, which may be more clearly expressed as the downward thrust of a knife held in the clinched hand, is still used by many tribes for the general idea of "kill," and illustrates the antiquity of the knife as a weapon. The actual employment of arrow, gun, or club in taking life, is, however, often specified by appropriate gesture.

Smite the sinister palm earthward with the dexter fist sharply, in sign of "Going down"; or strike out with the dexter fist toward the ground, meaning to "shut down;" or pass the dexter under the left forefinger, meaning to "go under." (*Burton.*)

Right hand cast down. (*Macgowan.*)

Right hand clinched, thumb lying along finger tips, elevated to near the shoulder, strike downward and outward vaguely in the direction of the object to be killed. The abstract sign for *Kill* is simply to clinch the right hand in the manner described and strike it down and out from the right side. (*Cheyenne* I.)

Both hands clinched, with the thumbs resting against the middle joints of the forefingers, hold the left transversely in front of and as high as the breast, then push the right, palm down, quickly over and down in front of the left. (*Absaroka* I; *Shoshoni and Banak* I.) "To force under—literally."

With the dexter fist brought in front of the body at the right side, strike downward and outward, with back of hand upward, thumb toward the left, several times. (*Dakota* I.) "Strike down."

With the first and second joints of the fingers of the right hand bent, end of thumb against the middle of the index, palm downward, move the hand energetically forward and downward from a foot in front of the right breast. (*Dakota* IV.) "Striking with a stone"—man's first weapon.

Hold the right fist palm down, knuckles forward, and make a thrust forward and downward. (*Dakota* VI, VII; *Hidatsa* I; *Arikara* I.)

The left hand, thumb up, back forward, not very signally extended, is held before the chest and struck in the palm with the outer edge of the right hand. (*Mandan and Hidatsa* I.) "To kill with a blow; to deal the death-blow."

Right hand, fingers open but slightly curved, palm to the left; move downward, describing a curve. (*Omaha* I.)

Another: Similar to the last, but the index-finger is extended, pointing in front of you, the other finger but half open. (*Omaha* I.)

Another: Close the right hand, extending the forefinger alone; point toward the breast, then throw from you forward, bringing the hand toward the ground. (*Omaha* I.)

Both hands, in positions (**AA**), with arms semiflexed toward the body, make the forward rotary sign for **Fighting** or **Battle;** the right hand is then raised from the left outward, as clutching a knife with the blade pointing downward and inward toward the left fist; the left fist being held *in situ*, is struck now by the right, edgewise as above described, and both suddenly fall together. (*Oto and Missouri* I.) "To strike down in battle with a knife. Indians seldom disagree or kill another in times of tribal peace."

Place the flat right hand, palm down, at arm's length to the right, bring it quickly, horizontally, to the side of the head, then make the sign for **Dead.** (*Wyandot* I.) "To strike with a club, dead."

Deaf-mute natural signs.—Strike a blow in the air with the clinched fist, and then incline the head to one side, and lower the open hand. (*Ballard.*)

Strike the other hand with the fist, or point a gun, and, having shot, suddenly point to your breast with the finger, and hold your head sidewise on the hand. (*Cross.*)

Use the closed hand as if to strike, and then move back the head with the eyes shut and the mouth opened. (*Hasenstab.*)

Put the head down over the breast, and then move down the stretched hand along the neck. (*Larson.*)

——— In battle, To.

Make the sign for **Battle, Fight,** then strike the back of the fingers of the right hand into the palm of the flat and slightly arched palm of the left, immediately afterward throwing the right outward and downward toward the right. (*Ute* I.) "Killed and falling over."

——— You; I will kill you.

Direct the right hand toward the offender and spring the finger from the thumb, as in the act of sprinkling water. (*Long.*) The conception is perhaps "causing blood to flow," or, perhaps, "sputtering away the life," though there is a strong similarity to the motion used for the *discharge of a gun or arrow.*

Kind. See **Good heart.**

Knife.

Hold the left hand clinched near the mouth, as if it held one end of a strip of meat, the other end of which was between the teeth, then pass the edge of the right hand as in the act of cutting obliquely a little upward from right to left between the other hand and mouth, so as to appear to divide the supposed meat. (*Long.*)

Cut past the mouth with the raised right hand. (*Wied.*) I have given you a different sign, which is the only one I have ever seen. (*Matthews.*) Although the signs (*Oto and Missouri* I, and *Wied's*) are different in their execution as applied to local parts, the same conception pervades each—"something used to sever or separate." (*Boteler.*) *Wied's* sign probably refers to the general practice of cutting off food, as much being crammed into the mouth as can be managed and then separated by a stroke of a knife from the remaining mass. This is specially the case with fat and entrails, the aboriginal delicacies.

Cut the sinister palm with the dexter ferient downward and toward one's self: if the cuts be made upward with the palm downward, meat is understood. (*Burton.*)

Right hand, palm outward, little finger representing the edge of a knife, drawn downward across palm or inside of left hand. (*Cheyenne* I.)

Left-hand fist (**B**) held six or eight inches in front of the mouth, back outward, as though holding a large piece of meat in the mouth, and then the right hand with the back outward, fingers extended, joined, upright (**S**), is passed from right to left on a curve between the mouth and the left hand as though cutting the piece of meat in two. (*Dakota* I.) "May have come from their first manner of using the knife in eating."

Hold the left fist, back outward, about eight inches in front of the mouth, and move the opened right hand, palm backward, fingers pointing obliquely upward toward the left, obliquely upward and downward from side to side behind it. (*Dakota* IV.) "Holding a piece of meat with the left hand and the teeth, and cutting a piece off with a knife."

The left hand, fully opened, with the fingers close together, palm upward and finger tips to the front, is held before the person. Then the right hand, also fully opened and with fingers not spread apart, with the palm inward and the thumb upward, is laid transversely on the left palm—the outer edge only touching the left palm. Sometimes the right hand is then drawn away once to the right with a motion representing a cut. (*Mandan and Hidatsa* I.)

The left arm is semi-extended, and the left hand, in position (**X** 1), modified by being edgewise up and down. The right arm is then brought before the body, and the hand, in position (**X**), horizontal, is made to exert a carving motion at the knuckle or metacarpo-phalangeal joints of the left hand, which is concluded by a scooping or carving movement. (*Oto and Missouri* I.) "That by which we open joints and cut or carve."

Deaf-mute natural signs.—Imitate the act of whittling with one forefinger upon the other. (*Ballard.*)

Pass one forefinger over the other several times. (*Hasenstab.*)

——— Skinning with a.

The hands are placed as in the sign for **Knife**, then the right hand is held a little obliquely, *i. e.*, with forearm semi-pronated, and drawn, but never pushed, across the left palm repeatedly, advancing a little toward the finger-ends with each strepe. (*Mandan and Hidatsa* I.)

Hold the left closed hand, palm down, a short distance before the body, and make repeated cuts in front of the knuckles with the flattened right hand. (*Pai-Ute* I.) "Represents holding a flap of skin and separating it from the body."

——— To kill with a.

Clinch the right hand and strike forcibly toward the ground before the breast from the height of the face. (*Ute* I.) "Appears to have originated when flint knives were still used."

Know. (Compare **Good ; Indecision ; Understand** and **Yes.**)

The forefinger of the right hand held up nearly opposite to the nose, and brought with a half turn to the right and carried a little outward. Place any of the articles [*sic*] before this sign, which will then signify, I know, you know, he knows. Both hands being made use of in the manner described implies to know much. (*Dunbar.*)

Spread the thumb and index-finger of the right hand, sweep toward the breast, moving them forward and outward, so that the palm turns up. (*Wied.*) The right arm is flexed and raised; the hand is then brought before the forehead between the eyes as in position (**I** 1, modified by being palm outward and the index more opened); the hand and forearm then describe a quadrant forward and downward. There is no conceivable similarity between this sign and *Wied's* as executed and it is probable that the conceptions are likewise of different source or association. This same sign is used for *Knowledge* in an abstract sense. (*Boteler.*) "To have in mind or utter from the mind."

The thumb and index-finger made into a ring and passed from the mouth. (*Macgowan.*)

Thumb, first and second fingers of right hand extended (others closed), horizontal, backs upward, are carried from the natural position close to and in front of the body as high as the mouth, where the hand is carried with a curved motion, first upward and outward, and then downward to the level of the stomach, backs of the fingers looking obliquely downward. (*Dakota* I.) "I have heard your talk and know what you say."

The right index and thumb fully extended and spread, the other fingers loosely closed, index pointing forward and a little to the left, back of hand upward; then supinate the hand, thrown on its back and held about a foot in front of the right breast. (*Dakota* IV.)

Strike the left breast with the thumb and forefinger, keeping the other fingers closed. (*Omaha* I.)

Another: Curve three fingers of the right hand, touch tip of middle finger with thumb, extend forefinger, and shake hand forward and down. (*Omaha* I.)

Another: Same as the preceding, but thumb and fingers closed instead of ring shape. (*Omaha* I.)

Another: Curve three fingers of the right hand, place the thumb over their middle joints, extend the index, and shake forward and downward from the right side of the face. (*Omaha* I.)

Spread the index and thumb of the right hand fully apart, remaining fingers closed, palm toward the body, and move the hand forward and slightly downward and a little to the right from below the chin to a distance of eight or ten inches. (*Kaiowa* I; *Comanche* III; *Apache* II; *Wichita* II.)

Deaf-mute natural signs.—Tap the forehead slightly with the hand. (*Ballard.*)

Nod and point to the forehead. (*Cross.*)

Put the open hand to the breast, and at the same time bend down the head. (*Larson.*)

Place the right forefinger on the forehead, at the same time nod the head as if to say "yes." (*Zeigler.*)

All the ascertained gestures of deaf-mutes relating to intelligence are connected with the forehead, on which we, also, rest the forefinger, for show of thought.

——— I don't know.

First place the fingers in the position for **Know**; then turn the right hand upward with spread fingers, so that they point outward toward the right side. (*Wied.*)

Is expressed by waving the right hand with the palm outward before the right breast, or by moving about the two forefingers before the breast, meaning "two hearts." (*Burton.*)

First make the sign for **Know,** and then that for **Not** or **No.** (*Dakota* I.) "Do not know what you say. Indecision, doubt."

Make the sign for **I Know;** then the sign for **No,** or while the fingers are in position for **I Know** throw the hand outward as for **No.** (*Dakota* IV.)

The right arm is elevated and the hand, in type-position (**F**), is twisted several times before the chest, then suddenly everted and expanded. (*Oto and Missouri* I.) "Not screwed up or posted unaware."

Make the sign for **Know, to,** and throw the hand to the right as in **No, Not.** (*Kaiowa* I; *Comanche* III; *Apache* II; *Wichita* II.)

Deaf-mute natural signs.—Point to the bosom, meaning the speaker, place the hand upon the forehead and then move the hand away. (*Ballard.*)

Having put the finger to your breast, point to the forehead, shaking your head. (*Cross.*)

Put the stretched hand to the breast, and at the same time shake the head. (*Larson.*)

Place the right forefinger on the forehead, at the same time shake the head as if to say *No.* (*Zeigler.*)

——— I don't know you.

Move the raised hand, with the palm in front, slowly to the right and left. (*Burton.*)

Lake, Pond.

Make the sign of **Drinking,** and form a basin with both hands. If a large body of water is in question, wave both palms outward as in denoting a plain. (*Burton.*)

Sign for **Water** followed by sign for **Big** in the sense of Broad, Wide. For **Pond,** make sign for **Lake** and **Little.** (*Cheyenne* I.)

First make the sign for **Water,** and then the sign for **Big** in the sense of Flat. (*Dakota* I.) "Water spread out or level—not running water."

Make the sign for **Water,** then spread and slightly flex the thumbs and forefingers, and hold the hands, palms inward, near together or far apart, according to the size of the lake referred to. (*Dakota* IV.)

After making the sign for **Water,** indicate a circle, by extending the hands horizontally, allowing the tips of the forefinger and thumb of one hand to join those of the other. A larger circle is made to indicate a large lake by making two horizontal semicircular air lines with the hands, the termini of the lines joining. (*Shoshoni and Banak* I.)

Deaf-mute natural signs.—Make a circle with the forefinger, and imitate the act of drinking to signify **Water.** (*Ballard.*)

Partly open your mouth with the head held back, place the fingers—arranged in such a manner that the hand looks like a cup—to it, and then suddenly move the hand horizontally along a line describing an ellipse. (*Cross.*)

Lame.

Right-hand fist (**B**, turned downward) in front of the body; make the forward arched movements in imitation of the walk of a lame person. (*Dakota* I.) "From a lame person's manner of walking."

Lance, or **Spear.**

Is shown by an imitation of darting it. (*Burton.*)

Hands elevated and closed as though grasping the shaft of a lance, left hand before right, sudden motion made from the left shoulder diagonally forward (hands being carefully retained in their relative positions). (*Cheyenne* I.)

Right hand extended in front on a level with the shoulder, as though holding a spear or lance in position to use it. (*Dakota* I.) "From the throwing of a spear."

(1) Point to tent-pole; (2) cut off left forefinger, with right index; (3) rub it with the latter toward its tip; (4) place tip of right at base of left index; (5) thrust both forward. (*Apache* III.) "(1) Shaft- (2) head; (3) sharpens it to point; (4) adjusted head to shaft; (5) put it in service."

Large. See **Great** and **Big,** in sense of Large.

Laugh, To. Laughter.

Place the hands as in **Heavy,** but forward from each side of the lower jaw, then move them up and down a short distance rapidly, the face expressing a smile. (*Kaiowa* I; *Comanche* III; *Apache* II; *Wichita* II.)

Lazy.

Lay the arched right hand (**H**) on the middle of the upper left arm partially extended in front of the body, back of hand upward, and in the same manner lay the left hand on the right arm, and then carry both hands upward on their respective sides in front to the level of the face, where both index-fingers are extended and point upright (**J**), from whence they are carried slowly downward in front of the body to the level of the stomach. (*Dakota* I.) "Lazy; no good with the arms and hands."

Deaf-mute natural sign.—Having extended your cheeks, shake your head, implying no, and then compress them with the hands. (*Cross.*)

Lean. See **Poor.**

Leaves (of trees).

The sign for **Tree** must be made first, and in this case with the left hand. Beginning from low down on left side, with fingers and thumb separated, pointed upward (**P**), move the hand upward till it reaches a little above the head; this is the tree. Right hand in position (**G**) is brought to touch the different fingers of the left (which are the branches); forefinger and thumb describe the leaf, and is made larger to describe different kinds by parting the thumb and finger more or less. The sign for **Tree** alone is generally made with the right hand. (*Cheyenne* II.)

Make a tree and its branches, and then with the thumb and forefinger of the right hand describe a semicircle with the free ends of the thumb and finger turned downward in front of the body. (*Dakota* I.) "In resemblance of the drooping leaves of a tree."

Arms are extended from body like limbs of a tree. The right hand, in position (**N**), is then brought to left in position (**S**) modified by being horizontal edgewise. From the left, arm and hand still extended, the right drops successively to the ground. Autumn is represented by this sign, following the sign for **Sun.** (*Oto* I.) "Something that drops from spreading limbs and the time for such."

——— On trees.

Same motion of right hand as in **Tree,** with the left hand and arm in front above head, looking up, spread the fingers which were bent downward. Thumbs nearly touching, shake both hands sidewise and up and down quickly in imitation of aspen-leaf motion. When the tree is near, point in both cases to the tree or trees with the finger. (*Ojibwa* IV.)

Left-handed.

The left hand clinched is held before the neck; the elbow is then brought in to the side, at the same time giving to the forearm a twist, so as to bring the closed palm opposite the breast. (*Long.*)

Simply point to the left hand with the extended forefinger of the right. (*Dakota* I.)

Leggings.

Separate the thumb and index-finger of each hand, and draw them upward along both legs. (*Wied.*) I have described his sign in essentially the same terms; but as for the sign for *Coat*, I say the fingers are closed. The same remarks apply in one case as in the other. (*Matthews.*) Notwithstanding the indefinite and inexplicit manner in which *Wied's* sign is expressed, there is evident similarity to that of (*Oto and Missouri* I), both in conception and movement. (*Boteler.*)

The tip of the thumb of each hand is opposed to the tip of one or more of its corresponding fingers, as if they grasped something lightly. The hands are then held a few inches apart on the anterior aspect of one of the thighs as low down as they can reach without bending the body (the finger-tips nearly or quite touching the limb), and are then simultaneously drawn rapidly upward to the waist to represent the motion as if drawing on a legging. The motion may be repeated on the opposite limb. (*Mandan and Hidatsa* I.)

The body is bent forward and the limbs flexed upward and adducted. Both hands, with the backs arched upward and the thumb points in contact, seemingly encircle the leg at the ankle, and are drawn toward the body over each leg severally, as in drawing the above article on. (*Oto and Missouri* I.) "That which is drawn over the legs."

Let alone.

Right-hand palm down, arm bent at elbow, move downward by degrees as low as the knees. (*Omaha* I.)

Lie, Falsehood.

The forefinger and middle fingers extended, passed two or three times from the mouth forward. They are joined at the mouth, but separate as they depart from it, indicating that the words go in different directions. (*Long.*)

Pass the second and third finger of the right hand toward the left side in front of the mouth. (*Wied.*) My description is much to the same effect, but I add that the hand is moved forward. (*Matthews.*) Though the description of *Wied's* sign is condensed, there is an evident similarity in the execution and conception of this with (*Oto* I.) (*Boteler.*) The author means the index and middle finger as appears from other parts of his list. He counts the thumb as the first finger.—[ED.

Extend the two first fingers from the mouth. (*Burton.*) "Double tongue—a significant gesture."

Pass the hand from right to left close by and across the mouth, with the first two fingers of the hand opened, thumb and other fingers closed. (*Dodge.*)

Thrust the fore and middle finger, extended and separated, from the mouth. Literally "the forked tongue." (*Arapaho* I.)

Right-hand fore and middle fingers placed on a level with the base of the chin, first knuckle against or near the mouth; thrust forward and to the left. (*Cheyenne* I.) "Speak double, with two tongues."

If the two forefingers are parted and moved from the mouth, like the split tongue of a snake, it signifies lying. This sign is adopted in the sign-language of all the Indians, as well as the figure from which it is derived. (*Ojibwa* I.) "Speak with the forked tongue, *i. e.*, lie."

Fore and second fingers of the right hand extended and forked (**L** 1, with thumb resting on third finger) passed from right to left directly in front of the mouth. (*Dakota* I.) "Double-tongued."

Place the right hand, palm inward (toward the left), just in front of the mouth; strongly extend the index; also extend the middle finger, but bend it toward the palm until it is at an angle of 45° with the index; half close the ring and little fingers, thumb against ring finger; move the hand straight forward about eight inches; or, having placed the hand with its back forward, move the hand to the left. (*Dakota* IV.) "Two tongues."

Touch the region of the heart with the right hand, then close the hand, extend the first two fingers, passing them from the mouth forward. (*Dakota* V.)

Spread the extended index and second finger of the right hand, and pass them, palm toward the body, quickly by and past the mouth to the left. (*Dakota* VI, VII.)

Close the right hand, leaving the index and second fingers extended and separated; then pass them before the breast from right to left, the fingers pointing in that direction and the palm toward the body. (*Hidatsa* I; *Arikara* I.)

The sign is like that for **True,** except that both the middle and index fingers are extended, and these are held together while the hand is at the mouth, but they diverge as the hand moves forward. (*Mandan and Hidatsa* I.)

The arm is flexed and elevated to a level with the mouth. The hand is in position (**N** 1), modified by being horizontal instead of vertical, with palm toward the face and first two fingers separated. From inception of sign at right side of face the hand is thrust from right to left across the mouth. It is then repeated with other hand identically the same *from opposite* side of mouth. Sometimes both index-fingers are used from corners of mouth, palm outward, and made to thus diverge as they recede. (*Oto* I.) "Duplicity or double-tongued."

Make the sign for **Talk,** then throw the right fist outward and downward toward the right side, and snap the fingers from the thumb, as in **Bad.** (*Wyandot* I.) "Talk, bad."

After pointing at person addressed, the hand is placed in front of mouth, back toward mouth and fingers projecting forward (**P** 1, with knuckles upward). Then with quick motions move hand two or three times to a point six or eight inches from mouth, as though casting something from the mouth. Then move hand to side of mouth, the two first fingers only extended and slightly separated (**N** 1, changed to horizontal position and thumb obliquely extended), and past the mouth to a point on the left. (*Sahaptin* I.) "Words double-tongued."

Pass the right hand to the left close by and across the mouth, with the first two fingers of the hand opened, thumb and other fingers closed. (*Pai-Ute* I.)

With the third and fourth fingers of the right hand closed, resting the tips of the first two fingers against the ball of the thumb, place the hand in front of the chin or mouth, and while moving it diagonally forward and to the left let the fingers snap forward from the thumb; repeat this two or three times. (*Ute* I.) "Double-tongued."

Another: Hold the index, pointing upward, in front of the mouth, and move it repeatedly and alternately obliquely forward toward the right and left. The index and second finger extended and separated, are sometimes used to represent extraordinary lying. (*Ute* I.) "Talk two ways."

Sign as for **Truth,** but make the motion obliquely and alternately toward the left and right. (*Apache* I.)

Run the index from each corner of the mouth. (*Zuñi* I.)

Deaf mutes gesture *Truth* by moving one finger straight from the lips, "straight-forward speaking," but distinguish *Lie* by moving the finger to one side, "sideways speaking."

Lie down.

Point to the ground, and make a motion as if lying down. (*Burton.*)

The sign for **Sleep,** the eyes remaining open. (*Arapaho* I.)

Only differs from the sign for **Sleep** in inclining the head and arms sidewise (to the right) toward the ground, with elbow out as though used as a support. (*Dakota* I.)

Wave the hands low down, palms up (**X**), horizontally and laterally; **Sleep.** (*Apache* III.) "Suitable place to lie upon."

Deaf-mute natural sign.—Place the hand upon the cheek, incline the head to one side, and then lower the hand. (*Ballard.*)

——— Flat where you are.

Like **Down** and **Alone,** but the arm is extended in the direction of the party addressed, with palm down, and moved downward several times. (*Omaha* I.)

Life, Living.

Right-hand forefinger straight upright, others closed (**J**), is slowly raised up in front of the right side close to the body, as high above the head as the arm can be extended. (*Dakota* I.) "Raised up; coming up; growing also."

Light, Daylight. (Compare **Clear.**)

Make the sign of the sun in the eastern horizon, and then extend the hands together, with the palm upward, and carry them from each other outward. (*Long.*)

Daylight is equivalent to **Sunrise.** See also **Glass.** (*Dakota* I.)

The left arm is extended from left side of the body with the hand in type position (**I** 1) modified by the index being a little more extended. In this manner the left hand indicates the rising sun. The hands are now approximated at fingers, palms before the face, and gradually diverge, as do hanging curtains. The hands are held with the edges inward and outward. (*Oto and Missouri* I.) "That which accompanies the sun; the curtains of night being unfolded."

Deaf-mute natural sign.—Move the outstretched hands apart from each other. (*Larson.*)

Light (in weight).

Right hand horizontal, back downward, fingers extended and partially curved upward, is carried upward by easy motion in front of the body, with arm nearly extended and as though the hand contained some light object. (*Dakota* I.) "Not heavy to lift."

Point at an imaginary object; lift it with one hand, easily and rapidly, high above head (**Y**). (*Apache* III.)

Lightning. (Compare **Thunder.**)

First the sign for **Thunder;** then open or separate the hands, and lastly bring the right hand down toward the earth, in the center of the opening just made. (*Dunbar.*)

Right hand elevated before and above the head, forefinger pointing upward, brought with great rapidity with a sinuous, undulating motion; finger still extended diagonally downward toward the right. (*Cheyenne* I.)

As thunder and lightning generally accompany each other, there is no separate sign for **Lightning.** (*Dakota* I.)

Extend the index straight and earthward, raise the hand to the height of the head, and pass it quickly down toward the ground making a rotary movement with the finger in doing so. (*Apache* I.)

Listen. See **Hear.**

Little. Small amount. See **Small,** also **None.**

Locomotive.

Place the right hand, with the fingers and thumb partially curved upward and separated, knuckles outward, in front of the breast, and push it up and down a short distance while moving it forward (puffing smoke and forward movement of engine); then place both hands edgewise before the chest, palms facing, and while moving the left but a few inches forward, pass the right quickly by it and to the front to arm's length. (*Kaiowa* I; *Comanche* III; *Apache* II; *Wichita* II.) "Rapid motion as compared with anything else."

Lodge, Tipi, Wigwam.

The two hands are reared together in the form of the roof of a house, the ends of the fingers upward. (*Long.*)

Place the opened thumb and forefinger of each hand opposite each other, as if to make a circle, but leaving between them a small interval; afterward move them from above downward simultaneously (which is the sign for *village*); then elevate the finger to indicate the number—one. (*Wied.*) Probably he refers to an earthen lodge. I think that the sign I have given you for "skin lodge" is the same with all the Upper Missouri Indians. (*Matthews.*)

Place the fingers of both hands ridge-fashion before the breast. (*Burton.*)

Indicate the outlines (an inverted **V**, thus **Λ**), with the forefingers touching or crossed near the tips, the other fingers closed. (*Arapaho* I.)

Both hands open, fingers upward, tips touching, brought downward, and at same time separated to describe outline of a cone, suddenly stopped. (*Cheyenne* I.)

Place the tips of the fingers of both hands together in front of the breast, with the wrists some distance apart, as in the outlines of the letter A. (*Dakota* V.)

With both hands flat and extended, pointing upward, palms facing, place the tips of both together, allowing the wrists to be about three or four inches apart. (*Shoshoni and Banak* I.) "Outline of lodge-poles with covering."

Both hands flat and extended, placing the tips of the fingers of one against those of the other, leaving the palms or wrists about four inches apart. (*Absaroka* I; *Wyandot* I; *Shoshoni and Banak* I.) "From its exterior outline."

Tipi is the preferred word with the Sioux. *Wigwam* is not known. Both hands carried to the front of the breast and placed V-shaped, with the palms looking toward each other, edge of fingers outward, thumbs inward. (*Dakota* I.) "From the tipi."

Cross the ends of the extended forefingers, the right one either in front or behind the left, or lay the ends together; rest the ends of the thumbs together side by side, the other fingers to be nearly closed, hands nearly upright, and the fingers resting against each other, palms inward. (*Dakota* IV.) "Represents the tipi poles and the shape of the tipi."

Fingers of both hands extended and separated; then interlace them so that the tips of one hand protrude beyond the backs of those of the opposing one; hold the hands in front of the breast, leaving the wrists about six inches apart. (*Dakota* VII; *Hidatsa* I; *Arikara* I.)

The extended hands, with finger tips upward and touching, the palms facing one another, and the wrists about two inches apart, are held before the chest. (*Mandan and Hidatsa* I.)

Place the tip of the index against the tip of the forefinger of the left hand, the remaining fingers and thumbs closed, before the chests, leaving the wrists about six inches apart. (*Kaiowa* I; *Comanche* III; *Apache* II; *Wichita* II.) "Outline of lodge. This is an abbreviated sign, and care must be taken to distinguish from *to meet*, in which the fingers are brought from their respective sides instead of upward to form gesture."

Place the tips of the fingers of the flat extended hands together before the breast, leaving the wrists about six inches apart. (*Kaiowa* I; *Comanche* III; *Apache* II; *Wichita* II.)

Both hands flat and extended, fingers slightly separated then place the fingers of the right hand between the fingers of the left as far as the second joints, so that the fingers of one hand protrude about an inch beyond those of the opposite; the wrist must be held about six inches apart. (*Kaiowa* I; *Comanche* III; *Apache* II; *Wichita* II.) "Outline of Indian lodge and crossing of tent-poles above the covering."

Place the tips of the spread fingers of both hands against one another, pointing upward before the body, leaving a space of from four to six inches between the wrists. (*Pai-Ute* I.) "Represents the boughs and branches used in the construction of a Pai-Ute wik-i-up."

Place the tips of the two flat hands together before the body, leaving a space of about six inches between the wrists. (*Ute* I.) "Outline of the shape of the lodge."

——— Coming out of a.

Same as the sign for **Lodge,** Entering a, only the fingers of the right hand point obliquely upward after passing under the left hand. (*Dakota* I.) "Coming out from under cover."

——— Entering a.

The left hand is held with the back upward, and the right hand also with the back up is passed in a curvilinear direction down under the other side of it. The left hand here represents the low door of the skin lodge and the right the man stooping to pass in. (*Long.*)

Pass the flat right hand in short curves under the left, which is held a short distance forward. (*Wied.*) I have described the same sign. It is not necessary to pass the hand more than once. By saying curves, he seems to imply many passes. If the hand is passed more than once it means repetition of the act. (*Matthews.*) The conception is of the stooping to pass through the low entrance, which is often covered by a flap of skin, sometimes stretched on a frame, and which must be shoved aside, and the subsequent rising when the entrance has been accomplished. In the same tribes now, if the intention is to speak of a person entering the gesturer's own lodge, the right hand is passed under the left and toward the body, near which the left hand is held; if of a person entering the lodge of another, the left hand is held further from the body and the right is passed under it and outward. In both cases both hands are slightly curved and compressed.

A gliding movement of the extended hand, fingers joined, backs up, downward, then ascending, indicative of the stooping and resumption of the upright position in entering the same. (*Arapaho* I.)

(1.) Sign for **Lodge,** the left hand being still in position used in making sign for **Lodge;** (2) forefinger and thumb of right hand brought to a point and thrust through the outline of an imaginary lodge represented by the left hand. (*Cheyenne* I.)

First make the sign for **Lodge,** then place the left hand, horizontal and slightly arched, before the body, and pass the right hand with extended index-finger underneath the left—forward and slightly upward beyond it. (*Dakota* V; *Absaroka* I; *Shoshoni and Banak* I; *Wyandot* I.)

Left hand (**W**), ends of fingers toward the right, stationary in front of the left breast; pass the right hand directly and quickly out from the breast under the stationary left hand, ending with the extended fingers of the right hand pointing outward and slightly downward, joined, palm downward flat, horizontal (**W**). (*Dakota* I.) "Gone under; covered."

The left hand palm downward, finger-tips forward, either quite extended or with the fingers slightly bent, is held before the body. Then the right hand nearly or quite extended, palm downward, finger-tips near the left thumb, and pointing towards it, is passed transversely under the left hand and one to four inches below it. The fingers of the right hand point slightly upward when the motion is completed. This sign usually, but not invariably, refers to entering a house. (*Mandan and Hidatsa* I.)

Place the slightly curved left hand, palm down, before the breast, pointing to the right, then pass the flat right hand, palm down, in a short curve forward, under and upward beyond the left. (*Ute* I.) "Evidently from the manner in which a person is obliged to stoop in entering an ordinary Indian lodge."

——— Moving away a.

Hold the hands as for **Lodge** and push them forward a foot or eighteen inches. (*Dakota* IV.)

——— Moving this way a.

Hold the hands as in the sign for **Lodge** and draw them toward you. (*Dakota* IV.)

——— Taking down a.

Make the sign for **Lodge** and quickly throw the hands outward, at the same time opening all of the fingers. (*Dakota* IV.)

——— Great council.

Make the sign for **Lodge;** then place both hands somewhat bent, palms facing about ten inches apart, and pass them upward from the waist as high as the face. (*Hidatsa* I; *Arikara* I.)

Loiter. To gad, dawdle.

The hand is held as in the sign for **White-tailed Deer,** but the tip of the index-finger is made to describe lines of two or more feet in length from side to side, and to do this the whole arm must be moved. (*Mandan and Hidatsa* I.)

Long, in extent. See **Big** in the sense of **Long.**

Long, in time. See **Time.**

Look! See!

Touch the right eye with the index and point it outward. (*Burton.*)

(1.) Fore and middle fingers of right hand extended, placed near to the eyes, pointing outward, and (2) thrust with a slight downward curved motion quickly forward toward the object to be seen or looked at. (*Cheyenne* I.)

Included in **To look, Seeing,** with the addition of the sign for **Attention** made first. (*Dakota* I.)

Look, to. Seeing.

Fore and second fingers right hand (**N**) brought to the level of the eyes, extended fingers pointing outward, back of hand upward, horizontal, is then carried directly outward on the same level with a slight to-and-fro or sidewise motion of the fingers. Many Sioux Indians use *both* hands in making this sign with only the forefinger of each extended. (*Dakota* I.) "Turning the eyes in looking."

Deaf-mute natural signs.—Point to something and strain the eye toward it, accompanied by an expression of command. (*Ballard.*)

Put the open hand on the shoulder, or the hand, or the arm of somebody (when this body asks what); point with the forefinger of the other hand to something valuable to look at, nodding the head, so as to cause him to look at it, and then to lift up the eyes toward it. (*Hasenstab.*)

Keep the eyelids more open than usual, and then put the forefinger toward one of the eyes and quickly stretch and keep it in a straight line from the same eye. (*Larson.*)

Place the forefinger on the eye and then point with it as if to point to something. (*Zeigler.*)

Looking-glass. See **Mirror.**

Lost.

The right hand is brought in front of breast, palm outward (**T**, with left hand palm near thumb of right), right is moved forward at the same time the left is moved back toward right shoulder in oblique upward position with palm to right. (*Cheyenne* II.) "The right is the object disappearing from view till lost. The left obstructs the sight."

Use the same hand and motions as in **Crazy,** describe rapidly enlarging circles, then reverse motions of circles. Point in different directions and again repeat the circles around the head from right to left, with right hand. (*Ojibwa* IV.)

Hold the left hand flat, with the palm downward, about twelve inches before the chest, then pass the right, flat and extended, forward under the left and upward beyond it, toward the left. (*Kaiowa* I; *Comanche* III; *Apache* II; *Wichita* II.)

Love, or **Affection.**

The clinched hand pressed hard upon the breast. (*Long.*)

Pantomimic embrace. (*Arapaho* I.)

Both hands closed, right slightly above the left and brought up in front of and a slight distance from the breast, and hugged to middle of breast, left hand below the right. (*Cheyenne* I.) "The embrace."

Same as **Admiration.** (*Dakota* I.)

Cross the forearms near the wrists on the upper part of the chest, hands closed, backs forward. (*Dakota* IV.)

Hug both hands to the bosom as if clasping something affectionately. (*Wichita* I.)

Deaf-mute natural sign.—Kiss your hand and point to the heart, with a happy smile. (*Cross.*)

Italian sign.—Place the open hand over the heart. (*Butler.*)

Male (applied to animals).

Make a fillip with forefinger of right hand on the cheek. (*Dunbar.*)

Right hand, back upward, forefinger pointing outward and upward, elevated to front of person, and motioned once or twice up and down. (*Cheyenne* I.)

Same as **Male**, applied to man. (*Dakota* I.) "From the male organ of generation."

——— Applied to man.

Right hand closed, thrust through the left hand, which then clasps the wrists, very slight up-and-down motion made. (*Cheyenne* I.)

Another: Right hand closed, held in obliquely erect position, left hand loosely clasping it and rubbed up and down from the knuckles to the upper part of the forearm. (*Cheyenne* I.)

Another: Right-hand fingers and thumb brought to a point and thrust through the left hand, which then clasps the wrist as before; slight upward motion made. (*Cheyenne* I.)

Left-hand forefinger straight, pointing backward and upward; forefinger of the right hand laid across the back of the left, seemingly to isolate and lengthen out the forefinger of the left hand. (*Cheyenne* I.)

Sign for **Squaw,** followed by that of **Negative.** Also, sign for **Male** applied to animals. (*Arapaho* I.)

The extended forefinger of the right hand (of which the others are closed) is laid in the crotch, finger pointing downward, back of hand upward. (*Dakota* I.)

——— Applied to man and animals.

Deaf-mute natural sign.—Take hold of the pants, at the same time shake them. (*Zeigler.*)

Man.

With the forefinger of the right hand extended, and the hand shut, describe a line, beginning at the pit of the stomach, and passing down the middle of the body as far as the hand conveniently reaches, holding the hand a moment between the lower extremities. (*Dunbar.*)

A finger held vertically. (*Long.*)

Elevate the index-finger and turn the hand hither and thither. (*Wied.*) I have seen only the sign of the erect finger without the motions to which he refers. (*Matthews.*) The turning of the hand hither and thither probably was to convey more than the simple idea of man. It might have meant only *one* man, or that a man was *alone.*

A finger directed toward the pubis. (*Macgowan.*)

Hold the index-finger erect before the face. (*Dodge.*)

Generally, any sign as a Sioux, a Cheyenne, etc., is understood to refer to the male, unless the sign for a *Squaw* or *Woman* follows. (*Arapaho* I.)

Right-hand palm inward, elevated to about the level of the breast, forefinger carelessly pointing upward, suddenly pointed straight upward, and the whole hand moved a little forward, at the same time taking care to keep the back of the hand toward the person addressed. (*Cheyenne* I.)

The right hand is held in front of the right breast with the forefinger extended, straight upright (**J**), with the back of the hand outward; move the hand upward and downward with finger extended, etc. This is general. White man has a special sign, also negro, and each tribe of Indians. (*Dakota* I.)

First, the extended right index, pointing forward, back upward, is to be placed horizontally in front of the privates, or a little higher, and the hand suddenly lowered about an inch. Then carry the hand (index still extended) to the right and upward through an arc, and bring it upright to a position a foot or so in front of the right shoulder, its back forward, and the index pointing upward, and suddenly move it forward an inch or so. While making these movements the middle, ring, and little fingers are to be closed and the thumb against the middle finger. (*Dakota* IV.) "Male one."

Elevate the extended index before the right cheek, and throw the hand forward, keeping the palm toward the body. (*Dakota* VI.)

Place the extended index, pointing upward and forward, before the lower portion of the abdomen. (*Dakota* VII.)

The right hand in the position of an index-hand, pointing upward, is held a few inches in front of the abdomen or chest, the outer edge of the hand being usually forward. (*Mandan and Hidatsa* I.)

The left arm is elevated and the hand, in type position (**S** 1, horizontal), is drawn across before the body on a level with the shoulder. The right arm is then raised and extended before the body with the hand, in position (**J** 1), more stiffly extended. (*Oto and Missouri* I.) "A being with projecting sexual organ."

Raise the closed hand, with the index only extended and elevated, pointing upward to the front of the right breast (cheek or shoulder) keeping the back of the hand to the front. (*Kaiowa* I; *Comanche* III; *Apache* II; *Wichita* II.)

The forefinger of either hand is brought before the body, pointing upward. (*Pai Ute* I.)

Pass the extended right hand downward, forward and upward from the hip, then lay the extended forefinger across the back of the right wrist. (*Ute* I.) "Male genital organ and length of."

Deaf-mute natural signs.—Put the hands on the legs and draw the hands up, in imitation of the act of putting on a pair of pantaloons. (*Ballard.*)

Stretch up the open hand over the head, indicating the general height of the man; next use both hands as if to stroke the beard or the mustache, and then nod the head. (*Hasenstab.*)

Make the motion of taking the hat from the head. (*Larson.*)

——— Old. (Compare **Old.**)

Hold the right hand, bent at elbow, fingers and thumb closed sidewise. (*Cheyenne* I.) "Old age dependent on a staff." It is made more emphatic by a tottering step.

Place the right extended index, pointing forward, back upward, horizontally in front of the privates, and suddenly lower the hand about an inch, as for **Man;** then move the right first, its back outward, from twelve inches in front of the right breast, forward and backward two or three times about a foot, describing an ellipsis perpendicular to the ground. (*Dakota* IV.) "Progression of a man with a staff."

Place the closed right hand in front of and as high as the elbow, leaving the index curved and pointing toward the ground. If the man is very aged, cause the hand to tremble at the same time the gesture is made. (*Dakota* VII.)

Right hand closed, forefinger slightly curved, hand held before the body or right hip, palm down, allawing the forefinger to droop toward the ground. (*Comanche* II.) "Form a position of the flaccid glans penis of an old man."

Close the right hand, pointing forward from the body, palm down, then partly extend the index, the tip pointing toward the ground; the lower the hand is held and the more the index is crooked, the greater the age of the individual. (*Ute* I.) "Curved and flaccid glans penis of an aged individual."

Touch teeth; make the sign for **Negation;** touch hair; touch white tent. [*sic*]. (*Apache* III.) "Toothless, and white haired."

——— Young.

With the right hand, index only extended, place the hand a short distance in front of the hip at the height of the elbow, pointing upward at an angle of about 20°, palm to the left. (*Comanche* II.) "Tendency of erection in a young man."

Marching. See **Going.**

Marriage.

There is no marriage ceremony with the Sioux Indians, and consequently no sign designating "marriage," and it can only be expressed by **Companion.** (*Dakota* I.)

Married, to marry. (Compare **Same, Similar.**)

The hands are placed in front, the arms pronated, and the extended forefingers (the others being flexed) are placed in contact side by side. (*Mandan and Hidatsa* I.)

Close both hands, except the two forefingers, and place them side by side, pointing forward, in front of the breast. (*Iroquois* I.) This sign expresses *mated*, also *husband* or *wife*.

Bring the two forefingers side by side, hands pronated. (*Zuñi* I.)

Italian sign.—Pretend to put a ring on the ring finger, or lay the two forefingers together side by side; yet this last is more commonly used of any union or harmony. (*Butler*.)

Match.

As though striking a match on the palm of the left hand held in front of breast, with the right hand. (*Dakota* I.) "From the act of striking a match."

Hold the left hand before the body, extended and pointing toward the right, palm down, then place the tip of the thumb of the right hand against the index and second finger and pass them quickly along the inner edge of the left hand from the wrist forward as if lighting a match. (*Kaiowa* I; *Comanche* III; *Shoshoni and Banak* I; *Ute* I; *Apache* I, II; *Wichita* II.)

Medicine. (Compare **Doctor.**)

Carry the right hand in position as though holding a bottle in front of the mouth, and then tip it up as though drinking from the bottle. (*Dakota* I.) "Drinking medicine from a bottle."

The left hand with the arm semi-extended is held with the back upward before the body. The extended index of the right hand then rubs on the back of the left, as the mixing of medicine on a pill-tile. Both hands with the fingers, as in (**Q**), are then held tremblingly before the body's sides. The extended indices next compress the temples and the countenance assumes an appearance of distress. For medicine to induce sleep, quiet cough, check flow of blood, or purge, the signs appropriate to the latter conditions or words are conjoined. (*Oto and Missouri* I.) "Something stirred up for inward distress."

Deaf-mute natural sign.—Use the sign for **Sick, Ill,** and then the sign for **Drink.** (*Zeigler.*)

——— In Indian sense. (Compare **Indian Doctor, Shaman.**)

Stir with the right hand into the left, and afterward blow into the latter. (*Wied.*) There is a similarity in the execution of the (*Oto and Missouri* I) sign and *Wied's* sign. The stirring in the left instead of on its back as in the former may be a matter of caprice. It is probable that the conclusive blowing into the hand in the latter is to add mystery, as in the magician's trick. (*Boteler.*) All persons familiar with the Indians will understand that the term "medicine," foolishly enough adopted by both French and English to express the aboriginal magic arts, has no therapeutic significance. Very few even pretended remedies were administered to the natives and probably never by the professional shaman, who worked by incantation, often pulverizing and mixing the substances mystically used, to prevent their detection. The same mixtures were employed in divination. The author particularly mentions Mandan ceremonies, in which a white "medicine" stone, as hard as pyrites, was produced by rubbing in the hand snow, or the white feathers of a bird. The blowing away of the disease, considered to be introduced by a supernatural power foreign to the body, was a common part of the juggling performance.

The right index is pointed toward the ground in several different directions in front of the body, and then the other fingers and thumb are extended, and the hand carried from the median line of the body with a gradually decreasing spiral motion to directly over the head, where the hand points upright, edge of hand toward the front. (*Dakota* I.) "Mysterious, hence sacred; power of herbs over disease."

Medicine Man. Shaman. (Compare **Doctor.**)

This double sign is made with the right hand, describing man first. Bring hand up to front of right shoulder, forefinger straight upright (**J,** palm inward); move forward, then bring it up front of face with first and second fingers straight and upward, separated (**N,** palm inward). Then make three or four moves in a circle with the hand in that position. The arm is not necessarily moved, only the hand from wrist up. (*Cheyenne* II.) "The medicine sign as made, supposed to represent singing and shaking the wand in incantations."

Make shaking and short jerking motions of the body, with arms and hands jerking and fingers pointing to and around head, neck, and body, with wild gestures and distorted features, also short quivering steps, the toes scarcely leaving the ground, and wild shakes of the head. (*Ojibwa* IV.)

First make the sign for *Dakota* Indians (See TRIBAL SIGNS), (or any other tribe, as may be desired), then the sign for **Man,** and then the right hand with fore and index fingers extended, pointing upward, others closed (**N**), is carried from the partially extended position of the hand on completing the sign for **Man,** upward, with a spiral motion, directly above the head. (*Dakota* I.)

With its index-finger extended and pointing upward, or all the fingers extended, back of hand outward, move the right hand from just in front of the forehead, spirally upward, nearly to arm's length, from left to right, in the opposite direction to the sign for **Fool** and **Crazy.** (*Dakota* IV.)

Elevate and rotate the extended index before the forehead. (*Dakota* VI; *Hidatsa* I; *Arikara* I.)

Rub the right cheek with the palmar surface of the extended fingers of the right hand, then rub the back of the closed left hand with the index and second fingers of the right, and conclude by holding the left hand before the face, the index and second fingers only extended and separated, pass it upward and forward before the face, rotating it in so doing, the rotation occurring at the wrist. (*Kaiowa* I; *Comanche* III; *Apache* II; *Wichita* II.)

Point to herbs or plants at a short distance from the body; imitate the pounding up of the same in a vessel with the right hand, using the left as if holding one, then make the sign of **to Eat.** (*Apache* I.) "The preparation of plants or herbs for internal administration."

Meet, To. (Compare **Lodge.**)

Bring the extended forefingers from either side, allow the tips to come together before the body, keeping the wrists about six inches apart. (*Kaiowa* I; *Comanche* III; *Apache* II; *Wichita* II.)

Melon.

Fingers of the right hand separated and curved; place the palm about ten or twelve inches from the ground, fingers pointing forward, and pass it forward in an upward curve, corresponding to the elongated and convex form of a melon. (*Kaiowa* I; *Comanche* III; *Ute* I; *Apache* I, II; *Wichita* II.)

Mercy.

Extend both forefingers, pointing upward, palms toward the breast, and hold the hands before the chest; then draw them inward toward their respective sides, and pass them upward as high as the sides of the head by either cheek. (*Kaiowa* I; *Comanche* III; *Apache* II; *Wichita* II.)

——— On another, To have.

Hold both hands nearly side by side before the chest, palms forward, forefinger, only extended and pointing upward; then move them forward and upward, as if passing them by the cheeks of another person from the breast to the sides of the head. (*Kaiowa* I; *Comanche* III; *Apache* II; *Wichita* II.)

Mexican. (See also TRIBAL SIGNS.)

Right hand (**V**), back outward, is held, with fingers pointing downward, at the chin. (*Dakota* I.) "From the wearing of a beard. The Sioux say the Mexicans are the only Indians that wear a beard."

Move the right hand in a small circle from right to left beneath the chin, palm upward, fingers semiflexed, thumb against index. (*Dakota* IV.) "From the beard on the chin of the first they saw."

Place the right hand about three inches below the chin, wrist toward the neck and knuckles forward, palm up, curve the spread fingers and thumb so that the tips are all directed to one point; then work the fingers and thumb at the second joints as if scratching, but keep the palm motionless. (*Kaiowa* I; *Comanche* III; *Apache* II; *Wichita* II.) "Whiskers."

Pass the flat right hand across the cheeks and chin from ear to ear and from left to right. (*Ute* I.)

Extend thumbs and forefingers widely about head, a few inches from the head. (*Apache* III.) "Who wear hats with moderately broad brims."

Mine, My property. See **Possession.**

Mirror.

The palm of the hand held before the face, the fingers pointing upward. (*Long.*)

Place both palms before the face, and admire your countenance in them. (*Burton.*)

Right hand (**S**) held in front of the face, hand and head turned as though looking at different parts of the face. (*Dakota* I.) "From the use of a looking-glass."

The erect extended right hand, palm backward, is held before the face, at about the distance a mirror would be held, and the gaze is directed toward the palm. (*Mandan and Hidatsa* I.)

The left arm is elevated and semi-extended; the hand is then in position (**W**), modified by being, palm up, held before the face and the eyes intently directed as looking therein. (*Oto and Missouri* I.) "The ordinary hand-mirror used among the Indians."

Deaf-mute natural sign.—Outline the shape of a mirror with the two the forefingers in the air, then place the hand before the face and fix eye upon it. (*Ballard.*)

Miser (a griping man). (Compare **Penurious.**)

Italian signs.—Italians express this idea by a doubled fist, with the fingers curled very close in the palm; also with the fist drawn firmly toward one, with fingers spread like claws and clutching. (*Butler.*)

Modesty.

Italian sign.—Cover the eyes with the fingers of one hand drawn apart. (*Butler.*)

Money (metallic).

With the right hand point downward toward a piece of metal, and then carry it to the left hand held in front of the body horizontally, with palm upward (**X**) as though putting the piece of metal in it, and then with the edge of the fingers of the closed right fist (**A** 1) pound in the palm of the left. (*Dakota* I.) "This is shaping the metal into coins."

——— Paper.

Point to a piece of paper with the right hand, and then with it make the sign for **Writing** over the left hand held in front of the left breast, with the thumb and forefinger describing a semicircle, with the free ends pointing downward, other fingers closed. (*Dakota* I.) "This indicates money with writing on it."

Moon or **Month.**

The thumb and fingers are elevated toward the right ear. This sign is generally preceded by the sign for **Night** or **Darkness.** (*Dunbar.*)

Make the sign for **Sun,** after having made that for **Night.** (*Wied.*)

Combine signs for **Night** and **Sun.** To distinguish from the stars or a star, indicate relative apparent sizes. (*Arapaho* I.)

The right hand closed, leaving the thumb and index extended, but curved to form a half circle, and hold the hand toward the sky nearer the eastern horizon than zenith. (*Absaroka* I; *Hidatsa* I; *Wyandot* I; *Arikara* I; *Shoshoni and Banak* I; *Ute* I.)

First make the sign for **Night,** and then the sign for the **Sun.** (*Dakota* I.) "Night sun or luminary."

First make the sign for **Moon,** and then the right index is held upright in front of the body to indicate one. (*Dakota* I.)

Same as for **Sun,** but instead of the sign for **Day** make the one for **Night.** For **Quarter-moon** and **Half-moon** the ends of the index and thumb are kept about four inches apart. (*Dakota* IV.)

Same as for **Sun,** except that the tips of the finger and thumb, instead of being opposed, are approximated so as to represent a crescent. (*Mandan and Hidatsa* I.)

Extend the curved index and thumb so as to form a crescent, close the remaining fingers, and raise the hand toward the sky, with the outer edge of the hand forward. (*Kaiowa* I; *Comanche* III; *Apache* II; *Wichita* II.)

The index and thumb are curved so as to form a half circle, the remaining fingers closed. (*Apache* I.)

Make a cresent with thumb and index, project it toward the western horizon, and by successive jerks carry the same to zenith. (*Apache* III.) "New moon first seen above western horizon, seen each night successively higher toward zenith."

Deaf-mute natural sign.—Move the hand in a curved line from the setting to the rising of the sun, and turn the forefinger, outstretched, around and around like a wheel. (*Larson.*)

——— New.

Close the right hand, extend and curve the index and thumb so that they form a half circle; then hold the hand toward the eastern horizon. (*Ute* I.)

——— Full.

Reach out both arms as if hugging a tree, then make the sign for **Moon.** (*Ute* I.)

Make a crescent with thumb and index, project same to western hori-horizon, and by several successive jerks from west to east change crescent to circle; hold at eastern horizon; retrograde toward the west with a few less distinct jerks. (*Apache* III.) "Apparent lunar course."

——— Month.

Sign for **Moon,** and passed across the heavens from east to west. (*Apache* I.)

Deaf-mute natural signs.—Make a zigzag motion in the air with the forefinger. (*Ballard.*)

Move the finger in a zigzag way. (*Larson.*)

More. See **Add.**

Morning. See **East.**

Mother. See **Relationship.**

Mountain. (Compare **Hill.**)

Outline its projection against the sky with the hand, to which may be added sign for **High**—as a hill. (*Arapaho* I.)

Right hand (**A** 1) to the left side on level of the face is drawn in front of the face to the right side on the same level, where it is held for a moment. (*Dakota* I.) "A height or mountain directly in front."

Strike the palm of the left hand with the back of the right fist, palm of left hand backward; close the left hand, turning its back outward, and raise the right fist two feet above it, and then make the sign for **Large.** (*Dakota* IV.)

Place both hands, flat and extended, thumb to thumb, with palms downward, in front of and as high as the head; then pass them outward and downward toward their respective sides, describing the upper half of a circle, and bring them back to the point of starting. (*Wyandot* I.)

Place the flat and extended left hand edgewise before the face, thumb resting on the forefinger, back forward, finger tips pointing toward the right. (*Ute* I.)

Hold both hands up before the body as if molding a mountain; thrust hand, on edge, downward over other hand in several directions. (*Apache* III.) "Cañons down its sides."

Deaf-mute natural signs.—Use one hand so as to represent a plane inclined upward, and move the other hand over and above it. (*Ballard.*)

Move the outstretched hand obliquely. (*Zeigler.*)

——— Divide.

Hold the left hand flat, and extended before the face, fingers pointing toward the right, the tip of the thumb slightly turned upward; then with the extended index indicate the gap over the left palm against the base of the upturned joint of the thumb, allowing the index to go down upon the back a short distance. (*Apache* I.)

——— Pass.

Place the flat and extended left hand edgewise before the body, elevate the thumb, thus forming a crotch; then pass the index, pointing over the left hand, between the thumb and forefinger, from the palmar side down over the back. (*Ute* I.)

——— Range.

Make the sign for **Mountain,** holding the hand at various points continuously toward the horizon. (*Ute* I.) "Mountain, and continuation of, along the horizon."

Move, To.

The only difference from **Marching, Traveling,** consists in reversing the direction of the arched movements of the hands, bringing them toward the body instead of carrying them from the body. (*Dakota* I.) "Moving toward you."

Much. See **Quantity.**

Mud.

Both fists pushed alternately downward to the ground several times. (*Kaiowa* I; *Comanche* III; *Apache* II; *Wichita* II.) "Horse's feet."

Mule, or **Ass.**

Hold the open hands high beside the head, and move them from back to front several times like wings. (*Wied.*) This sign is still in use. (*Matthews.*)

To denote the mule or ass the long ears are imitated by the indices on both sides and above the head. (*Burton.*)

Both hands, fingers and thumbs brought to a point, placed by the side of the head, hollows forward, moved slightly backward and forward. (*Cheyenne* I.)

The hands (**T**) at their respective sides of the head make movements resembling the flopping of the mule's ears. (*Dakota* I.) "From the mule's habit of flopping his ears."

Move the opened upright hands back and forth several times above the ears. (*Dakota* IV.)

Place both flat and extended hands to either side of the head, the wrists as high as the ears, then rock the hands to and fro several times, movement being made at the wrists. (*Ute* I; *Apache* I.) "Large ears."

Murder.

Italian sign.—Push the thumb against the heart with a stern look. (*Butler.*)

Mute.

Place the extended and joined fingers of the right hand over the closed lips. (*Kaiowa* I; *Comanche* III; *Apache* II; *Wichita* II.)

Near or **Soon.**

The hands are brought before the body both in type position (**H** 1), modified by the fist being a little more closed and the index a little more extended; the points of indices then touch and are rotated together. Speaking of near future, the rotary movement forward is executed. (*Oto* I.) "Approaching contact."

Deaf-mute natural signs.—Move the forefinger from here to there a short distance. (*Ballard.*)

Use the sign for **Small,** at the same time point with the forefinger as if to something at the feet and then to something at a short distance. (*Zeigler.*)

Negro. (Compare TRIBAL SIGNS.)

First make the sign for **White Man,** then rub the hair on the right side of the head with the flat hand. (*Wied.*) The present common sign for "black" is to rub or touch the hair, which, among Indians, is almost universally of that color.

Place the left forearm transversely before the body, hand extended, flat, palm down, then pass the flat right hand downward from the lower portion of the breast, forward and upward (forming a curve); then indicate any object that has a black color. (*Absaroka* I; *Shoshoni and Banak* I.) "Literally—born black."

Make the sign for **White Man,** then point at a black object, or rub along the back of the left hand with the fingers of the right (for black). (*Dakota* IV.)

Touch the hair with the fingers of the right hand, then rub the thumb and middle finger together as if snapping them. Kinky hair. (*Wyandot* I.)

Sweep the hand over the hair (**W**); gather it up in finger-tips (**U,** with thumb advanced to finger-tips). (*Apache* III.)

Another: Sweep the hand over the face; then touch some black object. (*Apache* III.) "(1) Kinky hair. (2) Black face."

Grasp the hair between the thumb and index, rolling it several times between them. (*Wichita* I.)

Night.

The two hands, open and extended, crossing one another horizontally. (*Dunbar.*)

The head, with the eyes closed, is laterally inclined for a moment upon the hand. As many times as this is repeated, so many nights are indicated. Very frequently the sign of the **Sun** is traced over the heavens, from east to west, to indicate the lapse of a day, and precedes the motion. (*Long.*)

Move both hands, open and flat—that is horizontal—backs up, and in small curves in front of the breast and over one another. (*Wied.*) This I believe to be primarily the sign for darkness, secondarily for night. Night, as a period of time, is more commonly, I think, indicated by the sign for sleep. (*Matthews.*) The *Prince of Wied's* sign differs from the (*Oto and Missouri* I) in execution and conception, one representing the course of *the cause* (sun), the other the effect, obscurity, or night. "The time or day that the sun moves beneath us when we sleep." (*Boteler.*) The conception is *covering* and consequent obscurity. In the sign for **Day** by the same author he probably means that the hands, palms up, were *moved* apart, to denote *openness.*

Make a closing movement as if of the *darkness* by bringing together both hands with the dorsa upward and the fingers to the fore. The motion is from right to left, and at the end the two indices are alongside and close to each other. This movement must be accompanied by bending forward with bowed head, otherwise it may be misunderstood for the freezing over of a lake or river. (*Burton.*)

The sign for **Day** reversed. (*Arapaho* I.) "Everything is closed."

Both hands outspread, palms facing, passed in front of the body and crossed, the right hand over the left. (*Cheyenne* I.)

Place the flat hands in front of and as high as the elbows, palms up, then throw them inward toward the middle of the breast, the right over the left, turning the palms downward while making the gesture. (*Absaroka* I; *Hidatsa* I; *Kaiowa* I; *Arikara* I; *Comanche* III; *Shoshoni and Banak* I; *Apache* II; *Wichita* II.) *Note.*—"For the plural, the sign is repeated two or three times and the number indicated by elevating the fingers of the left hand, and right if necessary. Time is reckoned by nights, and if but two or three nights (or days) are mentioned, the sign is made that number of times without elevating the fingers, the number of times the gesture being made referring to the number of days, without the accompanying numeration."

Extend the arms to full capacity to the sides of the body on the level of the shoulders, palms downward, fingers joined, etc. (as **W**), and then move them to the front of the body on the same level or on a slight curve until they have crossed each other at the wrists. (*Dakota* I.) "The coming of darkness—the closing in of night."

Extend both hands to the front at about the height of the elbow (as in **W**), and then slowly move both at the same time, the right hand to the left and the left hand to the right, passing one above the other, representing the darkness closing over the earth. (*Dakota* III.)

From advanced positions, two feet apart, about eighteen inches in front of the line of the body, pointing forward, palms downward, at the height of the lower part of the chest, carry the opened hands inward, each one describing an arc, until they point obliquely forward and outward, the right two or three inches above the left. (*Dakota* IV.) "Darkness covers all."

Place the flat hands, palm down, in front of and as high as the elbows, then move them horizontally toward and past each other until the forearms cross. (*Dakota* VII.)

The sign for **Day** is first completed; then the hand, in position (**I** 1), index more opened, describes an arc of the horizon inverted, as the sun beneath us; this is followed by the sign for **Sleep,** the eyes closed and the head inclining to the right, supported in the open hand. (*Oto and Missouri* I.)

Both hands placed in front of the eyes, palms facing, and slowly closed, the eyes being closed at the same time. (*Wyandot* I.)

Deaf-mute natural signs.—Move the forefinger downward from the west to the east. (*Ballard.*)

Use the shut hand as if to scratch a match. (*Hasenstab.*)

No, not. See also **Know, I don't.** (Compare **Nothing.**)

The hand held up before the face, with the palm outward and vibrated to and fro. (*Dunbar.*)

The hand waved outward with the thumb upward. (*Long.*)

Wave the right hand quickly by and in front of the face toward the right. (*Wied.*) Refusing to accept the idea or statement presented.

Move the hand from right to left, as if motioning away. This sign also means "I'll have nothing to do with you." (*Burton.*)

Right-hand fingers extended together, side of hand in front of and facing the face, in front of the mouth and waved suddenly to the right. (*Cheyenne* I.)

A deprecatory wave of the right hand from front to right, fingers extended and joined. (*Arapaho* I.)

Place the right hand extended before the body, fingers pointing upward, palm to the front, then throw the hand outward to the right, and slightly downward. (*Absaroka* I; *Hidatsa* I; *Arikara* I.)

The right hand, horizontal, flat, palm downward (**W**), is pushed sidewise outward and toward the right from the left breast. *No, none, I have none,* etc., are all expressed by this sign. Often these Indians for *No* will simply shake the head to the right and left. This sign, although it may have originally been introduced from the white people's habit of shaking the head to express *No*, has been in use among them for as long as the oldest people can remember, yet they do not use the variant to express *Yes*. (*Dakota* I.) "Dismissing the idea, etc."

Place the opened relaxed right hand, pointing toward the left, back forward, in front of the nose or as low as the breast, and throw it forward and outward about eighteen inches. Some at the same time turn the palm upward. Or make the sign at the height of the breast with both hands. (*Dakota* IV.) "Represents the shaking of the head. Our shaking of the head in denial is not so universal in the Old World as is popularly supposed, for the ancient Greeks, followed by the modern Turks and rustic Italians, threw the head back, instead of shaking it, for *No*.

Hold the flat hand pointing upward before the right side of the chest, then throw it outward and downward to the right. (*Dakota* VI, VII.)

The hand extended or slightly curved is held in front of the body a little to the right of the median line; it is then carried with a rapid sweep a foot or more farther to the right. (*Mandan and Hidatsa* I.)

Place the hand as in **Yes,** and move it from side to side. (*Iroquois* I.) "A shake of the head."

Throw the flat right hand forward and outward to the right, palm to the front. (*Kaiowa* I; *Comanche* III; *Apache* II; *Wichita* II.)

Quick motion of open hand from the mouth forward, palm toward the mouth. (*Sahaptin* I.)

Place hand in front of body, fingers relaxed, palm toward body (**Y** 1), then with easy motion move to point, say, a foot from body, a little to right, fingers same, but palm upward. (*Sahaptin* I.) "We don't agree." To express *All gone*, use a similar motion with both hands. "Empty."

Elevate the extended index and wave it quickly from side to side before the face. This is sometimes accompanied by shaking the head. (*Pai-Ute* I.)

Extend the index, holding it vertically before the face, remaining fingers and thumb closed; pass the finger quickly from side to side a foot or so before the face. (*Apache* I.) This sign, as also that of (*Pai-Ute* I), is substantially the same as that with the same significance reported from Naples by DE JORIO.

Wave extended hand before the face from side to side. (*Apache* III.)

Another: The right hand, naturally relaxed, is thrown outward and forward toward the right. (*Apache* I.)

Extend the palm of the right hand horizontally a foot from the waist, palm downward, then suddenly throw it half over from the body, as if tossing a chip from the back of the hand. (*Wichita* I.)

Deaf-mute natural signs.—Shake the head. (*Ballard.*)

Move both hands from each other, and, at the same time, shake the head so as to indicate "no." (*Hasenstab.*)

Our deaf-mutes for emphatic negative wave the right hand before the face.

The Egyptian negative linear hieroglyph is clearly the gesture of both hands, palm down, waved apart horizontally and apparently at the level of the elbow, between which and the Maya negative particle "*ma*" given by *Landa* there is a strong coincidence.

None, Nothing; I have none.

"Little" or "nothing" is signified by passing one hand over the other. (*Ojibwa* I.)

Motion of rubbing out. (*Macgowan.*)

May also be signified by smartly brushing the right hand across the left from the wrist toward the fingers, both hands extended, palms toward each other and fingers joined. (*Arapaho* I.)

Is included in **Gone, Destroyed.** (*Dakota* I.)

Another: Place the opened left hand about a foot in front of the navel, pointing forward and to right, palm obliquely upward and backward, and sweep the palm of the open right hand forward and to the right over it and about a foot through a curve. (*Dakota* IV.) "All bare."

Pass the ulnar side of the right index along the radial side of the left index from tip to base. Some roll the right index over on its back as they move it along the left. The hands are to be in front of the navel, backs forward and outward, the left index straight and pointing forward and to the right, the right index straight and pointing forward and to the left; the other fingers loosely closed. (*Dakota* IV.)

With the right hand pointing obliquely forward to the left, the left forward to the right, palms upward, move them alternately several times up and down, striking the ends of the fingers. Or, the left hand being in the above position, rub the right palm in a circle on the left two or three times, and then move it forward and to the right. (*Dakota* VI.) "Rubbed out. That is all. It is all gone."

Pass the palm of the flat right hand over the left from the wrist toward and off of the tips of the fingers. (*Dakota* VI, VII.)

Brush the palm of the left hand from wrist to finger tips with the palm of the right. (*Wyandot* I.)

Another: Throw both hands outward toward their respective sides from the breast. (*Wyandot* I.)

Pass the flat right palm over the palm of the left hand from the wrist forward over the fingers. (*Kaiowa* I; *Comanche* III; *Apache* II; *Wichita* II.) "Wiped out."

Hold the left hand open, with the palm upward, at the height of the elbow and before the body; pass the right quickly over the left, palms touching, from the wrist toward the tips of the left, as if brushing off dust. (*Apache* I.)

Deaf-mute natural sign.—Place the hands near each other, palms downward, and reverse and move them over in opposite directions. (*Ballard.*)

Australian sign.—Pannie (none or nothing). For instance, a native says *Bomako ingina* (give a tomahawk). I reply by shaking the hand, thumb, and all fingers, separated and loosely extended, palm down. (*Smyth.*)

——— Exhausted for the present.

Hold both hands naturally relaxed nearly at arm's length before the body, palms toward the face, move them alternately to and fro a few inches, allowing the fingers to strike those of the opposite hand each time as far as the second joint. (*Kaiowa* I; *Comanche* III; *Apache* II; *Wichita* II.)

——— I have none.

Deaf-mute natural signs.—Expressed by the signs for none, after pointing to one's self. (*Ballard.*)

Stretch the tongue and move it to and fro like a pendulum, then shake the head as if to say "no." (*Zeigler.*)

Noon. See **Day.**

Nose.

Right index crooked, turned, pointing downward (other fingers and thumb closed), is passed downward from the upper part of the nose to the level of the nostrils, with back outward and finger touching the nose. (*Dakota* I.)

——— Bleeding.

Lean the head slightly forward and continue the movement, as above given, downward from the level of the nostril, and repeat several times, which indicates the dropping of the blood from the nostrils. (*Dakota* I.)

Now, at once.

The two hands forming each a hollow and brought near each other and put into a tremulous motion upward and downward. (*Dunbar.*)

Clap both hands together sharply and repeatedly, or make the sign of **To-day.** (*Burton.*)

Forefinger of the right hand extended, upright, etc. (**J**), is carried upward in front of the right side of the body and above the head so that the extended finger points toward the center of the heavens, and then carried downward in front of the right breast, forefinger still pointing upright. (*Dakota* I.)

Place the extended index, pointing upward, palm to the left, as high as and before the top of the head; push the hand up and down a slight distance several times, the eyes being directed upward at the time. (*Hidatsa* I; *Kaiowa* I; *Arikara* I; *Comanche* III; *Apache* II; *Wichita* II.)

Number. (Compare **Counting; Quantity.**)

Deaf-mute natural sign.—Count the fingers until the number nine is reached, and beyond it count doubles of the hands, each denoting ten. (*Cross.*)

Objection.

Italian sign.—A finger placed on the lower lip is understood to mean some new and suddenly started objection to a previous plan. (*Butler.*)

Obtain. (Compare **Possession.**)

First make the gesture for **Mine,** then move the right hand right and left before the face, the thumb turned toward the face. (*Wied.*)

Make the sign for **Searching,** hunting for, and then take hold of the object with the right hand and draw it in toward the body, near which the hand is brought to a stop. (*Dakota* I.) "Have hunted for and got it."

Deaf-mute natural sign.—Extend the hand, and close and move it back. (*Ballard.*)

Ocean.

Make the sign for **Water,** then place the flat hands, palms downward, thumbs joining, before the breast, and move them horizontally outward to either side. (*Absaroka* I; *Shoshoni and Banak* I.) "Broad water."

Officer.

Turn the hands upward and inward, allowing the tips of the fingers to touch the top of each shoulder. (*Sac, Fox, and Kickapoo* I.) "Epaulets."

Offspring. See **Child.**

Old. (Compare **Old Man, Aged** and **Time, long.**)

With the right hand held in front of right side of body, as though grasping the head of a walking-stick, describe the forward arch movement as though a person walking was using it for support. (*Dakota* I.) "Decrepit age dependent on a staff."

Place the closed right hand in front of and as high as the shoulder, leaving the index partly extended and bent; then move it slowly forward and toward the left in an interrupted manner. (*Wyandot* I.) "Slow movement and bent form of an old man."

——— Person.

Grasp the cheeks with both hands, using the tips of the fingers and thumbs. Wrinkles. (*Apache* I.)

Deaf-mute natural sign.—A trembling motion of the head. (*Ballard.*)

Opposite.

A clinched hand held up on the side of the head, at the distance of a foot or more from it. (*Long.*)

Bring the ends of the outstretched forefingers in close proximity, removing them again perpendicularly, and repeating the process several times. (*Ojibwa* II.) "Face to face with."

Left hand stationary in front of face on level of the eyes, forefinger alone extended, horizontal, pointing toward the right, &c. (as in **M,** except back outward); then the right hand is carried to the right eye, all fingers except the index closed, which points outward, straight toward the end of the stationary extended forefinger of the left hand. (*Dakota* I.) "Forefingers opposite one another."

Otter.

Draw the nose slightly upward with the two first fingers of the right hand. (*Wied.*)

Rub the end of the nose round and round with the ends of the fingers of the right hand. (*Dakota* IV.) "White nose."

Out, Outward, Without (in position.)

The semicircle as made in the sign for **In, Within,** with the upright right index placed without the circle between it and the body. (*Dakota* I.) "The variant of **In, Within.**"

Over (on the other side).

Collect the fingers of both hands to a point, place the left horizontally before the breast, pointing to the right, and the right behind the left, pointing to the right, palms down, then pass the right forward, over and down a short distance beyond the left. (*Kaiowa* I; *Comanche* III; *Apache* II; *Wichita* II.) "This sign is abbreviated by merely using the extended forefingers instead of the whole hand."

Pack, carrying a.

The hands are placed each side of the head, as if they held the strap of the *hoppas*, which passes round the forehead, in order to relieve that part, by supporting a portion of the weight of the burden; with this motion, two or three slight inclinations of the head and corresponding movements of the hands are also made. (*Long.*)

Packing. See also **Horse** (packing a).

Paint.

Daub both the cheeks downward with the index-finger. (*Burton.*)

Left hand held up, back and thumb upward, first and middle fingers of right hand dipped forward as though touching something, and then rubbed against back of hand near the base of the thumb, as in sign for **Grease.** (*Cheyenne* I.)

Make the sign for **Color,** and then touch the cheek with the extended fore and second fingers of the right hand. (*Dakota* I.) "Mixing the paint and applying to the face."

(1) Rub the right forefinger in the left palm; (2) then rub it on the cheeks. (*Apache* III.) "(1) Mixing or grinding; (2) applying."

Paper.

The left arm is semi-extended, the hand open in position (**W**), palm up; the right hand then approaches the left and in position (**K**) seemingly writes in the left palm; both hands then assume position (**W**), approximate each other, then widely diverge. (*Oto and Missouri* I.) "Something extended upon which to write."

Parent. See, also, **Relationship.**

——— Generically.

Place the hand, bowl-shaped, over the right breast, as if grasping a pap. (*Dodge.*)

Make the sign for **Father** twice. (*Absaroka* I; *Shoshoni and Banak* I.)

Collect the fingers and thumb of the right hand nearly to a point and pretend to grasp the left breast and draw it out toward the front about twelve inches. (*Dakota* VI.) "When this sign is made once it means *father* (which may be more specifically designated by elevating the finger as for *Man*, *i. e.*, man or male parent); when it is made *twice* it means *Parents*, and is used generically; when mother is meant, the signs for *Parent* and *Woman*, *i. e.*, long-haired parent; woman parent."

The right arm is extended (with the hand in type-position **J**), and made to point to the object. The hands (in type-positions **A A**) next approach the mammary region and thus hold for a moment. The right hand is then opened (as in type-position **S**, modified by being horizontal) and made to describe a semicircle downward, inward, and outward from the lower part of the trunk of the body. (*Oto and Missouri* I.) "Him or her from whom comes the offspring."

Same sign as for **Father,** also made for mother with the addition of the sign for **Woman;** literally *Woman Parent.* (*Kaiowa* I; *Comanche* III; *Apache* II; *Wichita* II.)

Partisan. See **Chief, War.**

Patience. See **Quiet.**

Peace. (Compare **Friendship.**)

Intertwine the fingers of both hands. (*Burton.*)

Pantomimic.—Simulate shaking of hands. (*Arapaho* I.)

The extended fingers, separated (**R**), interlocked in front of the breast, hands horizontal, backs outward. (*Dakota* I.) "Let us be friends. Let us be at peace."

The left arm semiflexed, hand closed and elevated, then spirally rotated across the forehead for **Anger;** this is followed by the sign for **No,** indicating *no anger.* The arms and hands then fall to sides, right arm is now raised, and closed hand clasped in the left axilla—the sign for *A dear friend.* The arms are then extended and drawn inward from before the body. (*Oto and Missouri* I.) "What comes of mutual friendship and good feeling."

Penurious or stingy. (Compare **Miser.**)

Clinch both hands firmly, the right hand resting on the left, both drawn to the chest, held firmly with a slight shake, pressed against breast, back of right hand above, with compressed lips and light shake or quiver of head. (*Ojibwa* IV.)

Person, A. An Individual.

Place the half-closed hands over the front of the forehead, backs outward, then pass them outward, downward over the cheeks and forward toward the chin. Face; visage. (*Wyandot* I.)

Pills.

The right arm and hand is brought before the breast; the index-finger and thumb in position (**H**) rotate together; the hand then approaches the mouth as in the sign for **Eat;** the act of swallowing is then executed. The right hand then sweeps hurriedly from the anal region of body. (*Oto and Missouri* I.) "Something rolled between thumb and finger, then swallowed, that will evacuate the bowels."

Pipe.

Make the same sign as to **Smoke.** (*Absaroka* I; *Shoshoni and Banak* I.)

First make the sign of filling the pipe, in front of the stomach with the right hand, left hand held representing the bowl of the pipe, with fingers nearly closed, back outward, edge of fingers downward, insert the right index from above between the thumb and forefinger as though inserting the tobacco, then knock it down with the edge of the right fist (**B**), then extend the left hand to nearly full capacity, with back downward as though grasping the pipe-stem from the right side, and the right hand grasps the stem from the left side, with back downward near the mouth end. (*Dakota* I.) "From the filling and manner of holding the pipe."

The palm of the right hand being upward, the fingers nearly closed and thumb extended forward, move the hand from the mouth straight forward about four inches, three or four times. Or the hand may be held upright, palm toward the left, thumb pointing toward the mouth, fingers closed. (*Dakota* IV.) "Holding the pipe-stem and removing it from the mouth as in smoking."

——— My.

Make the sign for **Pipe** and then throw the hand forward as for **I have.** It is my pipe. (*Dakota* IV.)

Pistol.

The same movement as for gun made lower down. (*Burton.*)

Left hand placed in position as though holding a pistol, right-hand forefinger resting against lower of left hand, and motioned as though cocking pistol. (*Cheyenne* I.)

The right hand in its position near the right eye, as given in the sign for **Gun,** denotes a pistol. This is from the shortness of the barrel. (*Dakota* I.)

The right arm is semi-extended before the front of the body; the hand assumes the posture of type-position (**B** 1), modified by being more opened and index crooked as hooking the trigger. From the center of the body the semi-extended arm is elevated to a level with the face and suddenly the hand is expanded as it projects forward. The right hand then drops to an extended position from the side of the middle of the body and the left hand is drawn edgewise across it, as in sign for **Bad.** (*Oto and Missouri* I.) "Something to shoot down what is bad."

Place, At this. (Compare **Here.**)

Place the left hand, slightly curved, about eighteen inches before the heart, pointing toward the right, the palm toward the face; collect the fingers of the right hand to a point, and strike the palm of the right against that of the left, the axes being at about right angles. (*Absaroka* I; *Shoshoni and Banak* I.)

——— Place, To arrive at a.

The hands are placed as in the sign for **Place, at this,** but the right hand is brought from a point at arm's length backward or out from the right shoulder, and struck against the palm of the left. *Absaroka* I; *Shoshoni and Banak* I.) The left hand, representing locality, is held in any direction to indicate the speaker's meaning, when the right hand follows the course and strikes the left at that position.

Plain, Prairie. (Compare **Earth.**)

Lay the hands flat upon their backs and move them straight from one another in a horizontal line. (*Wied.*) There is no simillarity with *Wied's* sign in execution, nor is the conception of the latter as plain as that of (*Oto and Missouri* I). (*Boteler.*)

Wave both the palms outward and low down. (*Burton.*)

This is expressed in the sign for **Flat, Level.** (*Dakota* I.) "The flat or level prairie."

After placing the hands near together, palms upward, fingers pointing forward, separate them about two feet; carry the extended right index, back upward, pointing forward, through a forward curve from side to side in front of the body three or four times, then make the negative sign. (*Dakota* IV.) "Nothing there."

The arms are semiflexed and brought, hands together, before the body. Then stoop forward and touch a piece of wood, or, in its absence, execute the sign for the same. The hands are then approximated before the chest in type position (**T**), then made to diverge widely, finally the right hand openly sweeps negatively to the side (*Oto and Missouri* I.) "Extended space where there is no wood."

(1) Wave the hands horizontally and laterally, palms up (**X**), holding left hand still (**X**); (2) thrust the right hand up, fingers extended (**P**); (3) rub off left palm with right (**X** reversed); smooth off left palm with right. (*Apache* III.) "(1) Level surface; (2) trees; (3) destitute of; (4) very smooth."

Plant, To.

Collect the fingers and thumb of the right hand to a point, directed toward the ground, and as the hand is moved straight forward from the body, dip it toward the ground at regular intervals. (*Kaiowa* I; *Comanche* III; *Apache* II; *Wichita* II.)

Plants, Vegetation.

Close the right hand; extend the index, pointing vertically, and place the tip of the thumb against the second joint, then pass the hand, back down, toward one side, in repeated moves, slightly elevating it at each rest. (*Wyandot* I.)

Pleased. See **Glad.**

Pond. See **Lake.**

Poor. Lean. Indigent.

The two forefingers extended, with the right as if it was a knife, imitate the motion of cutting the flesh off the left finger, beginning toward the tip, and cutting with a quick motion directed toward the base; at the same time turn the finger a little round, so as to expose the different parts to the action of cutting; intimating that the flesh has diminished from starvation. (*Long.*)

Hold the flattened hands toward one another before the breast, separate them, moving all the fingers several times inward and outward toward and outward from the breast. (*Wied.*) The left forearm is elevated and semi-extended from center of side of body. Left hand is in type position (**J** 1, horizontal). The right hand, in a similar position, seemingly shaves the left index-finger toward the body. There is no apparent identity in execution or conception of this sign with that of the Prince of Wied. (*Boteler.*) "Reduced to small dimensions."

Pass one forefinger along the other, leaving it at the tip, both extended, as if paring or whittling it. (*Arapaho* I.)

With the hands about four inches in front of the chest, ends near together, pointing inward, palms backward and fingers relaxed, quickly move them a few inches outward several times, each time nearly closing the fingers. (*Dakota* IV.) "Because the ribs show and the fingers fit in between them."

Place both hands with fingers joined but hooked upon the middle of the chest, and pull them apart as if tearing open the flesh. (*Kaiowa* I; *Comanche* III; *Apache* II; *Wichita* II.)

With the right hand to the breast, imitate the grasping and tearing open thereof. (*Shoshoni and Banak* I.)

Deaf-mute natural sign.—Place the hands upon the cheeks, and draw in the cheeks. (*Ballard.*)

——— In property.

Extend the left forefinger in front of the left side, remaining fingers and thumb being closed; then with the extended index make several passes over the back of the left from tip to base. (*Absaroka* I; *Shoshoni and Banak* I.)

Left hand in front of body, forefinger horizontal (**M** palm of hand to the right), right hand same position excepting to have palm inward, stroke the finger of left with forefinger of right from end of the finger to the knuckle several times. (*Cheyenne* II.)

Left hand as in (**K**) is held about twelve inches in front of the breast, and the right hand (**M**), forefinger extended, horizontal, palm downward, the side of the right finger glided over the back of the extended left forefinger. (*Dakota* I.)

The extended forefinger of the left hand is stroked rather rapidly two to many times from point to base with the extended index of the right hand, the motion resembling that of whittling a stick. Sometimes the left hand is slightly rotated so as to present different parts of the finger to the stroking process. (*Mandan and Hidatsa* I.)

Place the point of the extended index upon the back of the basal joint of the middle finger of the left hand. (*Wyandot* I.)

Rub the extended index back and forth over the back of the extended forefinger of the left hand from tip to base. (*Kaiowa* I; *Comanche* III; *Apache* II; *Wichita* II.)

Another: Pass the extended index of the right hand alternately along the upper and lower sides of the extended forefinger as in whittling toward the hand. (*Kaiowa* I; *Comanche* III; *Apache* II; *Wichita* II.)

Same sign as for **Apache.** See TRIBAL SIGNS. (*Comanche* II.)

Rub the back of the left hand back and forth with the palmar surface of the extended index. (*Ute* I; *Apache* I.)

Deaf-mute natural sign.—Pass the thumb over the forefinger several times (indicating "money"), next move the open hand from side to side once, and, at the same time, shake the head (indicating "no"), and then, nodding the head, point with the forefinger to some person who is poor. (*Hasenstab.*)

Pony. See **Horse.**

Position, Changes of. See **Ahead.**

Possession, mine; my property; To belong to.

The hand shut and held up to the view. (*Dunbar.*)

With the fist, pass upward in front of the breast, then push it forward with a slight jerk. (*Wied.*) There is no appreciable similarity in the execution of the *Oto and Missouri* I sign and that of *Wied.* The conception of the latter is difficult to see. (*Boteler.*) It appears to be the grasping and display of property.

Touch the breast with the index-finger. (*Dodge.*)

Right hand closed as though holding something elevated to level of and in front of the chin, drawn quickly with a downward curved motion toward the neck. (*Cheyenne* I.)

Both hands clinched about twelve inches before the body, palms inward, the right about eight inches above the left; both are then forcibly pushed toward the ground. (*Absaroka* I; *Shoshoni and Banak* I.)

First make the sign for **I**, personal pronoun, then point to or make the sign for the particular article to which reference is made, and complete the sign by crossing the arms at the wrists, about a foot in front of the breast, with hands natural, relaxed (**Y**), palms inward, upright, draw the hands to the body so that they will cover the right and left breasts (the right hand the left breast and the left hand the right breast). (*Dakota* I.) "Possession—my property."

Another: First make the sign for **I**, first personal pronoun, and then the sign for the property, if it has one; if not, then the particular articles constituting the property must be pointed to. (*Dakota* I.) "These things are mine."

Strike the palms of the hands together, palms inward, and then make the sign for **Some, I have some.** (*Dakota* IV.)

Throw the clinched right hand edgewise toward the earth, before and as far as the lower part of the body. (*Dakota* VI, VII.)

The arms are crossed and the hands loosely collected as in type-position (**B**), are folded on the chest. (*Oto and Missouri* I.) "That pertains to me."

Throw the fist, edge downward toward the ground. When possession is elsewhere, the arm is extended in that direction, and the above sign made. (*Kaiowa* I; *Apache* II; *Wichita* II; *Comanche* III.)

Both fists, with palms forward, are held before the body, right above left, and forcibly pushed downward a short distance. (*Shoshoni and Banak* I.)

Deaf-mute natural signs.—Point to the object owned, and then point to the breast. (*Ballard.*)

Point to something with the forefinger of one hand, and, nod the head, then rest the other open hand on the breast. (*Hasenstab.*)

Slap the breast with the hand, and at the same time open the mouth as if to say "My." (*Zeigler.*)

Our instructed deaf-mutes press an imaginary object to the breast with the right hand.

——— It belongs to me.

Deaf-mute natural signs.—Point to the object possessed, and then to the bosom, meaning the speaker. (*Ballard.*)

First point to the object with the forefinger, next rest the forefinger on the breast-bone, and then nod the head. (*Hasenstab.*)

Point with the right forefinger as if to point to something, and then from the thing to myself. (*Zeigler.*)

——— I have.

First make the sign for **I**, personal pronoun, and then the back of the right hand which points obliquely upward and toward the left, with fingers extended and joined, is carried out from the breast about eighteen inches, and placed in the palm of the left, held pointing obliquely upward and toward the right, palm upward, fingers extended, joined, and then both hands drawn in to the body. (*Dakota* I.) "Possession—I have it right here in my hands."

Place the fists, backs outward, about a foot in front of the navel, the right just above the left, then move them straight forward a couple of inches. Some place the right a little in advance of the left. (*Dakota* IV.) "Holding fast to everything."

Make the signs for **I**, **Me**, and **Have.** (*Hidatsa* I; *Arikara* I.)

Deaf-mute natural sign.—Move to and fro the finger several times to the breast. (*Larson.*)

——— Another has.

Pass the right hand quickly before the face, as if to say "Go away," then make the gesture for **Mine.** (*Wied.*) The arms are raised and closed over the breast as in the sign for **Mine.** They are then suddenly thrown open from the breast toward another person, with the palms outward. There is no similarity in either execution or conception of this sign with that of *Wied;* they are evidently of different origin. (*Boteler.*) "Not mine."

First point to the person who has the article, or who has done it, with the right index (**M**), and then make the sign for **Have it.** (*Dakota* I.) "Reverse of I have it."

Rotate the right hand, back upward, from side to side, six or eight inches, describing an arc, fingers separated a little, slightly bent, and pointing forward; then make the sign for **Have.** (*Dakota* IV.)

Another: Point at the person and then make the sign for **Have.** It belongs to him. (*Dakota* IV.)

Make the sign for **Possession, Mine,** in the direction of the person, or if the person is named and not present, the gesture is made to one side. (*Dakota* VI.)

Same sign as for **His.** (*Kaiowa* I; *Comanche* III; *Apache* II; *Wichita* II.)

——— To have.

Both hands clinched, held edgewise, the right about six inches above the left, and struck downward toward the ground as far as the waist, retaining the same distance between the hands during the whole of the gesture. (*Hidatsa* I; *Arikara* I.)

——— His.

Indicate the person, and throw the clinched hand edgewise toward the ground a short distance (stopping suddenly as if striking a resisting body), the hand directed toward the person or his possessions. (*Kaiowa* I; *Comanche* III; *Apache* II; *Wichita* II.)

——— It does not belong to me.

First make the gesture for **Mine,** then wave the right hand quickly by and in front of the face toward the right. (*Wied.*)

Reference having been made to the particular article the rest is expressed by the sign for **None, I have none.** (*Dakota* I.) "Not mine."

Deaf-mute natural signs.—The same sign as **It belongs to me,** supplemented by a shake of the head. (*Ballard.*)

Use the sign for **It belongs to me,** at the same time shake the head as if to say "No." (*Zeigler.*)

Potato.

Collect the fingers and thumb of the left hand to a point, hold them upward before the body (size), then with the fingers and thumb of the right hand similarly collected, pointing downward, make several motions forward and toward the earth (planting), then with the fingers and thumb of the right hand pointing upward, curved and separated, make a motion upward at arm's length (growth). (*Kaiowa* I; *Comanche* III: *Apache* II; *Wichita* II.)

Pour, To.

With the left hand held in front of the stomach to represent *kettle, bucket, &c.*; then the right hand held (**A**) to the right, and below the left hand, makes the sign for **Cup,** and is so carried over the stationary left, and turned up as though pouring its contents into the left. (*Dakota* I.) "Dipping of water with a cup to fill a bucket."

Prairie. See **Plain.**

Praise.

Italian sign.—The forefinger raised, inasmuch as to say a thing is to be extolled to the skies. (*Butler.*)

Pray. I pray you.

The palm of the hand is held toward the person or persons addressed. Sometimes both hands are so held. The Omaha and Ponka Indians say "wí-bdha-ha^{n}," I pray to you, I petition you; or "wí-bdhi-stu-be," *I smooth you down with the hand.* In praying to the sun the hand is elevated and held with the palm up. Say (Long's Exped., i, 384) gives the meaning, 43: "Be quiet, or be not alarmed, or have patience." (*Omaha* I; *Ponka* I.) "Soothing."

With the face inclined upward, eyes looking toward the heavens, both hands are brought together from their natural positions at the sides of the body, describing a considerable sweep in front of the face or above or below it, with palmar surfaces looking toward each other, upright, fingers extended, thumbs inward (**T**); then the body and hands are lowered toward the ground (nearly our kneeling), with eyes looking upward. (*Dakota* I.) "This is much the same as with civilized people."

The countenance placid and turned upward, the arms elevated, and with opened palms the hands vibrate and diverge to and from the body in imitation of an angel flying and in execution of the sign for *Deity.* The hand now assumes position (**K**), and the semi-extended index-finger points in successive jerks from the mouth upward, a rude imitation of the sign for *Talk.* (*Oto and Missouri* I.) "Speaking to the Winged-one above."

Deaf-mute natural sign.—Clasp the hands across each other, shut the eyes, and move the lips rapidly. (*Ballard.*)

Prayer.

If the flat hand is pressed to the lips, and thence moved upward to the heavens, it indicates a prayer or address to Deity. (*Ojibwa* I.)

Elevate one hand high toward the sky, spread hands opposite face, palms up and backward (**X**). (*Apache* III.) "Reference to God—desire to receive.

Pregnancy.

Pantomimically expressed by passing both hands, slightly arched, palms toward the body, from the pubis in a curve upward and in toward the pit of the stomach. (*Ute* I.) "Corresponds to the rotundity of the abdomen."

Pretty. See also **Good** and **Handsome.**

The fingers and thumb, so opposed as to form a curve, are passed over the face, nearly touching it, from the forehead to the chin; then add the sign for **Good.** (*Long.*)

Another: Curve the forefinger of the right hand, and place the tip on the ridge of the nose between the eyes, so as to represent a high Roman nose; then bring down the hand in a curvilinear manner, until the wrist touches the breast; after which add the sign for **Good.** (*Long.*)

Pass the extended hand (right usually), fingers joined, palm toward the body, in a caressing or stroking manner, vertically downward in front of, and thence horizontally from, the body, in immediate juxtaposition to the face or that part of the person desired to be indicated. (*Arapaho* I.)

Same as **Handsome.** (*Dakota* I.)

The same sign as for **Good.** (*Dakota* IV.)

Deaf-mute natural sign.—Arch the eyebrows and smack the lips. (*Ballard.*)

Priest.

Italian sign.—The palms laid together before the breast and the eyes fixed either on heaven or earth. (*Butler.*)

Prisoner. (Compare **In, Within.**)

The forefinger and thumb of the left hand are held in the form of a semicircle, opening toward and near the breast, and the forefinger of the right representing the prisoner, is placed upright within the curve, and passed from one side to another, in order to show that it will not be permitted to pass out. (*Long.*)

Sign for **Take prisoner** made, and the hands clinched and crossed back to back as though bound tightly. (*Cheyenne* I.) "Taken and bound."

Cross the wrists a foot in front of the neck, hands closed, backs forward. (*Dakota* IV.) "Tied."

Same as the sign for **Capture,** with the addition of the sign for **You** or **I** when a present individual is concerned. (*Oto and Missouri* I.) "To seize when in battle."

Both hands clinched, the right laid transversely across the left at the wrists. (*Wyandot* I.) "Tied arms."

Make the sign for **Battle;** then with the right grasp an imaginary person from the right side, extend both forefingers toward the ground, the remaining fingers and thumbs closed; place them side by side and move them toward the left. (*Apache* I.) "One grasped in battle and led away by the captor."

Deaf-mute natural sign.—Place the fingers of one hand upon those of the other, indicating the cross-bars of a jail, and then point with the forefinger to some person who is put in jail. (*Hasenstab.*)

Italian signs.—The vulgarism "to look between bars" for to be in prison corresponds to the gesture which is made by crossing the fingers of both hands before the eyes so as to make a checkered grate, or by covering the eyes with the fingers of one hand drawn apart. (*Butler.*)

——— Specifically for *captive.*

The arms hanging down and wrists together or crossed. (*Butler.*)

——— To take.

Both hands, fingers slightly hooked, thumb lying against forefinger, suddenly thrust forward to the left and jerked back quickly toward the body. (*Cheyenne* I.)

The left hand held about 18 inches in front of the left breast, obliquely upward, edge of fingers outward (**B** 2), is quickly seized around the wrist by the right hand, passed from in front of the right breast, back upward, and drawn rapidly in toward and near the left breast. (*Dakota* I.) "He is come; I have taken him; leading him captive."

Both arms are flexed before breast, and hands made to execute the sign for fight; the left arm then, semiextended, is left a second; then siezed at wrist by the right hand in type position (**G**); palms of both face the ground. (*Oto* I.) "One taken in fight or misconduct."

Property. See **Possession.**

Prudent or **Cautious.** (Compare **Danger.**)

Hold right hand in front of right breast or partly to the right side and lower down, palm down, gently move it up and down two or three times, then hold it still a few moments, and gently depress the hand, with slight bow of the head to right (*Ojibwa* IV.)

See **Danger**. *Prudent, cautious*, are not the opposite of *fool.* (*Dakota* I.)

Place the tips of the extended fingers against the temples, then point them upward, the eyes following the same direction. "Superior judgment." (*Wyandot* I.)

Purchase. See **Trade.**

Put it back. See **Let alone.**

Quantity, large; many; much; number.

The flat of the right hand patting the back of the left hand, which is repeated in proportion to the greater or lesser quantity. (*Dunbar.*) Simple repetition.

The hands and arms are passed in a curvilinear direction outward and downward, as if showing the form of a large globe; then the hands are closed and elevated, as if something was grasped in each hand and held up about as high as the face. (*Long.*)

Bring the hands up in front of the body with the fingers carefully kept distinct. (*Cheyenne* sign. Report of Lieut. *J. W. Abert, loc. cit.*, p. 431.)

Both hands closed, brought up in a curved motion toward each other to the level of the neck or chin. (*Cheyenne* I.)

Clutch at the air several times with both hands. The motion greatly resembles those of danseuses playing the castanets. (*Ojibwa* I.)

——— Many.

A simultaneous movement of both hands, as if gathering or heaping up. (*Arapaho* I.) Literally "a heap."

Both hands, with spread and slightly curved fingers, are held pendent about two feet apart before the thighs; then draw them toward one another, horizontally, drawing them upward as they come together. (*Absaroka* I; *Shoshoni and Banak* I; *Kaiowa* I; *Comanche* III; *Apache* II; *Wichita* II.) "An accumulatoin of objects."

Hands about 18 inches from the ground in front and about the same distance apart, held scoop-fashion, palms looking toward each other, separated fingers, etc.; then, with a diving motion, as if scooping up corn from the ground, bring the hands nearly together, with fingers nearly closed, as though holding the corn, and carry upward to the height of the breast, where the hands are turned over, fingers pointing downward, separated, as though the contents were allowed to drop to the ground. (*Dakota* I.)

Open the fingers of both hands, and hold the two hands before the breast, with the fingers upward and a little apart, and the palms turned toward each other, as if grasping a number of things. (*Iroquois* I.)

Place the hands on either side of and as high as the head, then open and close the fingers rapidly four or five times. (*Wyandot* I.) "Counting 'tens' an indefinite number of times."

Deaf-mute natural signs.—Put the fingers of the two hands together, tip to tip, and rub them with a rapid motion. (*Ballard.*)

Make a rapid movement of the fingers and thumbs of both hands upward and downward, and at the same time cause both lips to touch each other in rapid succession, and both eyes to be half opened. (*Hasenstab.*)

Move the fingers of both hands forward and backward. (*Zeigler.*)

——— Horses.

Raise the right arm above the head, palm forward, and thrust forward forcibly on a line with the shoulder. (*Omaha* I.)

——— Persons.

Take up a bunch of grass or a clod of earth; place it in the hand of the person addressed, who looks down upon it. (*Omaha* I.) "Represents as many or more than the particles contained in the mass."

Hands and fingers interlaced. (*Macgowan.*)

——— Much.

Move both hands toward one another and slightly upward. (*Wied.*) I have seen this sign, but I think it is used only for articles that may be piled on the ground or formed into a heap. The sign most in use for the general idea of *much* or *many* I have given you before. (*Matthews.*)

Both hands flat and extended, placed before the breast, finger-tips touching, palms down; then separate them by passing outward and downward as if smoothing the outer surface of a globe. (*Absaroka* I) *Shoshoni and Banak* I; *Kaiowa* I; *Comanche* III; *Apache* II; *Wichita* II.; "A heap."

Much is included in **Many** or **Big**, as the case may require. (*Dakota* I.)

The hands, with fingers widely separated, slightly bent, pointing forward, and backs outward, are to be rapidly approximated through downward curves, from positions twelve to thirty-six inches apart, at the height of the navel, and quickly closed. Or the hands may be moved until the right is above the left. (*Dakota* IV.) "So much that it has to be gathered with both hands."

Both hands and arms are partly extended; each hand is then made to describe, simultaneously with the other, from the head downward, the arc of a circle curving outwards. This is used for **Large** in some senses. (*Mandan and Hidatsa* I.)

Hands open, palms turned in, held about three feet apart, and about two feet from the ground. Raise them about a foot, then bring in an upward curve toward each other. As they pass each other, palms down; the right hand is about three inches above the left. (*Omaha* I.)

Place both hands flat and extended, thumbs touching, palms downward, in front of and as high as the face; then move them outward and downward a short distance towward their respective sides, thus describing the upper half of a circle. (*Wyandot* I.) "A heap."

Sweep out both hands as if inclosing a large object; wave the hands forward and somewhat upward. (*Apache* III.) "Suggesting immensity."

——— And heavy.

Similar to **Man,** except when hands are raised close the fists, backs of hands down, as if lifting something heavy; then move a short distance up and down several times. (*Omaha* I.)

——— Plenty.

Raise the arms above the head, fingers interlaced to represent the lodge poles, separate and bring together again. (*Omaha* I.) "You see how I sit in a large lodge and how comfortable I am."

Another: Same as the preceding, then bring open hands together in front, palms down, extended; separate the hands to their respective sides. (*Omaha* I.) "I am sitting in a good lodge, or have a good tent-fire here."

Another: Right arm curved horizontally, with the tips of the fingers toward the breast; then slightly extend the arm and describe a circle by returning the hand and passing it inward past the breast toward the right shoulder. (*Omaha* I.) "A man has plenty in his tent; or I have plenty around here."

Deaf-mute natural sign.—Stretch the left arm at full length, and move the forefinger of the right hand along it to the shoulder. (*Ballard.*)

Question; inquiry; interrogation. (Compare **Fool; Indecision.**

The palm of the hand upward and carried circularly outward, and depressed. (*Dunbar.*)

The hand held up with the thumb near the face, and the palm directed toward the person of whom the inquiry is made; then rotated upon the wrist two or three times edgewise, to denote uncertainty. (*Long.*) The motion might be mistaken for the derisive, vulgar gesture called "taking a sight," "donner un pied de nez," descending to our small boys from antiquity. The separate motion of the fingers in the vulgar gesture as used in our eastern cities is, however, more nearly correlated with some of the Indian signs for **Fool.** It may be noted that the Latin "sagax," from which is derived "sagacity," was chiefly used to denote the keen scent of dogs, so there is a relation established between the nasal organ and wisdom or its absence, and that "suspendere naso" was a classic phrase for hoaxing. The Italian expressions "restare con un palmo di naso," "con tanto di naso," etc., mentioned by the Canon DE JORIO, refer to the same vulgar gesture in which the face is supposed to be thrust forward sillily.

Extend the open hand perpendicularly with the palm outward, and move it from side to side several times. (*Wied.*) This sign is still used. For "outward," however, I would substitute "forward." The hand is usually, but not always, held before the face. (*Matthews.*)

Right hand, fingers pointing upward, palm outward, elevated to the level of the shoulder, extended toward the person addressed, and slightly shaken from side to side. (*Cheyenne* I.)

Deaf-mute natural sign.—A quick motion of the lips with an inquiring look. (*Ballard.*)

Australian sign.—One is a sort of note of interrogation. For instance, if I were to meet a native and make the sign: hand flat, fingers and thumb extended, the two middle fingers touching, the two outer slightly separated from the middle by turning the hand palm upward as I met meet him, it would mean: "Where are you going?" In other words I should say "*Minna?*" (what name?). (*Smyth.*)

——— Has he?

Deaf-mute natural sign.—Move to and fro the finger several times toward the person spoken of. (*Larson.*)

——— Have you?

Deaf-mute natural sign.—Move the finger to and fro several times toward the person to whom the one is speaking. (*Larson.*)

——— When?

With its index extended and pointing forward, back upward, rotate the right hand several times to the right and left, describing an arc with the index. (*Dakota* IV.)

——— Are you?

Deaf-mute natural signs.—Point to the person spoken to and slightly nod the head, with an inquiring look. (*Ballard.*)

Point with the forefinger, as if to point toward the second person, at the same time nod the head as if to say "yes." (*Zeigler.*)

Quick, Quickly.

Same motion and position as **Soon,** arm slightly more raised and moved out faster and thrown back more rapidly by a quick motion. (*Ojibwa* IV.)

Both hands should be placed horizontal, palms upward, pointing forward, about three inches apart in front of the lower part of the chest, and then quickly raised about eight inches, at the same time shaking them a little from side to side. (*Dakota* VI.) "Idea of lightness, and therefore quick motion."

Quiet, be; be not alarmed; have patience.

The palm of the hand is held toward the person. This is also the sign for *Surrender*. (*Long.*)

Place the forefinger or the hand over the mouth. (*Arapaho* I.)

The right hand with palm downward, horizontal, flat, fingers extended and pointing forward (**W**) brought to the front median line of body, and about a foot from it, and then carried on the same level to the right side of the body, where two or three quick upward and downward movements of the hand and arm complete the sign. (*Dakota* I.) "Sit down; be quiet."

Deaf-mute natural signs.—Hold the hand still in the air. (*Ballard.*)

Fold the arms. (*Larson.*)

——— Patience.

The open flat right hand is laid, back outward, over the left breast, and then both hands, with fingers extended, separated, and somewhat curved, are held horizontally, with palmar surfaces looking toward one another, a few inches apart, in front of the body (**P**, turned horizontally) (this is *Expanded, large*); and then the hands are closed (fists **B**) and passed slowly over one another with a slight shaking or tremulous motion of the hands and arms. (*Dakota* I.) "A (large) patient, enduring heart."

Rabbit. See also **Hare.**

The fore and little finger of the right hand are extended, representing the ears of the animal; the hand is then bobbed forward to show the leaping motion of the animal. (*Long.*)

Rain. (Compare **Snow** and **Heat.**)

Begin with the sign of **Water,** then raise the hands even with the forehead, extending the fingers outward, and give a shaking motion as if to represent the dripping of water. (*Dunbar.*)

The sign for **Water** precedes that for **Snow.** (*Long.*)

Scatter the fingers downward. The same sign denotes **Snow.** (*Burton.*)

Imitate its fall with the hand, palm down, partially closed, fingers separated and pointing downward; then move the hand in a direct course toward the ground. (*Arapaho* I.)

Both hands, fingers and thumbs drooping, held to the level of the head; fingers slowly closed and opened, and motion downward made, as though flirting water from the ends of the fingers. (*Cheyenne* I.)

Carry both hands over the head, with fingers separated, curved downward (**Q**), palms inward, then make a sloping downward movement with the hands, flexing and extending all the fingers slightly and with considerable rapidity, as nearly as possible in imitation of the drops of rain falling on the face and body. (*Dakota* I.) "From the falling of rain."

Suspend the hands about a foot in front of the shoulders, backs forward, fingers separated and bent a little and pointing downward, and shake the fingers, or approximate and separate their ends once or several times. Some, each time while separating the ends of the fingers, throw the hands downward about eight inches. (*Dakota* IV.)

Hold the right hand pendent, with fingers separated and pointing downward, before the right side and on a level with the head; then thrust it downward and back to its first position, repeating the movement two or three times. (*Dakota* V, VI; *Hidatsa* I; *Arikara* I.)

The hand is held on a level with the top of the head, fingers separated and pendent; it is then moved downward rapidly a few inches and suddenly arrested. It may be restored to its original position and the motion may be repeated once or oftener. (*Mandan and Hidatsa* I.)

Both hands held to either side and in front of the head, palms down, fingers pendent and separated; then move the hands downward and back again, repeating several times. (*Wyandot* I.)

Raise right hand in front of breast, say fifteen inches from body, back upward, fingers hanging down (**P** 1, with fingers hanging down, hand horizontal), then make motion as though sprinkling water, moving hand up and down from wrist out. (*Sahaptin* I.) "Sprinkling water."

Hold the right hand in front of the side of the head, palm down, fingers pendent and separated; then move the hand up and down a short distance several times. (*Ute* I.)

Hold the right hand in front of or to one side of the face, palm down, fingers pendent and separated; then move the hand downward and back to its original position, several times, most of the movement being at the wrist. For a very heavy rain both hands are similarly employed, but with a more vigorous motion. (*Apache* I.)

Deaf-mute natural signs.—An up-and-down motion of the extended fingers. (*Ballard.*)

First, point out in some direction; next, move the open hand down fast and up slowly, successively, and then nod the head. (*Hasenstab.*)

Move the fingers upward and downward. (*Larson.*)

——— Drizzling.

Place the closed hand at the height of the face, palm down, leaving the index pointing downward and a little toward the left, then move the hand up and down several times. (*Apache* I.)

Rainbow.

The right hand with index-finger only extended, and pointing upright (**J**), is carried from left to right in front of the head, on a curve, and then downward toward the horizon with a gradually decreasing spiral motion, until finally, without this motion, hand carried on downward, with finger pointing upward. (*Dakota* I.) "To go across the heavens wrapped with different colors."

Rash, or **rashly.**

Indicate by quick, unsteady motion of hands and body, agitate both hands and body, move hands quickly to left in front and to right, palms down, with jerking motions up and down, most violently downward, head shaken a little. (*Ojibwa* IV.)

Recently. See **Time, Recently.**

Relationship.

——— Aunt (maternal).

Make the signs for **Mother, Hers** (to the right), **Brother and Sister,** and **Woman.** (*Kaiowa* I; *Comanche* III; *Apache* II; *Wichita* II.)

——— Aunt (paternal).

Make the signs for **Father, Hers** (to the right), **Brother and Sister,** and **Woman.** (*Kaiowa* I; *Comanche* III; *Apache* II; *Wichita* II.)

——— Brother, sister; brother and sister.

The sign for **Man** (and for sister, of a woman) succeeded by placing the ends of the fore and middle fingers of one hand together in the mouth. (*Long.*)

The two first finger-tips are put into the mouth, denoting that they fed from the same breast. (*Burton.*)

Place the fore and middle fingers in the mouth, thus implying nursing at the breast by a common mother. (*Arapaho* I.)

Tips of the fore and middle fingers of the right hand placed between nostrils and mouth; brought with a sudden curved motion forward, outward, and obliquely to the right expresses relationship between children of same father and mother; must be preceded by the sign for **Man** or **Woman,** to specify *brother* or *sister*. (*Cheyenne* I.)

Put the tips of the fore and second fingers of either hand between the teeth. (*Sac, Fox, and Kickapoo* I.) "Two persons sucking one breast."

First and second fingers of right hand extended (**N**), placed obliquely, backs outward, on the lips, or their tips in the mouth. (*Dakota* I.) "Nourishment from the same breast."

Place the tips of the extended fore and middle fingers of the right hand between or against the lips, and afterward draw them forward about a foot, the other fingers to be closed and the back of the hand upward; then make the first part of the sign for **Man,** *i. e.*, the index horizontally in front of the privates. Sister: The same as for brother, but instead of the sign for man use the sign for woman after drawing the fingers from the mouth. (*Dakota* IV.) "Sucking the same breast."

Another: With the right hand closed, leave the index and second fingers extended, palm down; place the tips near the mouth and jerk them forward about six inches. (*Dakota* VI.) "We derived nourishment from the same breast."

Bring the right hand to the lips, touching them with the index and middle fingers. (*Omaha* I.) "To suck from the same breast." Obtained from J. La Fleche, as a *Pani* and *Omaha* sign.

Thrust the first fingers into the mouth. (*Omaha* I.)

The left arm is semi-extended, with hand in position (**J** 1), modified by being held horizontal outward and index extended; the right arm, hand and finger now assume the same position above and behind left. Now, the right index is brought aside of left, each extended parallel. Finally the right index is brought to point to the cardiac region. (*Oto* I.) "Two separate births, alike in appearance and at heart."

Thrust the index and second fingers into the mouth. (*Ponka* I.) "We two sucked from the same breast."

——— Brother (said by male).

Bring the left arm and hand to the left breast, as if in embracing. (*Wyandot* I.)

——— Brother (said by sister or other brother).

Make the sign for **Brother and Sister,** followed by that for **Man.** (*Kaiowa* I; *Comanche* III; *Apache* II; *Wichita* II.)

Extend the first two fingers of the right hand; bring the hand, with fingers thus extended, opposite the mouth; then place these two fingers between the lips. (*Comanche* I.) "Both took nourishment from the same breast."

Right-hand fingers and thumb closed; bring in front of the right side, extending and separating the fore and second fingers, which are slowly brought together so as to lie side by side. (*Pai-Ute* I.)

Both hands closed, forefingers extended; bring them together, vertically, in front of the body, with the palms forward. (*Pai-Ute* I.)

Place the first two fingers of the right hand between the lips. (*Wichita* I.)

——— Brother's daughter.

Make the signs for **Brother, Woman,** and **Born** (quickly or continuously, with termination of last sign to the right of the body). (*Dakota* VI.)

——— Brother's son.

Make the signs for **Brother, Man,** and **Born** (to the right side of the body, and quickly after or continuously with termination of gesture for **Man**). (*Dakota* VI.)

——— Brother's wife.

Make the signs for **Brother, Man, Woman,** and **Possession** (**His**). (*Dakota* VI.)

Make the signs for **My, Brother** (**Brother and Sister** and **Man**), **His** (made to the right), **Same** or **similar,** and **Woman.** (*Kaiowa* I; *Comanche* III; *Apache* II; *Wichita* II.)

——— Brother and sister.

With the right hand closed, leaving the index and second fingers extended and slightly bent, bring the hand before the chin, palm down, finger tips nearly touching the mouth; then draw them downward and forward. This sign is made when the person alluded to is present, and has been referred to by either person, or an inquirer. (*Kaiowa* I; *Comanche* III; *Apache* II; *Wichita* II.) "Nursed from the same breast."

The first phalanges of the first and second fingers are placed between the lips and then withdrawn. This represents somewhat faultily that both have been nourished from the same source. The sign is for common gender. Masculine or feminine may afterward be indicated by the appropriate signs. (*Mandan and Hidatsa* I.)

——— Brother and sister (said by female).

Indicate the individual, then grasp the tip of the forefinger with the thumb and index. (*Wyandot* I.)

——— Daughter.

Make the same sign as for **Girl**; then indicate *Parent* or *Possession* by pointing to the person, or laying the index vertically against the breast or pointing to it. (*Absaroka* I; *Shoshoni and Banak* I.)

First make the sign for **Offspring,** then designate *Age* as described in the sign for **Child,** and complete by the sign for **Woman.** (*Dakota* I.)

Make the signs for **Birth** and **Woman.** (*Kaiowa* I; *Comanche* III; *Apache* II; *Wichita* II.)

Deaf-mute natural sign.—Point the finger to the ear (because of the ear-rings) and then put down the hand, when naturally stretched out, to the knee. (*Larson.*)

——— Daughter's daughter.

Make the signs for **Woman** and **Brother** (Sister), **Born** (to the right of the body), and **Woman.** (*Dakota* VI.)

——— Daughter's son.

Make the signs for **Woman** and **Brother** (Sister), **Born,** and **Man,** both off from the right side of the body. (*Dakota* VI.)

——— Family, members of one.

Grasp the tip of the forefinger with the thumb and fingers of the right hand. When more fingers are used than the index thumb, it indicates more individuals. (*Wyandot* I.)

——— Father, mother (parents).

The same sign of issue from the loins as **Offspring,** with additions: *e. g.*, for **Mother,** give **I** or **My,** next **Woman,** and then the symbol of parentage. For **Grandmother** add to the end clasped hands, closed eyes, and like an old woman's bent back. (*Burton.*)

With the right hand pretend to grasp the right breast with the extended fingers and thumb; then draw them outward about twelve inches Compare **Parentage.** (*Absaroka* I; *Shoshoni and Banak* I.)

The right hand, with fingers arched, separated tips, pointing inward toward the abdomen, is carried in and out two or three times with a sort of grabbing motion. (*Dakota* I.) "Part of the same body."

After making the sign in front of the privates for **Man,** make the sign for **Mother.** (*Dakota* IV.)

Another: Touch the right or left breast with the joined ends of the fingers of the right hand, and then make the motion in front of the right breast with the fist as for **Old man.** (*Dakota* IV.)

Same sign as for **Parent.** (*Dakota* VI; *Wyandot* I.)

Collect the fingers and thumb of the right hand to a point or nearly so, and pretend to grasp the breast, and draw the hand forward about eight inches. (*Kaiowa* I; *Comanche* III; *Apache* II; *Wichita* II.)

Deaf-mute natural signs.—Close the hand while the thumb is still up and then rest the thumb on the lips. (*Hasenstab.*)

Move the forefinger along the jaws—because of the beard. (*Larson.*)

——— Father's brother.

Make the signs for **Parent** (Father), **Possession** (**His**) (to the right), **Man,** and **Brother.**

——— Father's father.

Make the signs for **My, Father, Over,** and **Father.** Sometimes, for illustrating more clearly, the sign for **Aged** is added. (*Kaiowa* I; *Comanche* III; *Apache* II; *Wichita* II.)

——— Father's mother.

Make the signs for **My, Mother** (parent and woman), **Over,** and **Mother.** (*Kaiowa* I; *Comanche* III; *Apache* II; *Wichita* II.)

——— Father's sister.

Make the signs for **Parent** (father), **Possession** (**His**), **Man** (to the right), and **Sister.** (*Dakota* VI.)

——— Grandfather.

Make the signs for **Parent, Time long ago, Aged** and **Born.** (*Dakota* VI.)

——— Grandmother.

Ends of fingers of both hands touching the breasts on their respective sides, (this is *mother*), then make the sign for **Woman,** by drawing the hand downward at the right side of the head as though passing a comb through the long hair, and then complete by the sign for **Old,** by describing with the right hand in front of the right side of the body part of a circle after the manner of using a cane for support in walk ing. (*Dakota* I.) "Denotes an aged person. Decrepit age dependent on a staff."

Make the signs for **Parent, Woman, Time long ago, Aged** and **Born.** (*Dakota* VI.)

——— Husband.

Sign for **Companion,** (*Dakota* I.) "United."

——— Husband (said by wife).

Make the sign for **Same,** followed by that for **Man.** (*Kaiowa* I; *Comanche* III; *Apache* II; *Wichita* II.)

———Mother.

Touch the breast and place the forefinger in the mouth. (*Arapaho* I.)

Right-hand fingers and thumb closed inward as though clasping breast (mammæ) and drawn outward three or four times. (*Cheyenne* I.)

Make the sign for **Father,** followed by that for **Woman.** (*Absaroka* I; *Shoshoni and Banak* I.) "Woman parent."

From a foot in front of the right or left breast, move the right hand, its palm backward and its fingers semiflexed and spread, near to the breast, and after bringing the ends of the fingers together, move the hand forward again and half open the fingers and spread them. Make these motions two or three times, somewhat rapidly. (*Dakota* IV.) "In imitation of sucking the breast."

Ends of the fingers of both hands (**S**) touching the breasts of their respective sides. (*Dakota* I.) "Indicating the mammæ—one who has nursed a child."

Make the signs for **Parent** and **Woman.** (*Dakota* VI; *Wyandot* I; *Kaiowa* I; *Comanche* III; *Apache* II; *Wichita* II.)

Deaf-mute natural sign.—Close the hand except the little finger and then rest the finger on the lips. (*Hasenstab.*)

———Mother's brother.

Make the signs for **Woman** and **Parent** (mother), **Possession** (**Hers**) (to the right), **Man,** and **Brother.** (*Dakota* VI.)

——— Mother's father.

Make the signs for **My, Mother, Over,** and **Father.** (*Kaiowa* I; *Comanche* III; *Apache* II; *Wichita* II.)

——— Mother's mother.

Make the signs for **My, Mother, Over,** and **Mother.** (*Kaiowa* I; *Comanche* III; *Apache* II; *Wichita* II.)

—— Mother's sister.

Make the signs for **Woman** and **Parent** (mother), **Possession** (**Hers**) (to the right), and **Sister.** (*Dakota* VI.)

——— Nephew (brother's son).

Make the signs for **My, Brother, His, Born,** and **Woman.** (*Kaiowa* I; *Comanche* III; *Apache* II; *Wichita* II.)

——– Nephew (sister's son).

Make the signs for **My, Sister, Hers, Born,** and **Man.** The signs for *Hers, Born,* and *Man,* are made to the right of the body, nearly at arm's length, as belonging to another. (*Kaiowa* I; *Comanche* III; *Apache* II; *Wichita* II).

——— Niece (brother's daughter).

Make the signs for **My, Brother, His, Born,** and **Woman.**. (*Kaiowa* I; *Comanche* II; *Apache* II; *Wichita* II.)

——— Niece (sister's daughter).

Make the signs for **My, Sister, Hers, Born,** and **Woman.** The signs for *Hers* and *Born* are made to the right of the body nearly at arm's length, as belonging to another; although the sign must be made at the speaker's head, the sign is understood as referring to the preceding signs. (*Kaiowa* I; *Comanche* III; *Apache* II; *Wichita* II.)

———· Sister.

Sign for **Brother,** and, if necessary to distinguish gender, add that for **Squaw.** (*Arapaho* I.)

Same sign as for **Brother,** except designation of the sex. (*Dakota* I.) " We are from the same brother."

Make the signs for **Woman** and **Brother.** (*Dakota* VI.)

——— Sister.

The right arm is flexed upward, and hand, in position (**N** 1), modified by fingers being approximated, is then approached to the mouth and ends of fingers approximated and inserted between lips; the hand is then withdrawn, and the index-finger, extended, points to the cardiac region; the hands are then collected, as in type (**A**), and held in mammary region. (*Oto* I.) "A female dear to my heart and lips."

Pass the flat right hand, palm toward the body, from the pubis downward, forward and upward, then elevate the first two fingers of the right hand. (*Ute* I.)

Deaf-mute natural sign.—An uninstructed deaf-mute, as related by Mr. Denison, of the Columbian Institution, invented, to express *Sister*, first the sign for **Female,** made by the half-closed hands, with the ends of fingers touching the breasts, followed by the index in the mouth.

——— Sister (said by brother or other sister).

Bring the left arm and hand to the left breast, as if in an embrace, then elevate the forefinger. (*Wyandot* I.)

Make the sign for **Brother and Sister,** followed by that for **Woman.** (*Kaiowa* I; *Comanche* III; *Apache* II; *Wichita* II.)

——— Sister's daughter.

Make the signs for **Woman** and **Brother** (*sister*), **Woman** and **Born** (to the right side of the body) in a continuous movement. (*Dakota* VI.)

——— Sister's husband.

Make the signs for **Brother** and **Woman** (*sister*), **Man** and **Possession** (**Hers**), both to the right. (*Dakota* VI.)

Make the sign for **My, Sister,** (*brother and sister* and *woman*), **Hers** (made to the right), **Same** or **Similar,** and **Man**. (*Kaiowa* I; *Comanche* III; *Apache* II; *Wichita* II.)

——— Sister's son.

Make the signs for **Woman** and **Brother** (*sister*), **Man** and **Born** (to the right of the body). (*Dakota* VI.)

——— Son. See **Child** (**offspring**); **Male,** and **Man.** (*Arapaho* I.)

Same as the sign for **Daughter,** excepting designation of sex. (*Dakota* I.)

Make the signs for **Birth** and **Man.** (*Kaiowa* I; *Comanche* III; *Apache* II; *Wichita* II.)

——— Son's daughter.

Make the signs for **Born, Born,** and **Woman.** (*Dakota* VI.)

——— Son's son.

Make the signs for **Born, Born,** and **Man.** (*Dakota* VI.)

——— Uncle (maternal).

Make the signs for **Father, His** (to the right), **Brother and Sister,** and **Woman.** (*Kaiowa* I; *Comanche* III; *Apache* II; *Wichita* II.)

——— Uncle (paternal).

Make the signs for **Father, His** (to the right), **Brother and Sister,** and **Man.** (*Kaiowa* I; *Comanche* III; *Apache* II; *Wichita* II.)

——— Wife. (See also **Companion.**)

The dexter forefinger is passed between the extended thumb and index of the left. (*Burton.*)

Same as **Companion.** (*Dakota* I.)

Make the sign for **Woman,** and then lay the two forefingers together side by side, straight and pointing forward, the other fingers loosely closed. (*Dakota* IV.) "Two joined as one."

(1) Make the sign for **Woman;** (2) two fingers left hand extended (**N**, horizontal, forward, and fingers touching); (3) both fists to chest, *con amore;* (4) left arm circled before and drawn toward the body. (*Apache* III.) "(1) A woman (2) mated, (3) very dear to me, (4) and whom I embrace."

——— My wife.

Make the sign for **Same,** followed by that for **Woman.** (*Kaiowa* I; *Comanche* III; *Apache* II; *Wichita* II.)

Make the sign for **Woman,** and then move the right fist, back outward, forward a foot or eighteen inches from six inches in front of the navel. (*Dakota* IV.) "Woman I have."

——— Wife or mother.

The hands, in position (**A**), are brought to the chest and slightly passed along over the mammary prominence. The right hand then passes sweepingly downward and outward, palm toward the body, describing rudely the exit of the child from the loins in the obstetrical curve of Carus. The right arm is then raised and the extended index touches the præcordium. (*Oto and Missouri* I.) "Woman or mother of my heart."

Repeat; often. See also **Do it again.**

Extend the left arm, also the index-finger, and with the latter strike the arm at regular intervals, from front backward, several times. (*Wied.*) I have seen this sign. (*Matthews.*)

Deaf-mute natural sign.—Put the fingers of the two hands together (tip to tip) and rub them rapidly. (*Ballard.*)

Italian sign.—A man who puts his finger behind his ear, or who sticks out his chin and parts his lips, wishes to have something repeated which he has imperfectly heard. (*Butler.*)

Retreat, or to return through fear.

Begin with the sign for **Traveling moderately** or **Marching,** then draw the tips of the fingers and thumb together and retreat the hand to the body. (*Long.*)

Both hands closed (**B**) at the left breast on the same level and nearly joined; then carry them outward right to right, left to left, in front of the shoulders, with arms more than half extended, and in transit open the hands so that when brought to a stop the palms will be upward, fingers slightly separated, curved upward. (*Dakota* I.) "Heart was not brave; what was I to do?; would seem to indicate utter helplessness under the circumstances. In such cases the man would not be stamped a coward. See the conception of *Brave.*"

Ridge.

Right hand horizontal, back outward, fingers extended, edge of hand downward, is drawn from left to right about a foot in front of the face; if a jagged ridge, indicate by upward and downward sharp movements of the hand; if a level one, by drawing the hand on as nearly the same level as possible. (*Dakota* I.) "From the appearance of a ridge."

Riding (horseback). See **Horse** and **Going.**

River. (Compare **Broad.**)

The hand, in the form of a scoop or ladle, is carried to the mouth, as if conveying water, and drawn along in a horizontal line with the edge downward, about the height of the breast. (*Long.*)

Another: Hold up the fingers of the left hand, a little diverging from each other (representing a mountain range), and to convey the idea of the streams flowing from them, place the index-finger of the right hand alternately between each two of them and draw it away in a serpentine manner. (*Long.*)

Open the right hand and pass it before the mouth from above downward. (*Wied.*) If *Wied's* sign is complete there is a similarity in conception, but the (*Oto* I) sign represents the conception "water," and that which retains it at sides and directs the flow. "Something we drink, retained by banks at sides." (*Boteler.*)

The finger traces serpentine lines on the ground. (*Ojibwa* I.)

Make the sign for **Drinking,** and then wave both the palms outward. A rivulet, creek, or stream is shown by the drinking sign, and by holding the index tip between the thumb and medius; an arroyo (dry watercourse), by covering up the tip with the thumb and middle finger. (*Burton.*)

A movement of the extended hand, palm down and horizontal, fingers joined, indicative at once of the flowing of water, and the meandering of its current. (*Arapaho* I.)

Sign for **Water,** followed by the sign for **Snake.** (*Cheyenne* I.) "A river is flowing water.

Make the sign for **Water,** then place the extended flat hand, palm down, before the breast, and push it forward to arm's length. (*Absaroka* I; *Shoshoni and Banak* I.)

Right hand to the left side of body, level of shoulder, edge of fingers outward, extended, pointing obliquely downward toward the left (**S** turned downward), carry the hand downward on a double curve in front of the body and toward the right. (*Dakota* II.) "Running water."

Make the sign for **Water,** (*i. e.*, place the right hand, upright, six or eight inches in front of the mouth, back outward, index and thumb crooked and their ends about an inch apart, the other fingers nearly closed; move it toward the mouth, and then downward nearly to the top of the breast-bone, at the same time turning the hand over toward the mouth until the little finger is uppermost); then draw the right hand, its back forward, the index extended and pointing toward the left, and the other fingers closed, from about two feet in front of the left shoulder toward the right until it is a foot or so outside of the line of the right shoulder. (*Dakota* IV.) "Long water."

Right hand brought, cup-shaped, palm upward, to the mouth; hand, as in type (**F** 1), modified by being a little more relaxed; both hands are then extended and the edges held vertically, palms facing, but hands apart; the hands in same relative positions are then moved to and from the body. (*Oto* I.) "That we drink; flowing and retained between banks."

Collect the fingers of the right hand to a point and bring it to the mouth, palm first, then wave the flat hand, palm down, horizontally from right to left. (*Wyandot* I.) "Broad water."

Raise hands to sides, fingers extended, palms oblique (**X** 1, with palm oblique), then at same time move both on parallel lines as far as arms can reach, showing a trough; then place right hand three or four inches from mouth, palm upward and hollowed as though holding water, and move it quickly past the mouth, resting in last position just in front of chin. (*Sahaptin* I.) "Water running in a trough."

Put hand in front of mouth, palm upward and hollowed as though holding water, and move it past the mouth, resting in front of chin; then from a point in front of breast make winding movement to a point as far from the body as arm will reach, fingers naturally relaxed, (**Y** 1, palm vertical), as though tracing course of stream. (*Sahaptin* I.)

Hold the right hand flattened and extended, with palm down, to the side of the right hip, then pass it forward toward the left side in a serpentine movement. (*Comanche* II.) "The hand represents the flat surface of the water, the movement the serpentine course."

Make the sign for **Water,** then hold the extended forefinger of the left hand diagonally in front of the body; palm down, then pass the point of the index along the left from the base of the palm to the tip of the forefinger. (*Ute* I.)

Hold the right hand flat and extended at the height of the hip, and push it, palm downward, in a serpentine manner diagonally across toward the left. (*Apache* I.)

Deaf-mute natural signs.—An undulating motion of the hand. (*Ballard.*)

Move the forefinger forward in a circle just below the mouth, indicating the motion of rolling waves, and then point to the place of it. (*Hasenstab.*)

Raise the hand toward the mouth and then move the same hand in a line showing the flowing of the water. (*Larson.*)

——— Across a.

With the forefinger of the right hand describe near the ground a wavy line in the direction of the geographical course of the stream indicated, and then with the same finger describe a short, straight line across the former and from the direction of the journey. (*Dakota* II.)

——— Headwaters or source of a.

Hold the extended and flat left hand vertically before the body at the height of the elbow, then point to the palm with the index and make the sign for **River** away from the left hand with the right. (*Apache* I.)

Road.

Having the opened hands eight inches apart, pointing forward, palms upward, in front of the chest, move them, each one alternately, back and forth about eight inches. (*Dakota* IV.)

Both hands flat and extended, pointing forward from the chest, palms upward, thumbs an inch higher than the outer edges which are placed nearly together, in this position pass the hands forward nearly to arm's length. (*Kaiowa* I; *Comanche* III; *Apache* II; *Wichita* II.) "From the depression usually indicating a traveled trail."

——— On, or in the middle of a.

First make the sign for **Road**, extend the left forefinger pointing forward and to the right, then place the tip of the extended index, pointing downward, upon the second joint of the forefinger. (*Kaiowa* I; *Comanche* III; *Apache* II; *Wichita* II.)

——— Wagon.

Make the wagon sign, and then wave the hand along the ground. (*Burton.*)

With the right hand, forefinger extended and pointing downward, other fingers closed (**J** turned downward), describe from its natural position outward in front and to the left of the body as far as the arm can be extended the winding course of a prairie wagon-road. (*Dakota* I.) "From the winding course of roads."

Robe. See **Clothing.**

Rocky (as a hill).

An ascending motion of the extended right hand, fingers joined, palm down, toward and over the clinched left, which is constantly interposed as an obstacle. If impassable, the right hand should not pass over the left, or the sign may be completed in this manner and that of *negation* added. (*Arapaho* I.)

Round up Cattle, To. See **Cattle.**

Run, Running. (Compare **Walking.**)

The arm nearly doubled upon itself, and then the elbow thrown forward and backward, as in the act of running. (*Long.*)

Both hands, fists (**B**), carried upward on their respective sides to the level of the shoulders and then make the upward and downward motions from the shoulders in imitation of their movements held in this position when running. (*Dakota* I.) "From the movements of the arms when running."

Extend and point both forefingers inward, the right three or four inches behind the left, palms backward, at the height of the breast, then, while moving the hands forward alternately, throw the right index over the left and the left over the right. (*Dakota* IV.)

Both arms are flexed and fists brought before body at center, about four inches apart. The hands in position (**B**) are then moved forward successively and alternately as an animal galloping or trotting. (*Oto* I.) "Motion of limbs in movement."

With one or both fists placed near the side of the chest, move them forward and backward as in running; the motion being at the shoulder. (*Kaiowa* I; *Comanche* III; *Apache* II; *Wichita* II.)

Deaf-mute natural signs.—A rapid motion of the feet on the floor. (*Ballard.*)

Use both arms and both open hands in a way similar to that in which the legs are used to run. (*Hasenstab.*)

Move the hands up and down in the two parallel lines from the breast. (*Larson.*)

——— Rapidly, swiftly.

Lay both hands flat, palm downward, and pass the right rapidly high and far over the left, so that the body is somewhat raised. (*Wied.*)

Sacred.

Right hand upright (**S**), palm toward the left, is moved straight upward and downward in front of the face. (*Dakota* I.)

Sad, Sorry, Troubled. (Compare **Angry.**)

The right hand is partly, sometimes quite, closed, held in front of the chest, almost or quite in contact with it, and made to describe a circle of three or four inches radius, once or oftener. (*Mandan and Hidatsa* I.) "This indicates the various conflicting emotions which may be referred to by an Indian when he says 'My heart is bad.'"

Both fists placed before the breast, palms down, thumbs touching, move the outer edges downward as if breaking a stick, then place the palm of the hand (or the fingers) over the heart. (*Wyandot* I.) "Broken heart."

Saddle.

First make the sign for **Horse,** then turn the left hand outward to the left, and extend the flexed fingers so that the hand is horizontal, palm downward (**W**), at the same time the fingers of the right hand are likewise extended, and then on a curve, first to the right, then upward and to the left, the hand is brought, with palm downward, nearly crosswise over the back of the left, which it pats gently several times. (*Dakota* I.) "Indicating saddle from its position on the horse."

The sign for **Horse** is first made by drawing the open hand edgewise across before the face; the front and middle fingers of the right hand then straddle the index and middle fingers of the left. Finally, in representation of the hand-made and reclining saddle-tree of the Indian, the two front fingers of the right are made to stand inclined on the front and back of the left. (*Oto and Missouri* I.) "Something astride of a horse that inclines before and behind."

——— Pack, To.

The clinched fist is held before the chest at a variable distance, the second row of knuckles to the front, usually, the forearm being semi-pronated so as to make the metacarpo-phalangeal articulation of the index the highest point of the hand. The fist is then struck by the palmar surface of the extended fingers of the right, first in the back and then on the side, or *vice versa.* (*Mandan and Hidatsa* I.)

Salt.

The body is bent forward, and the palm print of the right fingers, in type-position (**W**), touch the earth before the body. The open hands are then approximated before the body, palms in contact, then diverge, the right index touching the tongue in type-position (**K**). The countenance assumes a mien of dislike. The motion to the ground would indicate the salt-licks of the plains. (*Oto and Missouri* I.) "Something from the ground—savory to the taste."

Salutation. Hand-shaking.

"He shook hands"—with the party greeted—"and then pressed his own open bosom." A *Kansas* sign. (Indian sketches by John T. Irving, Philadelphia, 1835, vol. I, p. 61.)

It is noticeable that while the ceremonial gesture of union or linking hands is common and ancient in token of peace, the practice of shaking hands on meeting, now the annoying etiquette of the Indians in their intercourse with whites, was not used by them between each other, and is clearly a foreign importation. Their fancy for affectionate greeting was in giving a pleasant bodily sensation by rubbing each other's breasts, arms, and stomachs. The senseless and inconvenient custom of shaking hands is, indeed, by no means general throughout the world, and in the extent to which it prevails in the United States is a subject of ridicule by foreigners. The Chinese, with a higher conception of politeness, shake their own hands. The account of a recent observer of the meeting of two polite Celestials is: "Each placed the fingers of one hand over the fist of the other, so that the thumbs met, and then standing a few feet apart raised his hands gently up and down in front of his breast. For special courtesy, after the foregoing gesture, they place the hand which had been the actor in it on the stomach of its owner, not on that part of the interlocutor, the whole proceeding being subjective, but perhaps a relic of objective performance."

Same; similar to what is mentioned before. (Compare **Companion.**)

Place the two forefingers parallel to each other, and push them forward a little. (*Dunbar.*)

The two forefingers opened forward, laid side by side as in sign for **Companion,** and gently pushed forward. (*Cheyenne* I.)

Same sign as for **Companion.** (*Dakota* I.)

The forefingers of both hands extended, joined, backs upward, are carried outward from the breast on the same level for a foot. (*Dakota* I.) "No difference—both the same."

The hands are placed in the same positions as in **Ahead** and **Behind,** except that the forefingers are placed exactly side by side. If it is to be shown that two things are exactly alike or constantly alike or beside one another, the hands are moved forward together for a short distance. (*Mandan and Hidatsa* I.)

Extend forefinger and middle finger of right hand, pointing upward, thumb crossed over the other fingers, which are closed; move hand downward and forward. (*Omaha* I.)

With the forefingers only extended, place the hands in front of the chest, palms down, so that the extended fingers lie side by side. (*Kaiowa* I; *Comanche* III; *Apache* II; *Wichita* II.) "One like the other."

The forefingers only of both hands extended, placed side by side before the body, palm down. (*Apache* I.)

Italian sign—Lay the two forefingers together side by side. (*Butler.*) "Union or harmony."

Satisfied. See **Glad.**

Saw.

Same as the sign for **Cheyenne Indian.** (See TRIBAL SIGNS.) (*Dakota* I.) "From the use of the saw."

Scalp.

Grasp the hair with the left hand, and with the right one flattened cut away over the left. (*Wied.*) Still in use. (*Matthews.*) Although *Wied's* sign seems inexplicit in description, there is a remarkable similarity in the execution and conception between that and the (*Oto and Missouri* I.) (*Boteler.*) "That part removed by the knife, as represented."

The left hand stationary, eighteen inches in front of stomach (**D**), as though grasping the scalp-lock, and then the right hand (**X**), with palm upward, fingers extended, pointing obliquely toward the left, is passed with a backward or inward motion under the left hand from in front of it, just as though drawing the knife inward in scalping. (*Dakota* I.) "From the act of scalping."

Rapidly carry the left hand to the front of the upper part of the chest and close it, back forward, as if grabbing the hair; then draw the right hand, palm downward, from left to right beneath it, as if cutting. (*Dakota* IV.)

The left hand is raised to the vertex of the head and seizes the hair called by the Indians the scalp-lock; thus firmly held, the right is raised and edgewise executes a severing sweep around the forehead. (*Oto and Missouri* I.) "That which is removed as represented."

Grasp the hair on the top or right side of the head with the left hand then draw the flat right hand with the edge toward and across the side of the head from behind forward. (*Pai-Ute* I.)

Scarce, Few.

Place the hand in the position given for **Come,** when it is moved from side to side, arrested in its motion at intervals, and where so arrested is depressed an inch or two. (*Mandan and Hidatsa* I.)

Scissors.

With the fore and middle fingers imitate the opening and shutting of the blades of the scissors. (*Long.*)

Search for. See **Hunting.**

Secret; To secrete. See **Hide.**

See; Seeing; Saw. (Compare **Look at.**)

The forefinger, in the attitude of pointing, is passed from the eye toward the real or imaginary object. (*Long.*)

Pass the extended index-finger forward from the eye. (*Wied.*) Same as my description, but briefer. (*Matthews.*)

Strike out the two forefingers forward from the eyes. (*Burton.*)

Two fingers projecting. (*Macgowan.*)

Place the fore and middle fingers (of the right hand usually), separated, extended, and pointing outward, in front of the eyes, indicating the direction of supposed lines of sight. (*Arapaho* I.)

Pass the extended index forward from the eye. (*Absaroka* I; *Shoshoni and Banak* I.)

Same as **Look, To.** (*Dakota* I.)

With the index and middle fingers of the right hand extended, and their ends separated about two inches, point forward at the height of the eyes, the other fingers to be closed and the thumb on them, back of hand upward. Hold the hand still or move it forward a few inches. (*Dakota* IV.) "Two eyes."

With the fingers of the right hand as for **Seeing,** move the hand from side to side several times at the wrist, describing a curve. (*Dakota* IV.)

Extend the index and second finger of the right hand, and move them horizontally forward from the eyes. (*Dakota* VI, VII.)

The right hand, held as an index, is placed near the right ear, its back almost or quite touching the cheek, and is then moved forward. (*Mandan and Hidatsa* I.)

Close the right hand, leaving the index (or both index and second fingers separated) extended, pass from the eye forward, the finger tip pointing in the same direction. (*Kaiowa* I; *Comanche* III; *Apache* II; *Wichita* II.)

Another: Draw a circle around the eye with the extended index, or with both index and second fingers. (*Kaiowa* I; *Comanche* III; *Apache* II; *Wichita* II.)

Close the third and little fingers of the right hand, lay the thumb over them, separate the extended index and second fingers as far apart as the eyes, bring the hand to the eyes, fingers pointing outward, and pass the hand outward. (*Wichita* I.)

Deaf mute natural signs.—Open the eyes wide and strain them at vacancy. (*Ballard.*)

Point the finger to the eye. (*Larson.*)

Move the open hand up and down successively in front of the eyes. (*Hasenstab.*)

Place the forefinger on the eye as if to see something. (*Zeigler.*)

——— One another.

Both hands closed with the palms facing, forefingers straight, flexed at metacarpal joint so that they are horizontal and pointing toward one another at a distance of eight or ten inches; sometimes slightly moved to and from one another. (*Absaroka* I; *Shoshoni and Banak* I.) "Sign of **To See** with *both* hands, as representing *two* individuals."

Seen, or **Discovered.**

The sign of a **Man** or other animal is made; after which the finger is pointed toward and approached to your own eyes. It is the sign for **Seeing** reversed. (*Long.*)

Same as **Found.** (*Dakota* I.)

This is made in a manner the reverse of **See.** (*Mandan and Hidatsa* I.)

Deaf-mute natural sign.—Nod, having touched the eye. (*Cross.*)

Shame.

Both hands to front of face, hand and fingers upward, back outward (**S**), pass the left hand slowly from left to right in front of the eyes, and the right in same way to the left. (*Cheyenne* II.)

Head inclined forward and downward, eyes looking directly downward; place the upright hands, with backs outward (**S**), about three or four inches in front of face so as to hide it from view as much as possible. (*Dakota* I.) "From covering the face to hide the shame."

Both hands flat, with extended fingers joined; place the left outward before the left cheek, pointing upward and backward toward the right side of the crown, and the right several inches from and before the left, pointing upward and forward, backs outward, the face at the same time being turned toward the left. (*Kaiowa* I; *Comanche* III; *Apache* II; *Wichita* II.)

Sheep.

Right hand held forward from the lower part of the right side of the abdomen, palm down, arched, fingers slightly separated, and make arched interrupted movements forward. (*Ute* I.) "Manner of the movement of the animal while grazing."

——— Ewe.

The right hand, extended and slightly arched, held before the body, about two feet from the ground; then push it forward over a slight arc once or twice. (*Apache* I.) "Illustrates the animal's height and interrupted manner of moving forward while browsing."

——— Mountain; Bighorn. (*Ovis montana*, Rich.)

Move the hands in the direction of the horns on both sides of the head by passing them backward and forward in the form of a half circle. (*Wied.*) This sign is still in use. (*Matthews.*)

Place the hands on a level with the ears, the palms facing backward and the fingers slightly reversed, to imitate the ammonite-shaped horns. (*Burton.*)

Move the opened hands backward, one on each side of the head above the ears, palms inward, fingers slightly flexed and pointing backward. (*Dakota* IV.)

Place the right hand at the height of and straight forward from the elbow, palm downward, close the two middle fingers, extend and crook the index and little fingers, at the same time extending the thumb so that it passes downward and below the closed fingers. (*Ute* I.) "Curved horns and nose is represented."

——— Wether.

Make the sign for **Sheep, ewe;** then place the arched left hand transversely in front of the body (pointing toward the right) and nearly close the right, and make a movement from the left backward and downward toward the body as if drawing a rope. (*Apache* I.) "Height of the animal, walking as it grazes, and the long tail."

Shield.

Is shown by pointing with the index over the left shoulder, where it is slung ready to be brought over the breast when required. (*Burton.*)

Both hands made to describe a circle slightly to one side and in front of the body. (*Cheyenne* I.)

Shoes, moccasins.

Raise the foot and stroke it from front to back with the index-finger of the hand on the same side. (*Wied.*) I have seen this sign. (*Matthews.*) The similarity between the (*Oto and Missouri* I) sign and *Wied's* exists more in the idea or conception than the motion exerted. It is not probable that they ever were identical in execution. (*Boteler.*)

Draw the foot upward and incline the body forward so that the hands can reach the foot when the drawing on of the shoe or moccasin is imitated. (*Dakota* I.) "From the drawing on of the shoe."

Stoop and, with the fingers in the same position as for **Dress, tunic**, excepting that the forefingers are to point downward and the thumbs nward, move the hands from the toes backward through slight curves, one on each side of one of the feet. (*Dakota* IV.)

Another: Make the same sign above the foot without stooping. (*Dakota* IV.)

Both hands in type-position (**W**) are approximated at the points of the index-fingers before the toes of either foot. Then the hands diverge and describe a curve around the sides of the foot to the heel, from which point both hands are pulled suddenly upward. The sign is perfect, representing the pointed oval of the moccasin and the use of both hands in drawing them over the heel. For **Boots** the last motion is extended up on both sides of the limb to midway the foreleg. (*Oto and Missouri* I.) "That which incloses the foot and is drawn on."

Shoot; shot. See, also, **Arrow** and **Gun.**

——— Discharge of a deadly missile.

The hand is clinched in such a way that the thumb covers the nails of the other digits; the forefingers are then suddenly extended as in the act of sprinkling. This is much like the sign for **Bad,** but here the arm is not moved and the fingers not strongly flexed. (*Mandan and Hidatsa* I.)

——— Struck by a deadly missile.

The left hand is held before the chest at a convenient distance, thumb upward, back outward, fingers slightly bent, and is struck in the palm with the back of the clinched right fist. (*Mandan and Hidatsa* I.)

Short, In stature.

A short person is described with right hand brought up as high as the head on right side, forefinger straight upright (**J**, back outward); move the hand down, keeping fingers upward till it reaches the waist or below. The body is usually bent to the right a little in the movement as the hand goes down. *Short distance* is described the same as **Close** or **Near.** (*Cheyenne* I.)

——— In extent.

Place the hands, palm to palm, a short or the required distance apart. (*Arapaho* I.)

——— Curtailed.

The arms are semiflexed before the body; the hands approximated at palms, then made to diverge to indicate some length. The right hand then approximates the left and, edgewise, imitates a cutting-off of a short piece of the finger-ends. The word does not seem to be well understood by the Indian unless applied to some object, in which case there is a compound sign. (*Oto and Missouri* I.) "Length reduced by cutting off."

——— A little, short extent, or time, according to connection.

Raise left hand to position in front of body, forefinger extended horizontal (**M** 1, changed to left); then raise right hand, first finger extended (**M** 2); place end of the finger near end of forefinger on left hand, and move it slowly up the finger, resting near its base or near base of thumb. (*Sahaptin* I.)

Only by less separation of hands. (*Apache* III.)

Sick; ill.

Hold the flattened hands toward one another before the breast, bring them, held stiff, in front of the breast, and move them forward and backward from and to the same. (*Wied.*) As is evident, no similarity of execution or design exists between *Wied's* sign and the (*Oto and Missouri* I.) (*Boteler.*)

Contract shoulders and chest, bring hands in front of throat and chin, with a shrinking, contracting motion and a shiver (if ague) or blow short breaths as if panting; then carry the left hand to the forehead and press, indicating headache. (*Ojibwa* IV.)

Touch the part that is the seat of the pain and then withdraw quickly the touched limb, or flinch at the pressure made on the part, at the same time the emotions of the face express suffering. (*Dakota* I.) "From the fear of increased pain by pressure."

The open, relaxed hands are to be quickly thrown about four inches forward and outward, several times, in front of the stomach; fingers spread a little, ends about four inches apart, palms backward. (*Dakota* IV.) "The pulsation of the heart."

Assume an appearance of distress, with general features relaxed. Both arms are then elevated, semiflexed, and the hands assume the type position (**Q** 1), modified by fingers being more curved and less rigid. The hands are brought tremblingly thus to the sides of the body, chest, etc., and then raised to the forehead and the extended indices made to compress the temples. (*Oto and Missouri* I.) "That which produces inward or bodily distress."

Place the tips or ends of the extended fingers and thumb gently over the heart, leaning the head slightly toward the left, accompanied by a drooping or closing of the eyelids. (*Ute* I.)

Another: Collect the fingers and thumb of the right (or left) hand to a point, and place the tips alternately to the right and left sides of the chest, accompanied by a simultaneous dropping of the head, with the eyes partially or entirely closed. (*Ute* I.) "Location of pain."

Deaf-mute natural signs.—Place the hand upon the breast and protrude the tongue. (*Ballard.*)

Place your palm on the forehead and shudder. (*Cross.*)

Place the open hand on the forehead, and then move the head down, with the mouth half opened. (*Hasenstab.*)

Put forth a part of the tongue out of the mouth and at the same time raise the hand to the breast. (*Larson.*)

Place the hand on the breast, at the same time open the mouth as if to vomit. (*Zeigler.*)

——— Very.

Both hands flat, extended, and fingers joined, place against the cheek bones and withdraw slowly. (*Wyandot* I.)

Sign language.

Tap the back of one hand with the palmar surface of the fingers of the other, alternately and repeatedly, then close both hands, leaving the forefingers and thumbs fully extended and separated; place them about four inches apart, palms facing, and rotate them in short vertical circles, in such a manner that when the right hand occupies the upper portion of its circle the left will be below. (*Kaiowa* II; *Comanche* III; *Apache* II; *Wichita* II.) "Hands and conversation."

Silence.

Lay the extended index, pointing upward, over the mouth so that the tip extends as far as the nose, or alongside the nose. (*Shoshoni and Banak* I; *Ute* I.)

Sing, to.

Right-hand fingers and thumb partially unclosed, placed in front of the mouth, shot upward, and slightly shaken. (*Cheyenne* I.)

Ball of the right hand resting on the chin, fingers extended obliquely upward and toward the left, as though catching the words thrown out of the mouth. (*Dakota* I.) "Catching the words."

Move the right hand through a small circle in front of the mouth, back forward, fore and middle fingers spread a little, extended and upright, other fingers closed, thumb on middle finger. (*Dakota* IV.) "Opening and closing the mouth and the sounds coming forth."

(1) Put thumbs and forefingers spread to make large circle; (2) beat tip of right hand several times towards it; (3) wave forefinger several times quickly from lips upward. (*Apache* III.) "(1) Drum; (2) beating accompaniment on it; (3) singing."

Deaf-mute natural sign.—Put one finger to your mouth, opening and shutting it alternately; set in motion your arms, after the manner in which a singer acts. (*Cross.*)

Singing, Sacred.

Move the upright right hand in a circle in front of the mouth, the fingers slightly bent and separated so as to form a circle, back outward. (*Dakota* IV.)

Sister. See **Relationship.**

Sit down.

The fist is clinched, and the motion of it is then the same as if it held a staff and gently stamped it upon the earth two or three times. (*Long.*)

Make a motion toward the ground, as if to pound it with the ferient of the closed hand. (*Burton.*)

Quickly lower the extended hand, palm down, indicating spot and action. (*Arapaho* I.)

Right hand held to one side, fingers and thumb drooping, struck downward to the ground or object to be sat upon. (*Cheyenne* I.)

Shut both hands, thumbs up (or above), raise hands a little, and lower at same time with a squatting or sitting motion of body if the person giving the sign is standing; if sitting, point to the place and make motion with the hands and arms. (*Ojibwa* IV.)

Right hand clinched, outer edge downward, and pushed toward the the ground. (*Absaroka* I; *Shoshoni and Banak* I.)

Another: Make the sign of **To sit,** but make it toward the spot indicated for the visitor to occupy. (*Absaroka* I; *Shoshoni and Banak* I.)

Right hand in natural position, fingers closed (**A**), palm upward, extend the forearm, with elbow fixed, straight toward the front (**L**), and carry it toward the ground or seat. (*Dakota* I.) "Sitting down on a chair or the ground."

Incline the body forward and move the right fist downward about eighteen inches from in front of the stomach, at arm's length forward, bent upward at the wrist, and back outward. (*Dakota* IV.) "Down in a bunch."

The clinched fist, thumb upward, is held outward, usually to the right, the elbow forming nearly a right angle, the hand is then depressed and suddenly arrested. This is a modification of the sign for **Stay,** or **Abide.** When the sign is made imperatively the arm is sometimes stretched toward the place where it is desired that the person addressed shall sit. Sometimes a particular spot, mat, or seat, if convenient, is struck with the fist in making the sign. (*Mandan and Hidatsa* I.)

Deaf-mute natural sign.—Point at the place where you wish the person (spoken to) to sit, and make the motion of sitting. (*Cross.*)

——— and smoke.

Used as an invitation to a visitor, and is made by carelessly pointing to the individual, to indicate person, then make the sign **Sit down** and **To smoke.** (*Absaroka* I; *Shoshoni and Banak* I.)

——— Australian sign. See **Wait.**

Slave.

Slave is described the same as a **Captive.** The only persons used as slaves, or so considered, in wild tribes of plains are captives. Mexican children have been often taken as well as young people of hostile tribes. The right hand clinched (**C** with palm forward), upright, on a level with and to the right of right shoulder, is clasped around the wrist by the fingers and thumb of the left hand with back of hand (left hand) to front, and pull the right hand to the front twelve or fifteen inches. (*Cheyenne* II.)

Sleep, sleeping. (Compare **Night.**)

Point to the ground and make a motion as if of lying down; then close the eyes. (*Burton.*)

Close the eyes and incline the head, the cheek resting upon or supported by the extended hand. Time may be indicated by this means; "one sleep" (the sign for sleep and one finger touched or held up alone) being the equivalent of twenty-four hours or a day. (*Arapaho* I.)

Right hand, palm inward, placed by the side of the head; head drooped to the right, as if to fall into the open palm, and eyes partially closed. This also means to go to bed. (*Cheyenne* I.)

Another: Forefinger of right hand crooked and placed against or near the upper lid of the eyes; very short motion downward and outward. (*Cheyenne* I.)

The head inclined sidewise toward the right, against the palm of the right hand with fingers separated (**P**). (*Dakota* I.) "Head supported by a pillow."

Close the eyes, incline the head toward the right, and lay it in the opened right hand. (*Dakota* IV.)

The arm is brought to the side of head, with hand in position (**T**), and head inclined to right shoulder, resting in palm, eyes closed. (*Oto* I.) "Rest."

Incline the head to one side, close or partly close the eyes, and place the flat hand to within about six inches of the ear. (*Pai-Ute* I.)

Deaf-mute natural signs.—Place the hand upon the cheek, inclining the head to one side, and closing the eyes. (*Ballard.*)

Close your eyes and bend your head sidewise on the open hand. (*Cross.*)

First place the open hand on one side of the head, next move the head, the eyes having been shut, down to the side, and then point to the place to sleep. (*Hasenstab.*)

Close the eyes. (*Larson.*)

Shut the eyes, and incline the head as if to sleep. (*Zeigler.*)

Italian sign.—Lay the open hand under the cheek. (*Butler.*)

——— To sleep with another.

The person is first indicated by pointing, then place the forefingers of each hand side by side in front of the breast, back upward, at the same time inclining the head a little to the left and partially closing the eyes. (*Dakota* V.)

Cross both closed hands and arms before the breast as if in an embrace, then lay the extended index and forefinger side by side, palms down, pointing forward, and move them over toward the right so that the backs of the hands point downward toward the right at the termination of the sign. (*Ute* I.)

Sleepless.

The head is held nearer the middle-line than in the sign for **Sleep.** The hand is then raised in position (**N**), and made to quiver with palm outward. (*Oto* I.) "Unrest."

Slow.

Extend the left arm, curving the forefinger and holding it still. The right arm does the same but is drawn back with several short and circular movements. (*Wied.*)

Both hands in front of breast with fingers extended, &c., as (**W**), pass the right hand forward over the back of the left *slowly*. (*Dakota* I.) "Slow in motion."

The hands, four to eight inches apart, about a foot in front of the lower part of the chest, with the forefingers extended, pointing forward and backs upward, should be slowly lowered about eight inches and at the same time separated by bringing the elbows to the sides. The other fingers are to be nearly closed, thumbs against the middle fingers or under them. (*Dakota* IV.) "Going backward, and therefore slow."

Deaf-mute natural signs.—A slow, horizontal movement of the hand (*Ballard.*)

In reference to walking, walk slowly for a little distance; to sewing, slowly copy such a manner as a dressmaker actually does, and so on. (*Cross.*)

Small; Little; a few; small amount. (Compare **Nothing.**)

Pass the nearly closed hands several times by jerks over one another, the right hand above. (*Wied.*) There are various signs for **Little,** depending on the nature of the object described. I have given you one. I do not remember this of the Prince of Wied. (*Matthews.*)

(1) Fingers and thumb of both hands closed, hands bent backward from the wrist, and thus (2) crosses right above the left before the breast. (*Cheyenne* I.)

First lay the open hands on the body, backs outward, and then make the sign for **Man,** or the animal or thing to which the sign is to be applied, and then close the hands, fists (**A** 1), left outside of the right and about a foot and a half in front of the left breast, and the right held just in front of the left breast; carry the left hand inward, and the right hand outward, to the body on a curve until the right fist is over the left. (*Dakota* I.) "Denotes small in body or stature."

Place the right fist or half-closed hand about three inches above the left, in front of the navel, radial side of the fists upward; then bend the hands backward as far as possible at the wrists, and move the right wrist over the left, at the same time turning the palms a little upward. (*Dakota* IV.) "So small or so little that it can be held in the closed hands."

The extended forefinger of the left hand (usually erected) is pinched near its extremity between the thumb and index-finger of the right hand. The degree of smallness is to some extent shown by the height of that portion of the left forefinger which appears above the right thumb-nail. For extra demonstration the eyes are often partly closed and the forefinger pinched tightly. (*Mandan and Hidatsa* I.)

The thumb and front fingers of the right hand are collected and, as in type-position (**G**), are made to grasp something; or both hands in like position are held parallel, facing each other. (*Oto and Missouri* I.) "That contained between the finger-ends."

With the forefinger only extended, place the inner edge of the extended index about half an inch from the tip of the forefinger. (*Ute* I.)

Extend the thumb and index, bringing their palmar surfaces to within half an inch of one another, the remaining fingers closed or nearly closed. (*Apache* I.) "The positions of the fingers are the same as if holding a very small body."

Hold imaginary object between left thumb and index; point (carrying right index close to tips) to the last. (*Apache* III.)

——— In size and also in quantity.

Right-hand in front of the body, mark off on the index-finger, with the thumb, a small portion of it, other fingers closed. (*Dakota* I.) "A portion of anything, a small amount."

Deaf-mute natural signs.—Put one forefinger upon the other a little way from the tip. (*Ballard.*)

Place the tip of one forefinger on the first joint of the other, and then half open the eyes, and move the lips from each other, while the upper and lower teeth are kept toward each other. (*Hasenstab.*)

Put the open hands together. (*Larson.*)

Use the teeth as if to press the end of the tongue between. (*Zeigler.*)

Smell.

Touch the nose tip. A bad smell is expressed by the same sign, ejaculating at the same time "Pooh!" and making the sign of **Bad.** (*Burton.*)

Fore and middle fingers of right hand placed at or near the nostrils, drawn downward and forward with slightly curved motion. (*Cheyenne* I.)

Fore and second fingers of right hand extended (others closed) (**N**), carried directly to the nose and then forward for a few inches in front of the nose, fingers pointing obliquely downward. (*Dakota* I.) "From the act of smelling."

Deaf-mute natural signs.—Hold the nose with thumb and forefinger; or imitate the act of sniffing. (*Ballard.*)

Point at the nostril with one finger, with a displeased or pleased expression. (*Cross.*)

Place the forefinger beneath the nostril, at the same time raising the upper lip several times in succession, as if to smell something. (*Zeigler.*)

Smoke.

Begin with the sign for **Fire,** then raise the hand upward, with the fingers open as if to represent smoke. (*Dunbar.*)

Snuffle the nose and raise the fingers of both hands several times, rubbing the fingers against each other. (*Wied.*) The rubbing suggests the old mode of obtaining fire by friction, and the wrinkling or snuffling of the nose indicates the effect of the smoke on that organ.

With the crooked index, describe a pipe in the air, beginning at the lips; then wave the open hand from the mouth to imitate curls of smoke. (*Burton.*)

Similar to the sign for **Fire,** the fingers still and the hand ascending by a constantly revolving motion. (*Arapaho* I.)

Sign for **Fire** made slowly. (*Cheyenne* I.)

Clinch the right hand, and hold it, palm toward the left and downward, about twelve inches in front of the lower portion of the chest. (*Absaroka* I; *Shoshoni and Banak* I.) "Holding the pipe."

Same as the sign for **Fire** with the hand carried up higher. (*Dakota* I.) "From the ascent of the smoke."

Make the sign for **Fire,** and then hold the opened upright hands, fingers a little spread, side by side, in front of the face. (*Dakota* IV.) "Hides everything."

Close both hands, place them side by side toward the ground, palms downward, then raise them quickly, extending the fingers and thumbs in doing so, and make spiral curves upward a short distance. (*Ute* I.)

Snake.

The forefinger is extended horizontally, and passed along forward in a serpentine line. (*Long.*)

A gliding movement of the extended hand, palm down, fingers joined, in imitation of reptilian locomotion. (*Arapaho* I.)

Right hand forefinger pointing, placed in front of and on a level with left shoulder, drawn along to the right with undulating sinuous motion, imitating the motion of a snake crawling. (*Cheyenne* I.)

Extended forefinger of right hand (others closed) (**J** pointing downward instead of upward) in front of the breast, move it in imitation of the movements of the snake in crawling. (*Dakota* I.) "From the crawling of a snake."

The hand, held as an index hand, pointing forward, is held near the body in front and usually to one side; it is then advanced rapidly and with a tortuous motion, like that of a snake crawling. (*Mandan and Hidatsa* I.)

Same sign as that for **Shoshoni Indian.** (See TRIBAL SIGNS.) (*Comanche* II; *Pai-Ute* I.)

With the index only extended, palm down and the hand at the right hip, pass it forward and toward the left, moving it from side to side (in a serpentine manner) in doing so, the motion being made at the wrist. (*Apache* I.)

Deaf-mute natural signs.—A zigzag motion forward with the forefinger. (*Ballard.*)

Move the arm in a serpentine form. (*Larson.*)

Point with the forefinger as if to point to something, at the same time move it crookedly, and also at the same time stretch the tongue and move it to and fro like a pendulum. (*Zeigler.*)

Snow. (Compare **Frost.**)

Begin with the sign for **Rain**, then the sign for **Air** or **Cold**, and conclude with the sign for **White.** (*Dunbar.*)

The hand is held up about as high as the head, with the fingers suffered to dangle downward; it is then bobbed a little up and down, as if to throw off drops from the ends of the fingers. (*Long.*)

Scatter the fingers downward. The same sign denotes rain. (*Burton.*)

Imitate its fall with the hand, palm down, partially closed, fingers separated and pointing downward. To indicate **Rain,** the hand is moved in a direct course toward the ground; **Snow,** the hand moves other than in a direct course to show drift, lighter fall, &c. (*Arapaho* I.)

Same sign as for **Rain,** though the hands are moved in and out more than in that sign, as if covering the body. (*Dakota* I.) "From the falling of the snow."

The same sign as for **Rain ;** but when it is necessary to distinguish it from rain, it must be preceded by the sign for **Cold,** which makes it the same as winter. Often the sign for **Rain** or **Snow** is made with one hand alone. (*Dakota* IV.)

Make the signs for **Rain** and **Deep.** (*Dakota* VI; *Hidatsa* I; *Arikara* I.)

The hand in position of sign for **Rain** is moved downward slowly and with a wavering motion. (*Mandan and Hidatsa* I.)

The face is cast inquisitively toward the sky and the arms and clothing collected around the body as when one is chilly. The right hand is then raised above the head with the fingers collected much as in type-position (**H** 1), modified by finger ends being held a little more curved. The hand then falls by jerks, opening and closing successively. (*Oto and Missouri* I.) "Something falling that makes us chilly."

The hands are held as in the sign for **Rain,** but are then moved down toward the ground and outward to either side. Literally, "deep rain;" rain being indicated, the depth is shown by passing the hands outward toward their respective sides. (*Wyandot* I.)

Place the right hand as high as the head, in front or toward the right side, palm down, moving it quickly up and down several times for a short distance, then indicate the depth upon the ground with the flat hand, palm earthward. (*Apache* I.)

Make the sign for **Clouds;** then the hand descends from above the head (**Q**), tips down; when near the earth wave the hand. To show depth of snow on earth spread both hands, palms down (**W**). (*Apache* III.) "Represents the varying motion of snow-flakes."

Deaf-mute natural signs.—Point to the shirt bosom, signifying the color white, and move up and down the extended fingers. (*Ballard.*)

Do the same as to say *rain*, except point with the forefinger to some object that is white, indicating the whiteness of snow. (*Hasenstab.*)

Put the hands toward the breast and shake the body, and then move the outstretched hands upward and downward. (*Larson.*)

Soap.

The right-hand clinched (**D**), is rubbed on the left forearm, just above the wrist. (*Dakota* I.) "From its use in washing clothes."

Soft.

Open the left hand and strike against it several times with the right (with the backs of the fingers) [which also means **Hard**]; then strike on the opposite side so as to indicate the reunion. (*Wied.*) The supposed yielding substance is restored by the second stroke to its former shape.

Take some soft body in the hand, and touch and handle it lightly, alternately with each hand, held as though molding it into a round ball. (*Dakota* I.) "Handling the substance gently, so as not to injure it."

With the hands three or four inches apart, pointing forward, palms downward, fingers relaxed, lower them about four inches slowly and raise them rather rapidly several times. (*Dakota* IV.) "Yields and springs back; therefore is soft."

With the finger and thumb of the right hand approximated to a point, pretend to pick some pulverulent substance from the palm of the left, keep working the tips of the right as if allowing the contents to fall slowly back again into the left. (*Kaiowa* I; *Comanche* III; *Apache* II; *Wichita* II.)

Made only by rubbing folds of cloth (flexibility), or imitating the crumbling of bread. (*Apache* III.)

Deaf-mute natural sign.—Squeeze softly the clinched hand. (*Cross.*)

Soil. See **Earth.**

Soldier (American).

Pass each hand down the outer seam of the pants. (*Sac, Fox, and Kickapoo* I.) "Stripes."

Sign for **White Man** and then for **Fort.** (*Dakota* I.) "From his fortified place of abode."

Extend the fingers of the right hand; place the thumb on the same plane close beside them, and then bring the thumb side of the hand horizontally against the middle of the forehead, palm downward and little finger to the front. (*Dakota* II.) "Visor of forage cap."

The nearly closed hands, thumbs against the middle of the forefingers, being placed with their thumbs near together in front of the body, palms forward, separate them about two feet. (*Dakota* IV.) "All in a line in front."

Another: First make the sign for soldier, then that for **White man.** (*Dakota* IV.)

Place the radial sides of the clinched hands together before the chest, then draw them horizontally apart. (*Dakota* VI.) "All in a line."

Place the flat and extended right hand, palm downward, horizontally against the forehead. (*Ute* I.) "Visor of the cap."

——— Arikara.

Make the sign for **Arikara,** and that for **Brave.** (*Arikara* I.)

——— Dakota.

Make the sign for **Dakota,** and that for **Soldier.** (*Dakota* VI.)

Soldiers coming,

Both hands extended, fingers spead, place obliquely upward and in front of the breast, right above left a short distance; moved alternately and successively from right to left. (*Ute* I.) "The movement of bayonets in a charge."

Some. (Part of a number of persons or objects.)

Extend the index, hold the palm down, and imitate the motion of indicating different individuals or articles from left to right. (*Kaiowa* I; *Comanche* III; *Apache* II; *Wichita* II.)

Son. See **Relationship.**

Soon. See **Time, Soon.** (Compare **Near.**)

Sorrow. (Occasioned by filial disrespect.)

Right hand next to the heart, palm in, fingers slightly curved; then make a circular movement forward and outward toward front. (*Omaha* I.)

Sorry. See **Sad.**

Soup.

Sign for **Kettle,** setting on the fire, and then that for **Drinking.** (*Dakota* I.)

Sour.

Simulate tasting anything sour, *i. e.*, act of tasting and expression of face. (*Arapaho* I.)

Tip of forefinger touched against the tip of the tongue; then make the sign for **Hard.** (*Cheyenne* I.)

Raise the right hand to the mouth, as though having the substance to be tasted in it, and then spit. (*Dakota* I.) "Not liking the taste."

Deaf-mute natural sign.—Make wry mouths. (*Ballard.*)

Touch the tongue, shaking the head, with a look expressive of displeasure. (*Cross.*)

Close the eyes a little and shake the head. (*Larson.*)

Space; extent.

The left arm and hand are extended. The right hand is then brought (as in **S** 1 modified by being horizontal) to left and drawn across left arm edgewise at successive points. (*Oto* I.) "That is composed of smaller parts; many added."

Speak; speech.

The motion is like sprinkling water from the mouth by springing the forefinger from the thumb, the hand following a short distance from the mouth at each resilience, to show the direction of the word, or to whom it is addressed; this motion is repeated three or four times. (*Long.*)

Place the flat hand, back downward, before the mouth, and move it forward two or three times. (*Wied.*) My description is the same as this, but more precise. I believe I said the thumb is held forward. A knowledge of this fact would be essential to one who wished to imitate the sign correctly. (*Matthews*) There is sufficient similarity to Wied's sign in the position of the hand and forward movement to justify a supposition of former identity between that and (*Oto and Missouri* I.) The curved position of the fingers in the latter sign is not invariable. (*Boteler.*)

Extend the open hand from the mouth. (*Burton.*)

Fingers used as if picking something from the mouth. (*Macgowan.*)

Point the extended forefinger as from the mouth. (*Arapaho* I.)

Forefingers of both hands crooked inward, as in making the sign for **Morning;** motion backward and forward from mouth. (*Cheyenne* I.)

The gestures by which "speaking" is described are made close to the mouth. If the hand is passed several times across the lips it means addressing the people, *Harangue.* If the fingers of both hands are crossed before the mouth like a pair of scissors, it means a *Dialogue.* (*Ojibwa* I.)

Same as the sign for **Sing,** excepting the hand is carried farther outward from the mouth. (*Dakota* I.) "Carrying the words out of the mouth."

Place the right hand just in front of the mouth, palm forward, index half flexed, other fingers closed, thumb against middle finger; move the hand at the wrist forward two or three times through an arc of about six inches, each time bringing the end of the index against the end of the thumb. (*Dakota* IV.) "Opening and closing the mouth, and the sounds coming forth."

Pass the tips of the fingers of the right hand forward from the mouth. (*Pai-Ute* I.)

Place the knuckles of the right hand against the lips, and make the motion of flipping water from the index, each flip casting the hand and arm from the mouth a foot or so, then bringing it back in the same position. (*Wichita* I.)

Place the flat right hand, palm up, fingers pointing to the left, a short distance before the chin, and move it forward. This is sometimes repeated three or four times. (*Dakota* VI, VII.)

The right hand, not very rigidly extended, palm upward, thumb forward, is held in contact with the lower lip; it is then moved forward a few inches, and restored to its original position. These motions are repeated once or oftener. (*Mandan and Hidatsa* I.)

The right hand is brought to the mouth, palm upward, index-finger crooked, the others somewhat collected, and hand slightly cup-shaped. The hand and arm is then extended from the mouth, opening and closing partly by successive and delicate jerks toward the person or object addressed. The position of the hand is not invariable, and the true origin of the sign seems to be more in the conception of something coming at intervals from the mouth. (*Oto and Missouri* I.) "Opening of the mouth and that which issues therefrom."

Close the hand, except the index-finger, and, first touching with this the mouth, move it forward, back upward, partly closing and opening the finger with a rapid motion. (*Iroquois* I.)

The right hand is held to the right side of the mouth, fingers pointing forward, palm down, when the fingers and thumb are slowly opened and closed, representing the opening and closing of the lips in speaking. (*Wyandot* I.)

Pass the right hand, palm up, forward from the chin. (*Shoshoni and Banak* I.)

Deaf-mute natural sign.—Move the lips as if to speak. (*Zeigler.*)

——— Another speaks.

Place the hand as in the sign for **Speak,** beginning farther from the mouth, drawing it nearer and nearer. (*Wied.*) I have seen this sign. (*Matthews.*)

——— Conversation.

Several repetitions of the sign for **Speak.** (*Arapaho* I.)

Make the same sign as **Tell,** but with both hands, and toward each other. (*Hidatsa* I; *Arikara* I.)

——— I will speak to you. An interview.

Right arm flexed at elbow, and hand collected as in type (**G** 1), modified by being inverted and palm turned up. The arm and fingers are then suddenly extended, after being brought to the position of the heart. (*Oto* I.) "Approach; I will open myself to you."

——— Talking (one person).

Throw the opened, relaxed, right hand, pointing forward, palm upward and inward, six or eight inches toward the left, several times. (*Dakota* IV.)

——— Two or more persons conversing.

Both hands being opened, relaxed, and pointing forward eighteen inches apart, palms upward and a little inward, move them inward until near together three or four times; or, having the hands near together, move them from side to side several times, turning the ends obliquely, first toward the right and then toward the left, moving them from the wrists alone, or moving forearms also. (*Dakota* IV.)

——— Or talk in council.

The right arm is raised, flexed at elbow, and the hand brought to the mouth in type-position (**G** 1, modified by being inverted), palm up, and the index-finger being more open. The hand then passes from the mouth in jerks, opening and closing successively; then the right hand in position (**S** 1), horizontal, marks off divisions on the left arm extended. (*Oto and Missouri* I.) "That which issues from the mouth continuously or in parts."

——— Tell me.

Place the flat right hand, palm upward, about fifteen inches in front of the right side of the face, fingers pointing to the left and front; then draw the hand inward toward and against the bottom of the chin. (*Absaroka* I; *Hidatsa* I; *Kaiowa* I; *Arikara* I; *Comanche* III; *Apache* II; *Wichita* I.)

Place right hand (**Y** 1), slightly strained at the wrist, as though holding something on it) at a point, say a foot from the mouth, and move it toward the mouth two or three times. All the motion by the forearm, the arm to the elbow lying against the side. (*Sahaptin* I). "Pouring in or being fed."

——— I have told you.

Move the opened relaxed right hand from the mouth straight forward about a foot, fingers pointing toward the left, palm upward. (*Dakota* IV.)

——— Told me, a person.

Reverse the movement of the right hand as given in the sign for **Talk,** *i. e.*, the hands drawn inward toward the face, as though catching the words as uttered by another person, and carrying it to your own mouth. (*Dakota* I.)

Deaf-mute natural sign.—A rapid motion of the lips as if in the act of speaking, and move the finger to the bosom after some sign for the person telling. (*Ballard.*)

Spear. See **Lance.**

Spoon.

Right hand in front of body with thumb and forefinger bent in resemblance to the shape of the bowl of a spoon as much as possible. (*Dakota* I.) "Bowl of a spoon."

The left arm is elevated and semi-extended, the index-finger and thumb are approximated at ends, as in position (**H**), other fingers are closed. The right hand is then made to scoop downward and inward, with the index and middle fingers approximated and curved, palm inward. The ring and little fingers are closed. The right hand then approaches the left in the above position, which is now taken to the mouth. (*Oto and Missouri* I.) "The shape, size, and use of the instrument is indicated."

Spotted.

With the extended index make repeated transverse cuts across the extended forefinger of the left hand. (*Absaroka* I; *Shoshoni and Banak* I.)

Hold the left hand with its palm inward and fingers pointing forward, and alternately draw the palms and the backs of the right fingers across its upper edge several times from left to right. Or draw them across the hand and arm at different places as if wiping off the fingers. (*Dakota* IV.)

Extend the left forearm horizontally, pointing forward, then pass the left palm alternately across it from below upward (but not touching it) on the inner and outer sides. (*Hidatsa* I; *Arikara* I.)

The sign for the animal or thing is made first, then the arms are flexed, hands brought together in front of body, opened in full, flat, palm of one on back of other—a cross duplicate of position (**W**). Flat surfaces then pass horizontally over each other. (*Oto* I.) "That which has been rubbed or blurred."

Spring (season).

The sign for **Cold**, to which add the sign for being **Done** or **Finished.** (*Dunbar.*)

Signs for **Day** (or **Daylight**) and **Grass.** The seasons may also be distinguished by indicating a greater or less meridional altitude of the sun. (*Arapaho* I.)

With the right-hand fingers and thumb curved upward and separated (**P** with knuckles and back downward) beginning with the hand in this position as low down on the right side as you can reach by bending the body a little, then bring the hand up a few inches, keeping fingers up. (*Cheyenne* II.) "Represents grass growing."

Make the sign for **Grass.** (*Dakota* I; *Kaiowa* I; *Comanche* III; *Apache* II; *Wichita* II.) "From the season the grass springs up."

Make the sign for **Horse,** (**Riding a horse**); and then hold the left hand, palm looking obliquely downward and backward, in front of the abdomen, and pass the right hand, back upward, underneath it from behind forward; or, make the sign for **Offspring,** (*Dakota* IV.) "The mares have colts."

Make the signs for **Rain, Grass,** and **Good.** (*Dakota* VI.)

The right hand is gradually drawn toward the body, then approaches the ground, in type-position (**Q** 1), fingers more collected at the ends and less rigid. From the ground the hand is made to rise slowly and successively in representation of the upward tendency of vegetation. (*Oto and Missouri* I.) "The time when grain and grass grow."

Make the sign for **Rain,** then with the curved index only pointing upward, hold the back of the right hand near the ground and elevate it, gradually and in an interrupted movement, upward. (*Ute* I.) "After the rains the sprouts appear."

Squaw. See **Woman.**

Stars. (Compare **Moon.**)

The right-hand, forefinger and thumb crooked, is pointed in various directions above the head toward the heavens, and a moderately quick under-and-over movement of the finger and thumb, forming a crescent, is made. (*Dakota* I.) "From the twinkling of the stars."

Make the sign for **Night,** and then, bringing the ends of the right thumb and forefinger together, or flexing the forefinger within the thumb, quickly move the upright hand four or five times forward, here and there above the head. For *star*, hold the hand above the head, its inner edge uppermost. (*Dakota* IV.)

Stay, abide. I live or stay here.

From a foot in front of the neck move the right hand, its back forward and index extended, several times through a curve toward the right shoulder, each time rotating it to turn the palm forward; then throw the fist forward in front of the lower part of the chest and move it a foot or eighteen inches up and down back outward. (*Dakota* IV.)

The clinched fist, back forward, thumb upward, is held before the chest, then depressed a few inches and suddenly arrested. If you wish to tell a person at some distance to stay where he is, stretch the arm out at full length toward him in making the sign, otherwise the hand is held near the body. (*Mandan and Hidatsa* I.)

Clinch the right hand as if holding a stick, and make a motion as if trying to strike something on the ground with the bottom of the stick. held in an upright position. (*Wichita* I.)

Steal, To.

The left forearm is held, horizontally, a little forward across the body, and the right hand passing under it with a quick motion seems to grasp something and is suddenly withdrawn. (*Long.*)

Seize an imaginary object with the right hand from under the left fist. (*Burton.*) This implies concealed action and the transportation forming part of the legal definition of larceny. Our instructed deaf-mutes make the same sign.

Left arm and hand held diagonally to the body on level with elbow, right-hand forefinger hooked, quickly drawn under left hand and back to the side (sometimes all the fingers are hooked as though grabbing something or tearing it away). (*Cheyenne* I.)

Left hand held about a foot in front of the breast, horizontal, back outward, fingers extended and pointing toward the right; then the right hand, with the fingers extended, hooked, tips outward, hand horizontal, is passed outward under the left hand, and quickly drawn backward again behind the left hand, as though seizing and subsequently concealing the article. (*Dakota* I.) "Stealing and concealment."

The left arm is partly extended and held horizontally so that the left hand will be, palm downward, a foot or so in front of the chest. Then, with the right hand in front, a motion is made as if something were grasped deftly in the fingers and carried rapidly along under the left arm to the axilla. (*Mandan and Hidatsa* I.)

No special sign for this unless the portrayal of a **Texan** (see TRIBAL SIGNS) be accepted as the Mescalero type for *a thief*, as these poor wretches are said to have been dreadfully harassed and plundered by Texans (tay-ha-nas) for many years. *Patricio* gave several narratives; in one the Texans came and *drove off* his horses; in another the Texans entered a house and *took* (shown by a quick grabbing) property. (*Apache* III.)

Deaf-mute natural signs.—Look around, put forward the hand, and close it as if to take something, and move it to the side. (*Ballard.*)

Bend forward your body and bring the hand, clinched, in the manner of taking something under your arm, at the same time looking around as if to see that no one has seen your deed. (*Cross.*)

Take anything spoken of and put the hand in the pocket, and turn and run away. (*Larson.*)

Use the hand as if to take something, at the same time look around as if to see if somebody comes. (*Zeigler.*)

Italian sign.—The open hand held before the face, and the fingers, beginning with the little one, turned round in a wheel, signifies a robbery. (*Butler.*)

——— A horse.

To express horse-stealing they saw with the right hand down upon the extended fingers of the left, thereby denoting rope-cutting. (*Burton.*)

Left hand horizontal, flat, in front and as high as the elbow. Right hand arched, joined, thumb resting near end of forefinger, downward (similar to **V**), and passed slowly under the left, backward toward the elbow and quickly across to its own side, to show crawling up to a horse, cutting its lariat and making off quickly. (*Dakota* III.)

Steamboat.

The sign for **Smoke** is made with the right hand extended upward at the side and above the head, and then with the mouth make the puffing sound in imitation of the sound from the escape-pipes. (*Dakota* I.) "From the puffing sound of the escape-pipes, which can be heard a considerable distance on a still day, and the smoke from the smoke-stack."

Make the sign for **Water,** by placing the flat right hand before the face, pointing upward and forward, the back forward, with the wrist as high as the nose; then draw it down and inward toward the chin; then with both hands indicate the outlines of a horizontal oval figure from before the body back to near the chest (being the outline of the deck); then place both flat hands, pointing forward, thumbs higher than the outer edges, and push them forward to arm's length (illustrating the forward motion of the vessel). (*Kaiowa* I; *Comanche* III; *Apache* II; *Wichita* II.)

Stingy, Covetous, Cowardly, &c.

First lay the palm of the right hand, horizontal, over the left breast: then make the sign for **Anger,** by carrying the fist (**B** 2) downward, in front of the body, from the face toward the left to the level of the heart, but not with any emphasis; and then the sign for **Good** is made by opening the hand, turning it palm downward (**S** 1), and carrying out from the breast for a foot or more, and then turning the hand, thumb downward, back toward the left, and carrying it out to the right side of the body on the same level, which is the sign for **No** or **Not.** (*Dakota* I.)

Make the sign for **Brave, Generous,** at the end of which sign the right hand is opened as in (**T** 1), modified by back of hand being more concave and swept semicircularly outward and downward from the right side of head. (*Oto* I.) "No good will, generosity, or courage."

Bring the left hand against the shoulder, with the elbow slightly before the hip, then tap the elbow with the knuckles of the right hand from below upward. (*Apache* I.)

Curve the fingers of the left hand so that their tips rest against the inner edge of the thumb, which should be about an inch from the palm; then bring the hand slowly to the pit of the stomach, back to the front. (*Ute* I.)

Stirrup.

Make first the sign for **Horse** and next the sign for **Saddle;** then catch the right hand, with its index hooked as in position (**I**), index more opened, by the left in similar position. Then raise the foot (either) and catch its sole by the hooked index of the left hand. Holding the leg thus, as half-mounted, throw the left arm into the air as the leg over a horse. (*Oto and Missouri* I.) "Something hooked to catch the foot when mounting a saddled horse."

Stone.

The right hand shut, give several small blows on the left. (*Dunbar.*)

Close the right hand, and strike the palm of the left hand two or three times with it. (*Long.*)

If light, act as if picking it up; if heavy, as if dropping it. (*Burton.*)

Fingers of right hand closed, thumb lying along the tips, struck once or twice into the palm of the left hand. (*Cheyenne* I.)

With the back of the arched right hand (**H**) strike repeatedly in the palm of the left, held horizontal, back outward, at the height of the breast and about a foot in front; the ends of the fingers point in opposite directions. (*Dakota* I.) From its use when the stone was the only hammer.

The right hand points to the earth with the extended index; then both hands, fingers divergent (as in **P** 1), inverted, approximate at the points of index-finger and thumb, then diverge until in descending the points of ring and little fingers touch the ground. The fingers then approach each other at their points uniformly and diverge three times. (*Oto and Missouri* I.) "Something, that would fill the hand, which lies scattered on the earth."

Deaf-mute natural sign.—Imitate the action of picking up and throwing a stone, and sometimes with indications of the size and form of the object by means of the left fist. (*Ballard.*)

Store.

First make the sign for **White man**, then for **Tipi,** and finally for **Trading.** (*Dakota* I.) "White man's house where we swap goods."

Make the sign for **White man's house,** and **To buy.** (*Kaiowa* I; *Comanche* III; *Apache* II; *Wichita* II.)

Stove.

First point to a piece of iron, and then with the hands in front of the body make the shape of a box-stove, and complete by the sign for **Fire.** (*Dakota* I.) "Iron of the stove and fire."

Study, To; to deliberate.

The arm is flexed and the hand assumes type-position (**O** 1) modified by the fingers being more curved, cup-shaped. The arm is then elevated and the hand twisted spirally from left to right upward before the center of the forehead. (*Oto and Missouri* I.) "To revolve in mind."

Stumble, To.

Hold the left hand flat, edgewise, extended before the breast, back to the front, fingers pointing to the right; then move the flat right hand, palm toward the body and fingers pointing downward, forward toward the left, and as the backs of fingers of the right strike the palm of the left drop the right hand over to the front and downward a short distance. (*Kaiowa* I; *Comanche* III; *Apache* II; *Wichita* II.) "To strike an object, and to trip or stumble."

Storm, Tempest, or **Hurricane.**

The three signs for **Wind, Big,** and **Fear,** in that order. (*Dunbar.*)

Make the **Rain** sign, then, if thunder and lightning are to be expressed, move, as if in anger, the body to and fro, to show the wrath of the elements. (*Burton.*)

Sign for **Clouds** is also used for storm. (*Dakota* I.) "Gathering of the clouds before a storm."

Deaf-mute natural sign.—*Rain* indicated by a repeated downward motion of the extended fingers. *Wind*, by a sidewise sweeping motion of the hands and blowing through the lips. (*Ballard.*)

Strong, Strength.

The hands are clinched; the left forearm is held almost perpendicularly near the breast, so that the fist is nearly opposite to the throat; the right arm is then carried up between the left and the breast, and continued on over the left fist to the outside of the latter; the right arm is then brought down so as to have the same direction with the other, and the fists rest opposite to each other in a line from the breast. This motion resembles the act of wringing a thick towel. If he would say "I am strong," he strikes himself upon the breast two or three times with his fist previously to the motion above described. If he would say "you are strong," he previously points to you, etc. (*Long.*)

Deaf-mute natural sign.—Imitate the action of a person exerting muscular force. (*Ballard.*)

——— Applied to man or animal.

Both arms raised on their respective sides to level with the shoulders, back of hands upward, fists (**A**) are quickly thrown downward to the level of the stomach on their respective sides, and brought to a sudden stop with a rebounding motion. The muscle of the arms, chest, and back are all brought into action in making this sign. (*Dakota* I.) "Exhibiting muscular power."

——— As a cord, rope, etc.

With both hands in front of the breast, fists (**B**), hands separated a few inches, make movements as though pulling on a cord or rope that would not yield. In addition to the muscles of the arms, etc., those of the face are brought more into action than in the above sign. (*Dakota* I.) "Cannot break it. It is strong."

Submission,

With both hands in front of face, open (**W**, palms oblique, downward, with the little-finger edge of the hands lowest), the fingers close to and pointing together, the head is slightly inclined forward and eyes cast down, hands are moved obliquely inward and downward till they come close to or reach the breast. Generally repeated two or three times (*Cheyenne* II.)

The right hand, with fingers extended (**S**), is carried to the right and to the left in front of the body and back to in front of the right shoulder, where all the fingers are closed excepting the index, which points upright, back of hand outward, and then the hand is thrown slowly forward in front of the body so that it is horizontal, back downward, index-finger pointing obliquely forward and downward. (*Dakota* I.) "The first part of this sign means everything is clear; nothing of this matter to come up hereafter; and the latter part, 'I accept, I yield, submit.'"

Sugar. (Compare **Sweet.**)

The right arm is bent at a right angle, and the hand, in type-position, (**K** 1, modified by the palm facing the mouth), is made to slowly and gently touch the tongue with the palm point of the index-finger. The hand is then dropped and approaches the tongue a *second time* in a semi-circle, the countenance and mouth indicating pleasure. (*Oto and Missouri* I.) "Something that can be tasted twice with pleasure."

The right hand, back outward, fingers as in (**U**), but turned downward, is carried from in front of the body upward to the lips, and a sound made by sucking in air. (*Dakota* I.) "It is sweet; I like it."

Summer.

Both hands, fingers and thumbs separated (**Q**, fingers downward), are moved outward to front and upward as far as arms will reach. The hands need not be in shape till they are out at arm's length. The sign is stationary. (*Cheyenne* II.) "Supposed to represent rays and heat of sun striking down."

Make the sign for **Grass** in front of the body, carrying the hand upward two or three feet from the ground, indicating that the grass is long; and then the left hand, representing a **Tree,** is held in front of the breast, and with the right hand make movements as though picking something from it and putting in the mouth. (*Dakota* I.) "The time when the grass is long and the cherries are ripe; hence, summer."

Make the sign for **Grass growing,** *i. e.*, move the right hand from the ground upward three or four inches at a time. (*Dakota* IV.) "The grass getting higher and higher."

Point to the sky, then pass the palms, turned upward, to the right and left, horizontally, before the breast. (*Dakota* VI.)

The countenance assumes an oppressive mien; the right arm is elevated and the index-finger in type-position (**J**), points to the sun in the zenith; both hands then wave above the head, in type-position (**P** 1) modified by being inverted; the hands, thus resembling the direct rays of the sun, approach the head. (*Oto and Missouri* I.) "The time when the rays of the sun descend direct and oppress us."

Same sign as for **Hot.** (*Kaiowa* I; *Comanche* III; *Apache* II; *Wichita* II.)

Made in the same manner as that for **Warm.** (*Apache* I.)

Sun. (Compare **Day.**)

The thumb and finger, forming a circle, elevated in front toward the face. (*Dunbar.*)

The forefinger and thumb are brought together at tips so as to form a circle, and held up toward the sun's track. (*Long.*)

Form a small circle with the forefingers and hold them toward heaven. (*Wied.*) I have given you this sign. (*Matthews*) There is no visible identity in the execution of the (*Oto* I) sign and *Wied's*, although a seeming similarity in conception exists; the similarity in the signs for day explains the practice of speaking of a day *as after one or more suns.* (*Boteler.*)

Join the tips of the thumb and forefinger of the same hand, the interior outline approximating a circle, and indicate thus the projection of its disk against the sky. (*Arapaho* I.)

Right-hand finger crooked, elevated, and held toward the east. (*Cheyenne* I.) The crook is an abbreviation of the circle representing the orb.

Right hand closed, the index and thumb curved, with tips touching, thus approximating a circle, and held toward the sky. (*Absaroka* I; *Shoshoni and Banak* I; *Ute* I; *Wyandot* I.)

Right hand extended at side of body on a level with the head; with the forefinger and thumb describe a crescent, other fingers closed. (*Dakota* I.)

Make the sign for **Day,** and then flex the right index and thumb until their ends are about four inches apart; or, as some do, until they are an inch and a half apart; or, as most do, bring the ends together; nearly close the other fingers and raise the hand in front of the forehead. The ulnar (inner) edge of the hand is usually turned toward the part of the sky where the sun is supposed to be: for sunrise, toward the east; for noon, toward the zenith; for sunset, toward the west. (*Dakota* IV.)

Close the right hand, curve the index-finger in the form of a half-circle, and in this position hold the hand upward toward the sun's track. (*Dakota* V.)

Close the right hand, forming a circle with the thumb and index, then hold the hand toward the sky. (*Dakota* VII.)

The partly bent index and thumb of the right hand are brought together at their tips, so as to represent a circle; and with these digits next to the face the hand is held up toward the sky, from one to two feet from the eye and in such a manner that the glance may be directed through the opening. (*Mandan and Hidatsa* I.)

The right arm is elevated, then extended to the left on a level with the left deltoid prominence. The hand is in type position (**I** 1) modified by being horizontal. The hand and arm thus pointing to the Orient, describes next the arc of the vault of the heavens and slowly sinks, wavering, extended from shoulder, pointing to the west. The sign for **Light** is next executed. (*Oto and Missouri* I.) "That which passes through the heaven's vault, shedding light."

Raise the right hand above the head, holding the open palm toward the sky (sun in prayer). (*Ponka* I.) "Wakanda—Praying to the sun."

Join the tips of the index and thumb so as to form a circle, close the remaining fingers and hold the hand toward the sky, with the outer edge forward. (*Kaiowa* I; *Comanche* III; *Apache* II; *Wichita* II.)

Form a circle with the index and thumb, tips touching, the remaining fingers closed, and hold them toward the sky. (*Apache* I.)

Deaf-mute natural sign.—Point toward the sky, make a circle with the forefinger, and wink as if dazzled by the sun's rays. (*Ballard.*)

——— Eclipse of.

First make the sign for the **Sun,** and then the sign for **Dead, Death.** (*Dakota* I.) "The sun is dead."

Sun-dogs. (Compare **Aurora Borealis.**)

First make the sign for the **Sun,** directly in front of the body with the right hand, and then the sign for **Fire,** on the same level and at both sides of it at the same time. (*Dakota* I.) "Fire built to heat the winter sun.

Sunrise.

Make the sign for **Day,** at the same time indicating position of the sun, just above the horizon, as in sign for **Sun.** (*Arapaho* I.)

Make the sign for the **Sun,** but point the crescent in the direction of the rising sun in the horizon, and then carry it slightly upward. (*Dakota* I.) "The coming up of the sun."

Make the sign for **Morning,** and then the sign for **Sun,** holding the inner edge of the hand toward the east and raising it a little. (*Dakota* IV.) "Uncovering the sun."

Deaf-mute natural sign.—The same sign as **Sun,** with the addition of pointing to the eastern horizon. (*Ballard.*)

Sunset.

Make the sign for **Night,** at the same time indicating position of the sun, just below horizon, as in sign for **Sun.** (*Arapaho* I.)

Right-hand forefinger crooked, as in sign for **Morning,** lowered toward the western horizon. (*Cheyenne* I.)

Point the crescent sign for **Sun,** in the direction of the setting sun in the horizon and below it. (*Dakota* I.) "Sun has disappeared from view."

Make the sign for **Sun,** holding the inner edge of the hand toward the west and lowering it a little, then make the sign for **Night.** (*Dakota* IV.)

Deaf-mute natural sign.—The same sign as for the **Sun,** and pointing to the western horizon. (*Ballard.*)

Superior. See **Ahead.**

Supplication.

Italian sign.—Falling upon the knees and clasping the hands or laying the palms together shows the supplication of a beggar. (*Butler.*)

Surprise. (Compare **Admiration** and **Wonder.**)

Throw the head and body backward with a quick motion and express surprise by facial emotions and the eyes. See connection with *Horror* under that word. *Wonder* is included in the sign for *Surprise.* (*Dakota* I.)

The right hand, palm inward, with the fingers slightly bent, is placed over the mouth in such a way as to leave the lips free to articulate. The index rests on the upper lip, but the palm does not touch the mouth. The thumb commonly rests against the right side of the nose, and one or more finger-tips on the face to the left of the mouth. While the hand is thus held, low groans, exclamations, or expressions of surprise are uttered. (*Mandan and Hidatsa* I.)

Clinch the fists and shrink away. Fists must be near waist and not at chin, as in **Fear.** (*Apache* III.)

Deaf-mute natural sign.—Part the lips, arch the eyebrows, and raise the hand. (*Ballard.*)

Surrender. See **Quiet.**

Surround.

At the height of the breast, backs of hands obliquely upward, thumb and forefinger of each extended, curved, and brought nearly together; other fingers of both hands closed. (*Dakota* I.) "Closing in on or surrounding anything."

——— Surrounded.

Form a circle about eight inches in diameter by extending and separating both thumbs and forefingers, and holding the hands opposite each other with palms inward; then move the hands about six inches from side to side. (*Dakota* IV.)

——— Surrounding the bison.

The sign for **Bison** is first made; the hand, with the forefingers and thumbs in a semicircle, are then brought two or three times together. (*Long.*)

First make the sign for **Bison** and then the sign for **Surrounding**. (*Dakota* I.)

Suspicion.

Italian sign.—Draw down one lower eyelid, which is as much as to say, "Let me open my eyes a little wider." A man convinced that others wish to impose upon him, and wishing to let them know that he is not imposed upon, points a finger at his eye as if to say, "My eye is wide open and sees what you are about." (*Butler.*)

Swallow, To. Swallowing.

Slightly flex the fingers of the right hand and place the thumb against the side of the index, the hand directed forward, palm upward, in front of the right breast, and, while turning the hand over, move it first upward, then backward through a curve to the mouth, and then downward to the top of the breastbone. (*Dakota* IV.)

Sweet. (Compare **Sugar** and **Sour.**)

Tip of forefinger touched against the tip of tongue; sign for **Good.** (*Cheyenne* I.)

Same as the sign for **Sour,** omitting the spitting, and smacking the lips instead. (*Dakota* I.) "Good; I like it."

Deaf-mute natural sign.—Any agreeable taste would be indicated by smacking the lips. (*Ballard.*)

Swift, swiftness.

The two index-fingers are held parallel together and pointing forward; the right one is then passed rapidly forward. (*Long.*)

Left hand held horizontal, with palm downward, fingers extended, joined, pointing outward (**W**), about 12 inches in front of breast; pass the right hand, carried outward from the right breast, by the stationary left, with a rapid motion. (*Dakota* I.) "The swift passing the slow."

Deaf-mute natural sign.—A slight moving of the body from side to side in rapid succession, and a slight movement of the feet on the floor. (*Ballard.*)

Italian sign.—The colloquial phrase, "hand over hand," exactly describes the Italian motion to express the same idea, namely, to do anything rapidly. (*Butler.*)

Swim, swimming.

The forefinger of the right hand extended outward and moved to and fro. (*Dunbar.*)

Hands brought together in front of the body about a foot (**W**), with fingers pointing outward; make a series of sidewise movements of the hands toward the right and left, on a curve, in imitation of the movements of the hands and arms in swimming. (*Dakota* I.) "From the act of swimming."

Sword.

Make the motion of drawing it. (*Burton.*)

Right hand flattened, fingers pointing upward, little finger front; motion made forward to imitate cutting. (*Cheyenne* I.)

Syphilis.

The left hand is closed, allowing the forefinger to be extended and pointing forward before the body; then, with the thumb and index of the right, pretend to pick off small particles of imaginary foreign bodies from various sides of the forefinger. (*Absaroka* I; *Shoshoni and Banak* I.) "From the ulcerating or 'eating' nature of the disease."

Talk. See **Speak.**

Taste. (Compare **Sweet** and **Sour.**)

Touch the tongue-tip. (*Burton.*)

Right-hand fore and middle fingers, pointed upward, touched to tip of tongue. (*Cheyenne* I.)

Simply touch the forefinger of the right hand to the tongue. (*Dakota* I.) "From the act of tasting."

Put one forefinger in the other palm, then to tongue. (*Apache* III.)

Telegraph.

Left index extended and held in front of the body, horizontal and pointing toward the right, back outward, is struck smartly crosswise several times by the right index, edge of hand downward, and then the sign for **Talking** or **Speaking** is made to complete it. (*Dakota* I.) "The first part of this sign denotes the striking of the key."

Tell. See **Speak.**

Texan. (Compare **Steal.**)

Place widely extended thumbs and forefingers as if inclosing a very large hat brim, out by sides of head. (*Apache* III.) "Such being esteemed by Texans."

Thanks. (Compare **Glad.**)

Thank you, or, more strictly, **Invoking a blessing.**

The right hand upright, opened and relaxed, fingers separated a lit tle, palm forward, is placed near the person's forehead and then moved downward in front of the face to the sternum, the hand being at the same time bent at the wrist until it becomes horizontal. (*Dakota* IV.) "Both hands are frequently drawn downward in front of the face."

Theft. See **Steal.**

There, I have been.

Hold the open left hand, its palm obliquely backward and upward, a foot in front of the chest; then, the right hand being closed excepting the index, which is to be extended upward, strike its palm and fingers against the palm of the left and hold the two hands still for a few seconds. (*Dakota* IV.)

Thick.

First make the sign for **Thin,** and then the sign for **No,** or **Not,** and then the two hands, with fingers extended and joined, are held horizontal, six or seven inches apart, in front of the breast, with their palmar surfaces toward one another. (*Dakota* I.) "Not thin."

Thin. See **Poor.**

Think; Guess. (Compare **Study.**)

Pass the forefinger sharply across the breast from right to left. (*Burton.*)

(1) Right-hand fingers and thumb loosely closed, forefinger crooked, slightly extended; (2) dipped over toward and suddenly forward from left shoulder or upper arm. (*Cheyenne* I.)

Right hand carried to the left breast, with the fore and second fingers extended, pointing downward, obliquely toward the left, back outward (**N,** turned obliquely downward), make several outward and inward movements of the extended fingers only. (*Dakota* I.) "'Stop! let me think.' The heart is regarded as the seat of all the functions of life, hence the sign of thinking from that organ."

Hold the left hand, pointing toward the right palm, backward, a foot in front of the neck; then move the right hand, palm toward the left, from an upright position just below the mouth over the left to arm's length, turning the end of the right hand downward until it points forward. (*Dakota* IV.) "The mind going straight forward."

Clinch the right hand and place the radial side (either the thumb or the middle joint of the index) against the lower portion of the forehead; the fist is usually placed between the eyes. At the same time the head, with eyes to the ground, is inclined and rested against the fist, as if in meditation. (*Dakota* VI, VII.)

Hit the chest with closed fist, thumb over the fist. (*Omaha* I.)

Deaf-mute natural sign.—In the sense of *Suppose* or *Presume* the sign was made by nodding the head slightly, accompanied by a steady fixing of the eye. (*Ballard.*)

Italian sign.—The forefinger on the forehead denotes either effort of thought or force of talent. (*Butler.*)

Thunder.

The sign of **Rain** accompanied by the voice imitating the rumbling sound of thunder. (*Dunbar.*)

Hands partially closed, backs outward, elevated to the ears; moved slightly out and in; face expressing annoyance or pain. (*Cheyenne* I.)

Another: The sign for **To Sing** exaggerated. (*Cheyenne* I.) "Great voice or big sing."

Right hand raised as high above the head as possible (with the hand as **T**), bring it down in front of the body with a quick motion, snapping the fingers, and separating them (as **Q**), the fingers pointing downward, back of hand outward. Same sign includes **Lightning.** Thunder and lightning so frequently accompany each other as to suggest to the Indian the idea of constancy; hence no separate sign for lightning. (*Dakota* I.)

From positions near together in front of the face, palms forward, separate the upright clinched hands about eighteen inches, and then, turning the palms inward, move the hands backward, one on each side of the head. (*Dakota* IV.) "Spreads and goes away."

Tie, To.

Make a circular motion around, over, and above the left hand—held in front of breast (fist, **A** 1)—with the right hand, with thumb and forefinger extended, crooked and meeting (other fingers closed), back of hand upward, of closed fingers outward, and then pass the right hand under the left, with thumb and forefinger separated and drawn inward or back ward again as though having seized hold of something and pulling hard on it, after which the right hand is dropped downward. (*Dakota* I.) "Putting a lariat around the pole and making it secure. Securing the horse."

Time.

The seasons, corresponding with our divisions of winter, spring, summer, and autumn, are denoted by their appropriate signs—*Winter*, by **Cold** or **Snow; ** *Spring*, by the **Springing up of the Grass;** *Summer*, by **Long Grass**, the **Time Cherries are Ripe,** etc.; and *Autumn*, by the **Falling of the Leaves.** *Hour* of the day is approximately denoted by the **Position of the Sun.** A *Month* (one moon) is also denoted by its appropriate sign. Days and nights can also be so denoted. (*Dakota* I.)

Deaf-mute natural sign.—No general sign. A *day* is indicated by moving the forefinger across the sky; *parts* of the *day* by portions of this movement; *days* numbered by sleeps, that is, by inclining the head on the hand repeatedly; *noon*, by the index-finger of the right hand applied to that of the left, as for the time when the hands of the clock meet and both point to the hour twelve. (*Ballard.*)

——— Future.

The arms are flexed and hands brought together in front of body as in type-position (**W**). The hands are made to move in wave-like motion up and down together and from side to side. (*Oto* I.) "Floating on the tide of time."

Count off fingers, then shut all the fingers of both hands several times, and touch the hair and tent. (*Apache* III.) "Many years; when I am old (white-haired)."

Deaf-mute natural sign.—To denote a future time, the sign is made by putting the hand on the cheek with the head slightly inclined, meaning *days*, and counting on the fingers to denote *how many*. There is no specific sign to distinguish the past from the future. (*Ballard.*)

——— Long.

Place the hands close together and then move them slowly asunder, so slowly that they seem as if they would never complete the gesture. (*Cheyenne* sign. Report of *Lieut. J. W. Abert*, *loc. cit.*, p. 426.) "This was used in narrating a tradition and referring to great antiquity in time; also applied to great, indefinite distance."

Signs for **Sleep** and **Many.** (*Arapaho* I.) Literally, "many sleeps."

Fingers of both hands clasped as though holding a string, left hand remaining stationary, right hand drawn along the imaginary string in proportion to the length of time to be represented. It also means *old* in the abstract. (*Cheyenne* I.)

Place the hands as in **Time** (**Short**); then draw them apart any distance thought necessary by the talker to convey the idea. (*Cheyenne* II.)

Both hands in front of the breast, thumb and forefinger of each extended, curved, and meeting at tips (other fingers closed), hands horizontal, backs outward, second phalanges of little fingers joined, then the hands are separated by slowly carrying right to right, left to left, still horizontal, and on the same level, by a series of short stops, as though passing a string between the thumb and forefinger of each and tightening on it, arms carried to full extent at sides of body. (*Dakota* I.) "Making time."

Throw the upright opened right hand forward three times from the wrist just in front of the right ear, the palm inward, fingers joined, thumb separated a little from the index. (*Dakota* IV.)

Hold the left hand, closed, about a foot in front of the left shoulder, the forefinger extended and pointing upward; then close the right hand, index only extended, horizontal; touch the tip of the left forefinger with the tip of the index, and draw the right hand backward to the right shoulder. (*Shoshoni and Banak* I.)

Place the left hand in front of the chest, the tips of the thumb and forefinger touching, with remaining fingers tightly closed; with the fingers and thumb of the right hand similarly placed; bring the tips of thumb and index of the right against those of the left, and draw them slowly apart, the left hand forward and outward from the left side, and the right backward over the front of the right shoulder. (*Kaiowa* I; *Comanche* III; *Apache* II; *Wichita* II.)

Another: Hold the left hand about twelve inches in front of the left shoulder, tips of forefinger and thumb touching; then bring the tip of the index against that of the thumb, the right touching those of the left, and draw them slowly apart, bringing the right hand toward the right shoulder as if drawing out a long thread. (*Wyandot* I; *Kaiowa* I; *Comanche* III; *Apache* II; *Wichita* II.)

Place the thumb and forefinger of each hand as if holding a small pin, place the two hands (in this position) as if holding a thread in each hand, and between the thumb and forefinger of each hand close together, and let the hands recede from each other, still holding the fingers in the same position, as if letting a thread slip between them, until the hands are two feet apart. (*Wichita* I.)

——— Lately, recently.

Right-hand fingers and thumb extended straight upward, separated

(**R**), is brought up to side of face (right) with palm toward face, and moved backward and forward two or three times. (*Cheyenne* II.)

Hold the left hand at arm's length, closed, with forefinger only extended and pointing in the direction of the place where the event occurred; then hold the right hand against the right shoulder, closed, but with index extended and pointing in the direction of the left. The hands may be exchanged, the right extended and the left retained, as the case may require for ease in description. (*Absaroka* I; *Shoshoni and Banak* I.)

The flat open right hand, turned back toward the right, fingers extended, pointing upward (**S**), is carried backward and forward at the right side of the head, and then the right hand is passed by the left hand, held horizontal, back toward the left (**S** turned horizontal instead of upright), about a foot and a half in front of the face. (*Dakota* I.) "Gone by in time."

Extend the right index, half close the other fingers, thumb against the middle finger, and after placing the hand, back outward and well-extended, on the upright forearm, four or six inches in front of the right ear, throw it forward about four inches three times, by jerks, from the wrist. (*Dakota* IV.)

——— Long ago.

Both hands closed, forefingers extended and straight; place one hand at arm's length, pointing horizontally, the other against the shoulder or near it, pointing in the same direction as the opposite one. Frequently the tips of the forefingers are placed together, and the hands drawn apart, until they reach the positions described. (*Absaroka* I; *Shoshoni and Banak* I.)

Place the flat right hand, palm forward, near the side of the head, and wave it by interrupted movements outward toward the right, gradually turning the palm more and more to the right. (*Kaiowa* I; *Comanche* III; *Apache* II; *Wichita* II.)

Another: Pass the right hand, flat and extended, edgewise and pointing upward from over the shoulder, outward toward the right in a waving motion, so that at each movement the hand is farther from the head, and at last the palm is turned nearly to the right. (*Kaiowa* I; *Comanche* III; *Apache* II; *Wichita* II.)

——— Short.

The sign for **Time** (**Long**) followed by that of negation. (*Arapaho* I.)

Both hands in front of breast, about six inches apart, arched (**H**, back outward), thumbs and forefingers horizontal, and pointed toward each other; move slowly together till thumbs and fingers of each hand touch, if a very short time is meant. (*Cheyenne* II.)

Indicate by pointing to the sun or above, as at the sun at high meridian, and move right hand to right a short distance. Or, if sun or moon is seen, point at, with same indication, a slow motion and short distance of rotation or change. (*Ojibwa* IV.)

The right index extended and pointing obliquely upward (**K**), is held ten or twelve inches in front of the breast, then the hand is turned horizontal, back upward, and drawn slowly inward to the body, fingers pointing toward the left and obliquely downward. (*Dakota* I.) "A short distance in time."

With the tips of the index and thumb of the right hand touching, pretend to draw a short fiber held by the forefinger and thumb of the left. (*Kaiowa* I; *Comanche* III; *Apache* II; *Wichita* II.)

Another: Place the tips of the forefingers and thumbs together as in **Time** (**Long**); then draw them about an inch apart. (*Kaiowa* I; *Comanche* III; *Apache* II; *Wichita* II.)

——— Some time ago. From a certain time mentioned.

Having placed the nearly closed left hand, back outward, about two feet in front of the lower part of the chest, and the right hand, back outward, about six inches back of it and a little to the right, fingers relaxed and separated a little, push the left hand a very little forward and toward the left, and draw the right backward and toward the right until it is about six inches in front of the right side; then drop the left hand and move the right one from the wrist up and down about eight inches two or three times. (*Dakota* IV.)

——— Soon.

Raise left hand and arm partly, palm toward the body, arm bent at right angle, hand and forearm drawn forward toward the body slowly, with slight bow of head toward body. (*Ojibwa* IV.)

——— Very long ago.

Wave the extended flat right hand in an interrupted manner outward and slightly backward from the right side of the head. (*Wyandot* I.)

——— Of day. See **Hour.**

——— To-day. See **Day.** (Compare **Now.**)

Tipi (**tepee**). See **Lodge.**

Tired, weary.

Strike the palmar surfaces of both hands (**W**) against the legs about midway between the thighs and knees, and carry out to the sides for a few inches with both hands as in (**W**), with extended fingers pointing forward, carrying the hands downward for nine or ten inches with a quick motion and coming to a sudden stop. (*Dakota* I.) "Legs have given out."

The left arm is partly extended forward and is gently struck near the bend of the elbow, usually above it, with the palm of the right hand; at the same time the head is usually inclined to the left side; then, in similar manner, the right arm is extended and struck by the left hand, and the head, in turn, inclined to the right. If the sign-maker aims to be particularly expressive, he assumes an appearance of weariness. (*Mandan and Hidatsa* I.)

Pass the hands down the legs, hands trembling, gather arms to side, fists before chin, and settle elbows down in the lap; facial expression corroborating. (*Apache* III.) "Action of an exhausted man."

Tomahawk, ax, hatchet.

Cross the arms, and slide the edge of the right hand, held vertically, down over the left arm. (*Wied.*) Still employed, at least for a small hatchet, or "dress tomahawk," as I might call it. The essential point is laying the extended right hand in the bend of the left elbow. The sliding down over the left arm is an almost unavoidable but quite unnecessary accompaniment to the sign. The sign indicates the way in which the hatchet is usually carried This is illustrated in Catlin's North American Indians by no less than fourteen portraits. In seven of these portraits the hatchet is represented in different positions. In one of these the position approximates that of this sign; in others the subject is so loaded down with weapons that he cannot give his ax the usual position, and in others there are some evidences of "posing" by the artist. Pipes, whips, bows and arrows, fans, and other dress or emblematic articles of the "buck" are seldom or never carried in the bend of the left elbow as is the ax. The pipe is usually held in the left hand. (*Matthews.*) There is not the least similarity in execution or conception between *Wied's* and the (*Oto* I) signs, the former being also very obscure. Something with a long handle and wide blade, used for chopping. (*Boteler.*)

Is denoted by chopping the left hand with the right. (*Burton.*)

Right hand elevated to level of chin, fingers open and flattened, thumb lying close to and along the forefinger, whole hand bent in the direction of the little finger and at a right angle to the wrist. (*Cheyenne* I.) "The motion of chopping imitated, using the forearm as the handle of the ax."

Right hand in front of the body as though grasping the handle of a tomahawk, and at the same time a slight upward and downward movement of the hand is made. (*Dakota* I.) "From the manner of holding the tomahawk."

With right hand closed or opened, and the palm obliquely upward toward the left, and the left hand opened, palm obliquely downward toward the right, and fingers forward, move them downward toward the left several times to imitate chopping with an ax. (*Dakota* IV.)

Place the extended flat right hand edgewise above the left, similarly held, both pointing toward the left and downward, and make a simultaneous cut in that direction with both. (*Dakota* VI; *Hidatsa* I; *Arikara* I.)

The left arm is extended, the hand edgewise, thumb up and fingers inclined downward, much in position (**L** 1), fingers opened. The extended right index is then brought to touch the lower thick part of the left hand, and then slowly drawn downward and backward to about the length of the handle. Both hands then, *in statu quo*, exert a uniform and simultaneous up-and-down motion, as in chopping wood. (*Oto and Missouri* I.) "Something with a long handle by which we chop."

The right hand, with extended index only, is brought to the mouth and the finger inserted; the act of smoking is then imitated as the pole of this instrument is hollowed and handle perforated to be used as a pipe. The right hand is now extended in position (**L** 1), modified by fingers being opened and inclined downward. The left hand is then superimposed to the left in position (**L** 1), modified by index being closed. (*Oto and Missouri* I.) "An ax through which one smokes."

To-morrow. See **Day.**

Trade; barter.

First make the sign of **Exchange,** then pat the left arm with the right finger, with a rapid motion from the hand, passing it toward the shoulder. (*Long.*)

Strike the extended index-finger of the right hand several times upon that of the left. (*Wied.*) I have described the same sign in different terms and at greater length. It is only necessary, however, to place the fingers in contact once. The person whom the Prince saw making this sign may have meant to indicate something more than the simple idea of trade, *i. e.*, trade often or habitually. The idea of frequency is often conveyed by the repetition of a sign (as in some Indian languages by repetition of the root). Or the sign-maker may have repeated the sign to demonstrate it more clearly. (*Matthews.*) Though some difference exists in the motions executed in *Wied's* sign, and that of (*Oto and Missouri* I), there is sufficient similarity to justify a probable identity of conception and to make them easily understood. (*Boteler.*) In the author's mind *Exchange* was probably intended for one transaction, in which each of two articles took the place before occupied by the other, and *Trade* was intended for a more general and systematic barter, indicated by the repetition of strokes, which the index-fingers mutually changed positions.

Cross the forefingers of both hands before the breast. (*Burton.*) "Diamond cut diamond." This conception of one smart trader cutting into the profits of another is a mistake arising from the rough resemblance of the sign to that for **Cutting.**

Cross the index-fingers. (*Macgowan.*)

Cross the forefingers at right angles. (*Arapaho* I.)

Both hands, palms facing each other, forefingers extended, crossed right above left before the breast. (*Cheyenne* I.)

The left hand, with forefinger extended, pointing toward the right (rest of fingers closed) horizontal, back outward, otherwise as (**M**), is held in front of left breast about a foot; and the right hand, with forefinger extended (**J**), in front of and near the right breast, is carried outward and struck over the top of the stationary left (+) crosswise, where it remains for a moment. (*Dakota* I.)

The sign should be made at the height of the breast. Raise the right index about a foot above the left before crossing them. (*Dakota* IV.) "Yours is there and mine is there; take either."

Place the first two fingers of the right hand across those of the left, both being slightly spread. The hands are sometimes used, but are placed edgewise. (*Dakota* V.)

Another: The index of the right hand is laid across the forefinger of the left when the transaction includes but two persons trading single article for article. (*Dakota* V.)

Strike the back of the extended index at right angle against the radial side of the extended forefinger of the left hand. (*Dakota* VI, VII.)

The forefingers are extended, held obliquely upward, and crossed at right angles to one another, usually in front of the chest. (*Mandan and Hidatsa* I.)

The palm point of the right index extended touches the chest; it is then turned toward the second individual interested, then touches the object. The arms are now drawn toward the body, semiflexed, with the hands, in type-positions (**W W**), crossed, the right superposed to the left. The individual then casts an interrogating glance at the second person. (*Oto and Missouri* I.) "To cross something from one to another."

Close the hands, except the index-fingers and the thumbs; with them open, move the hands several times past one another at the height of the breast, the index-fingers pointing upward and the thumbs outward. (*Iroquois* I.) "The movement indicates 'exchanging.'"

Hold the left hand horizontally before the body, with the forefinger only extended and pointing to the right, palm downward; then, with the right hand closed, index only extended, palm to the right, place the index at right angles on the forefinger of the left, touching at the second joints. (*Kaiowa* I; *Comanche* III; *Apache* II; *Wichita* II.)

Pass the hands in front of the body, all the fingers closed except the forefingers. (*Sahaptin* I.)

Close the fingers of both hands (**K**); bring them opposite each shoulder; then bring the hands across each other's pathway, without permitting them to touch. At the close of the sign the left hand will be near and pointing at the right shoulder; right hand will be near and pointing at the left shoulder. (*Comanche* I.)

Close both hands, leaving the forefingers only extended; place the right before and several inches above the left, then pass the right hand toward the left elbow and the left hand toward the right elbow, each hand following the course made by a flourishing cut with a short sword. This sign, according to the informant, is also employed by the Banak and Umatilla Indians. (*Comanche* II; *Pai-Ute* I.)

The forefingers of both hands only extended, pass the left from left to right, and the right at the same time crossing its course from the tip toward the wrist of the left, stopping when the wrists cross. (*Ute* I.) "Exchange of articles."

Hands pronated and forefinger crossed. (*Zuñi* I.)

Deaf-mute natural sign.—Close the hand slightly, as if taking something, and move it forward and open the hand as if to drop or give away the thing, and again close and withdraw the hand as if to take something else. (*Ballard.*)

Our instructed deaf-mutes use substantially the sign described in (*Mandan and Hidatsa* I.)

——— To buy.

Hold the left hand about twelve inches before the breast, the thumb resting on the closed third and fourth fingers; the fore and second fingers separated and extended, palm toward the breast; then pass the extended index into the crotch formed by the separated fingers of the left hand. This is an invented sign, and was given to illustrate the difference between buying and trading. (*Ute* I.)

Deaf-mute natural sign.—Make a circle on the palm of the left hand with the forefinger of the right hand, to denote *coin*, and close the thumb and finger as if to take the money, and put the hand forward to signify giving it to some one, and move the hand a little apart from the place where it left the money, and then close and withdraw the hand. as if to take the thing purchased. (*Ballard.*)

Italian sign.—To indicate paying, in the language of the fingers, one makes as though he put something, piece after piece, from one hand into the other—and gesture, however, far less expressive than that when a man lacks money and yet cannot make up a face to beg it; or simply to indicate want of money, which is to rub together the thumb and forefinger, at the same time stretching out the hand. (*Butler.*)

——— Exchange.

The two forefingers are extended perpendicularly, and the hands are then passed by each other transversely in front of the breast so as nearly to exchange positions. (*Long.*)

Pass both hands, with extended forefingers, across each other before the breast. (*Wied.*)

Hands brought up to front of breast, forefingers extended and other fingers slightly closed; hands suddenly drawn toward and past each other until forearms are crossed in front of breast. (*Cheyenne* I.) "Exchange; right hand exchanging position with the left."

Left hand, with forefinger extended, others closed (**M**, except back of hand outward), is brought, arm extended, in front of the left breast, and the extended forefinger of the right hand, obliquely upward, others closed, is placed crosswise over the left and maintained in that position for a moment, when the fingers of the right hand are relaxed (as in **Y**), brought near the breast with hand horizontal, palm inward, and then carried out again in front of right breast twenty inches, with palm looking toward the left, fingers pointing forward, hand horizontal, and then the left hand performs the same movements on the left side of the body. (*Dakota* I.) "You give me, I give you."

The hands, backs forward, are held as index hands, pointing upward, the elbows being fully bent; each hand is then, simultaneously with the other, moved to the opposite shoulder, so that the forearms cross one another almost at right angles. (*Mandan and Hidatsa* I.)

Trap (beaver.)

The two forefingers brought suddenly together in a parallel manner. so as to represent the snapping of the steel trap. (*Long.*)

Travail; plural, **Travaux** or **Travois.** (The corrupt French expression for the sledge used by Indians, probably from *traîneau.*)

The same sign as for **Dog.** (*Dakota* IV.)

Traveling. See **Going.**

——— Moderately; marching. See **Going.**

——— With great rapidity. See **Swift.**

Tree, trees. (Compare **Forest.**)

Vertically raise the forefinger, pointed upward, other fingers and thumb closed, back of hand down. (*Arapaho* I.)

Point with forefinger extended in front obliquely toward the ground, and with an extending motion of arm raise the hand and arm quickly to an angle of over 45°; extend arm at full length, then with fingers and thumb extended, shake the hand once or twice to indicate the branches; look up as if following motion of hand. (*Ojibwa* IV.)

First hold the right index in front of the breast, upright (**J**, back outward), for a moment, and then open the second and third fingers, separate them and let them point upward in different directions, thumb resting on the closed little finger. (*Dakota* I.) "The trunk of a tree and its branches."

With the hands upright, backs forward, fingers a little separated and slightly bent, the right behind the left and a foot in front of the chin, move the left a foot or so obliquely forward toward the left, and the right obliquely backward toward the right until it is in front of and near the right shoulder. (*Dakota* IV.)

Hold the right hand before the body, back forward, fingers and thumb extended and separated; then push the hand slightly upward. Made more than once in succession and at different points of the horizon, means trees or groves. (*Dakota* V, VI; *Hidatsa* I; *Arikara* I.)

Move the right hand, fingers loosely extended, separated and pointing upward, back to the front, upward from the height of the waist to the front of the face. For trees, not referring to a dense grove or a forest, the same sign is repeated several times toward different points in front of the body. (*Kaiowa* I; *Comanche* III; *Apache* II; *Wichita* II.) "Trunk and branches."

——— Grove of.

See **Trees.** (*Dakota* V, VI; *Hidatsa* I; *Arikara* I.)

Raise the right arm vertically, with fingers and thumb spread, then grasp the arm near the shoulder with the left hand. (*Wyandot* I.)

"Trunk and branches, the left hand representing the earth inclosing the base of the trunk."

True, truth.

The forefinger passed in the attitude of pointing, from the mouth forward in a line curving a little upward, the other fingers being carefully closed. (*Long.*)

Lower the hand in front of the breast, then extend the index-finger, raise and move it straight forward before the person. (*Wied.*) I have described the sign for this in much the same way. I think "lower the hand" refers simply to a preparatory motion; if the hand were hanging by the side, "raise the hand." I have usually seen the index-finger held horizontally, not perpendicularly, if that is what he means by raised. (*Matthews.*) The right arm is flexed at the elbow and the hand drawn up to the mouth. The index-finger is extended palm downward and made to pass steadily forward, describing an arc of a quadrant downward. Though *Wied's* sign is very inexplicit, there is much similarity between it and the (*Oto and Missouri* I) both as to conception and movement. In the former, the stress is on what comes from the mouth; in the latter what comes from the breast. (*Boteler.*) "That which comes straight from the mouth or breast."

If one finger is thrust forward in a straight line from the mouth, it means a straight speech, or speaking the truth. (*Ojibwa* I.)

Thrust the forefinger from the mouth direct to the front, *i. e.*, "straight," not "crooked speech." Also, the sign for **Lie, Falsehood,** followed by that of **Negation.** (*Arapaho* I.)

Right-hand fingers and thumb drooping, hold thumb inward against the heart; brought up to the level of the mouth, middle, third, and little finger closed, forefinger extended pointing forward, thrust suddenly, with a curved motion, straight forward from the mouth. (*Cheyenne* I.)

The extended forefinger of the horizontal right hand (**M**), other fingers closed, is carried straight outward from the mouth. This is also the sign for **Yes.** (*Dakota* I.) "One tongue; straight-forward talking."

Place the right hand in front of the mouth, back upward, index extended and pointing forward, other fingers half closed, thumb as you please, move the hand forward about eight inches. Some point the index forward and upward. (*Dakota* IV.) "One tongue."

Touch the breast over the heart with the fingers of the right hand; then with the extended index-finger of the right hand pass it forward from the mouth, elevate and hold it a moment. (*Dakota* V.) "This signifies 'one-tongued,' and coming from the heart as 'sincerity of thought.'" The breast, being the initial point, the sign nearly corresponds with the French deaf-mute sign for **Sincere.**

Pass the extended index, pointing upward and forward, forward from the mouth. (*Dakota* VI, VII.)

The sign is the same as that for **Yes,** except that the hand is held before and often in contact with the mouth and the motion made from that point. (*Mandan and Hidatsa* I.)

The right hand is gradually brought to the mouth which is in motion of talking. The hand is in position (**I** 1) modified by the index-finger being more extended. The hand and index then describe the arc of a quadrant, the index-finger pointing forward, outward and downward. (*Oto* I.) "What comes straight or unvarying."

Make the sign for **Speak,** then point upward with the extended index. (*Wyandot* I.) "Talk good."

Pass the extended index, pointing upward and forward, to the front several times. (*Ute* I.) "But one tongue; only one way in talking—to the front."

With the index only extended, pointing forward, push it forward from the mouth in a slightly downward direction and terminating as high as in the beginning. (*Apache* I.)

Strike with right index, erect, from lips forward; repeat the movement with emphasis, not returning to lips each time. (*Apache* III.) "That is so."

Run the finger straight out from the center or middle of the mouth. (*Zuñi* I.)

Deaf-mute natural sign.—Nod the head several times with an earnest look, in answer to an indication of doubt on another person's face. (*Ballard.*)

Deaf-mutes generally give the gesture of moving one finger straight from the lips. "Straight-forward speaking."

Try, To; To Attempt.

With both fists (**A,** knuckles outward) in front of breast, the left a little in rear of the right, move outward briskly and repeat the motion two or three times. (*Cheyenne* II.) "Keep pushing."

Right index, as (**J**), carried to the right and to the left, and in front of the body, when the hand is turned horizontal, finger pointing straight outward, and then the hand is drawn inward toward the body and slightly upward and then thrown forward and downward, on a curve, with a quick strong movement, so that the arm is fully extended in front of the body, with finger horizontal and pointing outward. (*Dakota* I.) "Anything it is I will try to do."

Turkey.

The open hands brought up opposite to the shoulders and imitating slowly the motion of the wings of a bird, to which add the sign for **Chicken.** (*Dunbar.*)

Understand. (Compare **Hear** and **Know.**)

The fingers and thumb of the right hand brought together near the tips, and then approached and receded, to and from the ear two or three times, with a quick motion, made within the distance of two or three inches The motion of the fingers is designed to represent the sound entering the ear. (*Long.*)

Vertically lower the hand (right usually employed), forefinger and thumb extended, other fingers closed and nails up, in a decisive or emphatic manner. This is often preceded by the sign of affirmation, *i. e.*, **Yes, I understand.** (*Arapaho* I.)

Right hand, middle, third, and little fingers closed, forefinger extended, thumb cocked upward, held a short distance in front of the mouth; sign for **Yes.** (*Cheyenne* I.)

To point with the forefinger to the ear means "I have heard and understand." (*Ojibwa* I.)

Make the sign **To Hear,** then place the hand quickly before the chin, the index pointing to the left; then move the hand forward and downward a short distance until the palm comes uppérmost. The motion takes place at the wrist. When the motion is quickly made at the termination of the sign **Hear** to the beginning of that for **Understand,** it is equivalent to the conjunction *and.* (*Shoshoni and Banak* I.)

Forefinger of the right hand extended and crooked, other fingers closed, thumb resting on the second, is carried behind the right ear, and then in the same position in front of the left breast, where it is held for a moment with hand upright, edge of fingers outward, back of hand toward the right. (*Dakota* I.) "I hear; I understand."

After making the sign for **To hear,** throw the back of the hand forward (retaining the position of the fingers), and move it forward and downward. (*Absaroka* I; *Hidatsa* I; *Arikara* I.)

Both arms are flexed and folded on the chest; the fingers are closed except the index, which is hooked much as in position (**I**), index more opened and hand horizontal. The hands thus are made to touch the sides of the chest and then passed uniformly forward toward the object; the same several times repeated. (*Oto and Missouri* I.) "Something known between you and me."

Make the sign for **To Hear,** and by merely reversing the palm conclude by that for **To Know.** Sometimes the sign for **To Know** is made only, as an abbreviation. (*Kaiowa* I; *Comanche* III; *Apache* II; *Wichita* II.)

Forefinger of right hand moved quickly from behind the ear to the front. (*Sahaptin* I.)

Another: Move right hand, palm toward head, all fingers extended (**T** 1), to a position behind ear; then move it past the ear to a point in front of breast; then turn the hand, palm down (**W** 1), and move to a point, say a foot from body, and a little to the right. This last is sign for **Good.** (*Sahaptin* I.) "Cutting off the sound or words."

Index to breast, then to lips with a vigorous thrust upward and forward, with an affirmative nod. (*Apache* III.)

Deaf-mute natural sign.—Look down at vacancy, with the eyebrows knit, and placing the hands on the forehead and then raising the head, slightly arch the eyebrows. (*Ballard.*)

——— Do not. See also **Hear, do not.**

Move the flat hand quickly past the ears means "I have not heard" and may mean that he *will* not understand, or that the request passes his ears unheeded. According to circumstances it may mean that it passes his ears because he considers it untrue. Slightly modified, it will indicate, "You are trying to take me in." (*Ojibwa* IV.)

Sign for **Understand,** followed by that for **No.** (*Arapaho* I;) (*Cheyenne* II.)

Point to the ear with the right index, slightly curved and remaining fingers closed; then place the tips of the fingers against the ball of the thumb, and snap them off—as if sprinkling water, from the ear outward and forward from the ear. (*Shoshoni and Banak* I.)

First make the sign for **Understand,** and then the sign for **No.** (*Dakota* I.) "Do not hear you, understand you."

The sign for **Hear,** followed by that for **No,** made to the side of the head. (*Apache* I.)

Pass one or two fingers from lips to the ear and make the sign for **No.** (*Apache* III.)

Deaf-mute natural sign.—Look down at vacancy, knit the eyebrows, putting the hand on the forehead and shake the head. (*Ballard.*)

Unready, unprepared, etc.

The arms are raised and extended parallel before the body. The hands assume position (**K** 1) modified by being horizontal. The index-fingers are then approximated and rubbed together at palmar points. (*Oto and Missouri* I.)

Vest. See **Clothing.** (*Dakota* IV.)

Vain.

Cannot be separated from **Proud, Pride.** (*Dakota* I.)

Deaf-mute natural sign.—Move the fingers of both hands up and down. (*Zeigler.*)

——— Purse-proud.

Italian sign.—Both hands stuck in the pocket. (*Ballard.*)

Vermillion. See **Color.**

Village (Indian). (Compare **Kettle.**)

Place the open thumb and forefinger of each hand opposite to each other, as if to make a circle, but leaving between them a small interval; afterward move them from above downward simultaneously. (*Wied.*) There is no similarity in execution of the (*Oto and Missouri* I) sign and that of *Wied's*, nor in their conception, as the village is not surrounded by a stockade. (*Boteler.*) The villages of the tribes with which the author was longest resident, particularly the Mandans and Arikaras, were surrounded by a strong circular stockade, spaces or breaks in the circle being left for entrance or exit.

Repetitions of the sign for **Lodge,** or that sign and that for **Many.** (*Arapaho* I.)

Partly extend the tips of fingers of both hands, slightly cross the hands perpendicular in front of breast, then describe a circle by a slight circular move of the hands and wrists, palms inside, and drop the hands a little, and in both descriptions point to the direction of the village, and repeat several times the form sign within reach of the arms and hand when a village is described, and but once if only one house is to be described, saying *Wig-wam!* (*Ojibwa* IV.)

First make the sign for **Lodge, tipi,** and then the sign for **Many.** (*Dakota* I, IV.) "Many tipis."

The arms are elevated and the hands approximated at the finger tips before the face; the hands and arms then diverge from points of contact to form the triangular representation of the wigwam door; the sign for wigwam or house being thus completed, the right fist, in type-position (**A**), marks the same successively around the subject's position on the ground. (*Oto and Missouri* I.) "Many houses collected in one locality."

Raise both hands to a position in front, a little to right of the face, fingers extended, pointing upward, palms facing each other (**R** 1, right and left). Then, with zigzag movement, up and down, pass them in front of face to left, hands say five inches apart. (*Sahaptin* I.) "Village—things standing on ground."

——— White man's.

Repetition of sign for **House,** also that sign and the sign for **Many.** (*Arapaho* I.)

Move both hands with two motions, first back of left toward palm right, about twelve inches apart, then forming a right angle forming a square representing the four sides of a house; then place the hands, thumbs under so as to show a cover, as a roof of a house, and pronouncing *Wig-wam.* (*Ojibwa* I.)

The sign for **White man** is prefixed to that for **Village.** (*Dakota* I.) "Many white men's houses."

Make the sign for **House,** and then the sign for **Many.** (*Dakota* IV.)

Make the sign for **Village,** followed by that for **White man.** (*Sahaptin* I.)

Outline with extended hands (**T** on edge) the vertical walls and peaked roof; also between earth and roof pass the hand (**X** reversed), horizontally, indicating house divided into two stories; dimensions large; wave hands about horizontally, level of chin, palms down (**W**), great extent of town. (*Apache* III.)

Volley.

The two hands as in sign for *discharge of a deadly missile,* are held in front, a few inches apart and directed toward one another, then the fingers are suddenly straightened as in the same sign; this may be repeated to represent the volleys of contending forces, and each hand may make its sign simultaneously or alternately with the other. (*Mandan and Hidatsa* I.)

Wagon.

Roll hand over hand, imitating a wheel. (*Burton.*)

The right hand, with fingers closed (fist **B**), is rotated at the right side of the body. (*Dakota* I.) "From the motion of the wheels."

Both hands held in front of the body, the fingers extended, the right fingers pointing to the left, the left fingers to the right, the tips of the fingers opposite the wrist of the opposite hand, the hands about six inches apart, both palms toward the body. By a movement of the elbows, rotate the hands over each other to the front, like a revolving wheel. (*Dakota* III.)

Place both hands, palms backward, at the height of the shoulders and a little in front of them, or place them near the sides of the body, flex both forefingers and thumbs until their ends are about an inch apart, the other fingers nearly closed, then throw the hands forward several times, each time bringing the ends of the thumbs and forefingers together, to imitate the rotation of wheels. (*Dakota* IV.)

Both arms are flexed at a right angle before the chest; the hands then assume type-position (**L**), modified by the index-finger being hooked and middle finger partly opened and hooked similarly; the hands are held horizontally and rotated forward side by side to imitate two wheels, palms upward. (*Oto and Missouri* I.) "The erect wagon-standard and curved rolling wheel."

Both hands closed in front of the body, about four inches apart, with forefingers and thumbs approximating half circles, palms toward the ground, move forward slowly in short circles. (*Kaiowa* I; *Comanche* III; *Apache* II; *Wichita* II.) "Wheels and revolving motion forward."

Swing the forefingers of each hand around each other, representing the wheel running. (*Sahaptin* I.)

(1) Circle both thumbs and indices, and hold them parallel; (2) place these circles with slight emphasis in two places; (3) seize left circle with right thumb and index and move right hand as if removing it; (4) dip right index downward; (5) sweep it around extended left index; (6) remake left circle. (*Apache* III.) "(1) Two wheels; (2) capable of progression; (3) took off one wheel; (4) dipped up grease; (5) greased axle; (6) put on wheel again." This probably means more than the simple idea of "wagon."

Deaf-mute natural sign.—An up-and-down motion of the bent arms in imitation of a man riding on horseback, and moving the fingers in circles to denote the motion of the wheels. (*Ballard.*)

Wait.

Australian sign.—"Minnie-minnie." (Wait a little)—Hand with fingers half clinched, between type-positions (**A**) and (**D**), thumb straight, shaken downward rapidly two or three times. Done more slowly, toward the ground, it means **Sit down.** (*Smyth.*)

Want, To.

Curve the index, and bring it in a curve downward toward the mouth, past it, and forward from the breast a short distance. (*Kaiowa* II; *Comanche* III; *Apache* II; *Wichita* II.)

War. See **Battle.**

——— To declare.

First make the sign for **Battle,** and then the sign for **Marching** or **Traveling.** Of course Indians do not make any formal declaration of war, and the above sign would be rendered *going to battle, going to fight.* (*Dakota* I.) "From fighting, battle."

War-path, On the.

With its index at a right angle with the palm, and pointing toward the left, its thumb extended and upright, back of hand forward and outward, the other fingers closed, move the right hand rapidly forward about a foot from just in front of the right shoulder. (*Dakota* IV.) "The thumb chasing the index."

Warm. (Compare **Hot.**)

Draw the hand across the forehead as if wiping off the perspiration. (*Apache* II.)

Wash.

Rub the hand as with invisible soap in imperceptible water. (*Burton.*)

Back of left hand briskly rubbed with palm of right. (*Cheyenne* I.)

Precisely the same as though washing the face with both hands with water contained in a wash-basin, about the height of the stomach. (*Dakota* I.) "From the act of washing."

Water. (Compare **Drinking.**)

The hand formed into a bowl and brought up to the mouth, passing a little upward without touching the mouth. (*Dunbar.*)

The hand is partially clinched, so as to have something of a cup-shape, and the opening between the thumb and finger is raised to the mouth and continued above it. (*Long.*)

Open the right hand and pass it before the mouth from above downward. (*Wied.*)

Wave the right hand, held open, palm to the mouth, as if about to hold the mouth shut, lick the palm of the hand with the tongue, moving the hand from above downward on the tongue. (*Burton.*)

Present the hollowed hand, cup-shaped, other fingers and thumb closed, back of the hand down. (*Arapaho* I.)

A hollow hand, with the motion of drawing water. (*Ojibwa* I.)

Right hand flat and slightly arched or curved; pass it downward before the face from the forehead to the chin, palm inward. (*Absaroka* I; *Shoshoni and Banak* I.) "To wash the face."

Same as the sign for **Drinking** or for **River,** which of course includes water. (*Dakota* I.)

Place the right hand upright six or eight inches in front of the mouth, back outward, index and thumb crooked, and their ends about an inch apart, the other fingers nearly closed; move it toward the mouth, and then downward nearly to the top of the breastbone, at the same time turn the hand over toward the mouth until the little finger is uppermost. (*Dakota* IV.) "Carrying a cup to the mouth and emptying it."

Collect the fingers and thumb of the right hand to a point, and bring them to the mouth, palm up. (*Wyandot* I.)

Place the flat right hand before the face, pointing upward and forward, the back forward with the wrist as high as the nose; then draw it downward and inward toward the chin. (*Kaiowa* I; *Comanche* III; *Apache* II; *Wichita* II.) "From the former custom of drinking with the bowl-shaped hand."

Pass the extended flat right hand downward before the face, fingers to the left and back forward. (*Shoshoni and Banak* I.) "From the manner in which it is used in washing the face."

Deaf-mute natural sign.—Raise and depress the hand in imitation of a man handling a pump, and move the hand to the mouth and raise the head in imitation of the act of drinking. (*Ballard.*)

——— A drink of.

Make sign for **Bring;** fingers still crooked as in sign for **Many,** brought over forward from the mouth with sudden downward curve. (*Cheyenne* I.)

——— Spring of.

(1) Hold the hands down, thumbs and indices widely separated, as if inclosing a round object twenty inches across; (2) trace a serpentine line from it with finger tip. (*Apache* III.) "(1) A spring; (2) a rivulet."

Watermelons, Squashes, and Muskmelons.

Pantomimically expressed by illustrating their form on the ground, and according to size for specific designation. (*Ute* I.)

Weak, Weakness.

Left hand is held in front of the body about a foot, hand horizontal, fingers extended and pointing toward the right, back of hand outward (**S** I), and the right hand (**S** 1) is passed from the right breast forward by the left, with the fingers pointing straight outward, back of hand toward the right, and then the right hand carried directly out toward the right side of the body. (*Dakota* I.) "No; no go; cannot go; am weak, sickly."

Well, Good Health.

Have a smiling countenance, raise both hands quickly to lower part of face and mouth as if in the act of eating, shake and gently touch the breast and body, as hands descend in front, separate over the abdomen with quivering motion, then move both hands outward, raise hands quickly, back of hands above, stand erect and throw chest forward. (*Ojibwa* IV.)

Palms of both hands, fingers extended, touch the body over the breast, stomach, abdomen, &c., and then make the sign for **Good.** (*Dakota* I.) "Body is all good; I am well."

The arms are passed tremblingly up the sides of the body, then the extended indices are made to press the temples, the countenance assuming all the while a visage of distress; the sign for **Sick** being thus completed the hand is thrown open negatively from the body to indicate emphatically—no sickness. (*Oto and Missouri* I.) "Not sick or in distress."

Weep, To; To Cry.

The forefinger of each hand extended, carried to its respective eye, back of hand outward, all fingers but the index closed, carry the fingers down the face as though following the course of the tear-drops. (*Dakota* I.) "From the dropping of the tears."

Make the sign for **Rain,** but in so doing hold the backs of the pendent fingers toward the face, and drop the hand repeatedly from the eyes downward. (*Ute* I.) "Literally, 'eye-rain;' drops of water from the eyes."

Sign for **Water** made from eyes. (*Cheyenne* I.)

Deaf-mute natural sign.—Rub the eye with the back of the hand as children do when crying. (*Ballard.*)

What? What do you say? See **Question.**

Wheat. (Compare **Grass.**)

Same sign as for **Grass,** begun near the ground, and gradually and interruptedly elevated to the height of about three feet. (*Ute* I.)

When? See **Question.**

Whence come you?

First the sign for **You,** then the hand extended open and drawn to the breast, and lastly the sign for **Bringing.** (*Dunbar.*)

Where?

With its back upward and index pointing forward, carry the right hand from left to right about eight inches, raising and lowering it several times while so doing, as if quickly pointing at different objects. (*Dakota* IV.)

Whiskey.

Make the sign for **Bad** and **Drink,** for "bad water." (*Burton.*)

Sign for **Water** and **Fire.** (*Cheyenne* I.)

Another: Sign for **Bring,** and right-hand fingers outspread, tips pointing upward, shaken before forehead with wave of head to indicate unsteadiness. (*Cheyenne* I.)

First make the sign for **Drinking,** and then the sign for **Fool.** (*Dakota* I.) "Very expressive."

The right arm is flexed and elevated, the right hand approaches the chest in type-position (**G** 1), modified by being held edge up. The hand thus seemingly grasping a bottle is passed semicircularly upward toward, then to the mouth and from it, the head receding. The hand then falls to the side, and the head inclining to either side is swayed to and fro, indicative of lost equilibrium. (*Oto and Missouri* I.) "Something drunk that stupefies the senses."

Deaf-mute natural sign.—Raise the hand, fingers placed together, toward the mouth, and shake the body to and fro sidewise. (*Larson.*)

White. See **Color.**

White Man; American.

Place the open index-finger and thumb of the right hand toward the face, then pass it to the right in front of the forehead to indicate the hat. (*Wied.*) Still used to some extent. (*Matthews.*) There is a plain and evident similarity in both execution and conception in the (*Oto* I) sign and *Wied's.* (*Boteler.*)

The sign for **Trade** also denotes the Americans, and, indeed, any white men, who are generally called by the Indians west of the Rocky Mountains "Shwop," from our "swap" or "swop." (*Burton.*) This is a legacy from the traders who were the first representatives of what used to be called the Caucasian race, met by the Indians.

A finger passed across the forehead. (*Macgowan.*)

Indicate upon the forehead with the hand the supposed line of contact of a hat. (*Arapaho* I.) "Literally, 'the hat-wearers.'"

Hold one hand horizontally over the forehead. (*Sac, Fox and Kickapoo* I.)

The extended forefinger of the right hand (**M**, turned inward) is drawn from the left side of the head around in front to the right side, about on a line with the brim of the hat, with back of hand outward. (*Dakota* I.) "From the wearing of a hat."

Draw the opened right hand horizontally from left to right across the forehead a little above the eyebrows, the back of the hand to be upward and the fingers pointing toward the left. Or, close all the fingers except the index, and draw it across the forehead in the same manner. (*Dakota* IV.) "From the hats worn by the whites."

Close the right hand, and draw the back of the thumb horizontally across the forehead from left to right. (*Hidatsa* I; *Arikara* I.)

The right arm is raised and the hand assuming position (**I** 1), index not entirely closed but loosely hooked, is then drawn across the forehead in a line corresponding to the hat mark. The Indians wearing generally a rudely constructed turban would most likely select the sign for hat to distinguish the white man, from its being his universal head-dress. (*Oto and Missouri* I.) "The man that wears the hat."

Point to the eye with the index, then place the half-closed hands, palms toward the forehead, over the eyes and pass them downward over the cheeks and forward toward the chin. (*Wyandot* I.) "Stated to be based upon the fact of the generally gray or light eyes of Americans first seen, followed by the sign for **Person.**"

Pass the palmar surface of the fully extended and separated thumb and index of the right hand across the forehead from left to right. Although this is the essence of the gesture, numerous slight variations or abbreviations occur. Frequently the extended index only is drawn across the forehead; sometimes the thumb is placed against the right temple as a support for the hand, as the index is drawn across to it. The hand and arm may be placed in almost any position, as it does not form an essential feature in expressing the idea. The left hand has also been used in instances when the right was engaged, as in holding a pipe while smoking, or from other causes. (*Kaiowa* I; *Comanche* III; *Apache* II; *Wichita* II.) "The line at which the hat rests against the forehead."

Raise right hand to side of head, arched, thumb and fingers horizontal, pointing to temple (**H** 1) then pass hand in that position over the forehead, at same time turning head to right. (*Sahaptin* I.) "Man with hat."

Place the extended and separated index and thumb of the right hand, palm downward, across the right side of the forehead. (*Pai-Ute* I.)

Imitate the stroking of chin whiskers, then place the back of the wrist against the chin, allowing the spread fingers and thumb to point forward and downward. (*Ute* I.)

Another: Draw the radial side of the extended index across the forehead from left to right. (*Ute* I.)

Make the sign for **Man;** place the hands on face; touch tent-cloth or some other white object, point to the hands. (*Apache* III.) "Man with white face or hands."

Another: Hand turned, tips down under chin (**Q**). (*Apache* III.) 'Beard; the Mescaleros being beardless."

Deaf-mutes generally, in especial the French, make the "hat" sign for *man* as distinguished from *woman*.

Wicked. See **Bad heart.**

Wide. See **Big** in the sense of wide.

Wife. See **Husband, Relationship,** and **Comparison.**

Wigwam. See **Lodge.**

Wild.

The hands in front of their respected breasts, with fingers and thumbs extended, separated and pointing downward, backs of hands outward, are quickly carried outward with a tremulous motion of all the fingers, and as the hands are carried outward increase the distance between them, *i. e.*, carry the hands out obliquely. (*Dakota* I.) "Probably from the movements of some kinds of game on being surprised."

Wind (air in motion). **Air.**

Right hand held perpendicularly upward and brought forward with a tremulous or vibratory motion until it passes beyond the face. (*Dunbar.*)

Stretch the fingers of both hands outward, puffing violently the while. (*Burton.*)

Indicate with the extended hand its direction and force, and emit a whistling sound. (*Arapaho* I.)

Both hands held up carelessly before the body, fingers naturally extended, swept suddenly with downward and upward curve to the left. (*Cheyenne* I.)

Right hand with the fingers slightly separated, upright palm forward or outward (**R**), is carried, held in this position, from behind the body, by the right side of the head, to the front of the body, on the same level as far as the arm can be extended, and at the same make with the mouth a sound in resemblance to the whistling of the wind. (*Dakota* I.) "From the whistling of air in motion swiftly by a person."

Make short, rapid, vertical vibrations of the hand, spread and palm down (**W**), sweeping it from behind forward, about height of waist; accompanied by a whizzing sound from the mouth. (*Apache* III.)

Deaf mute natural signs.—Blow through the lips and move the uplifted hand horizontally. (*Ballard.*)

Blow the air from the mouth, and then move the stretched hand in a line before the breast. (*Larson.*)

Winter. (Compare **Cold.**)

A shrinking, shivering condition; move as if drawing a blanket around the head and shoulders, then move both hands above the head, shaking the hand as in the case of **Falling leaves**; bring the hands (fingers down) toward the ground and undulate them near the ground; then with the right hand indicate the depth and level of the snow; pronounce the name of snow, *Occone.* (*Ojibwa* IV.)

Make the sign for **Cold,** and then for **Snow.** (*Dakota* I.) "Cold weather—the season of snow."

Shake the upright nearly closed hands back and forth several times in front of the shoulders as if shivering, palms inward; then suspend the hands about a foot in front of the shoulders, backs forward, fingers separated and bent a little, and pointing downward, and move all the fingers as if shaking something from them, or approximate and separate the ends once or several times, and while doing so, each time throw the hands downward about eight inches. (*Dakota* IV.) "Cold and snow."

Make the signs for **Rain, Deep,** and **Cold.** (*Dakota* VI.)

Make the sign for **Rain** with both hands, then pass the hands with palms down horizontally to the right and left before the lower part of the body. (*Dakota* VI; *Hidatsa* I; *Arikara* I.) "Precipitation and depth."

Both hands in position (**A**) tremble before the breast. This being enacted, the hands of the subject are extended, the arms likewise on a level with shoulders pointing to the horizon, both hands now describe the arc subtending the quadrant of the horizon until they meet over the head. (*Oto and Missouri* I.) "When cold days spread over us."

Same sign as for **Cold.** (*Kaiowa* I; *Comanche* III; *Apache* II; *Wichita* II.)

Bring both fists together in front of the breast as if drawing together the edges of a blanket. (*Apache* I.)

Wise.

Raise the right hand and fingers, gently tap the forehead over the right eye, and pass backward alongside of head with three or four taps of fingers on head. (*Ojibwa* IV.) "Probably intimating a *level head.*"

Touch the forehead with the right index and then make the sign for **Big** directly in front of it. (*Dakota* I.) "Big brain."

Touch the side of the head with the flat right hand, then elevate the hand toward the sky. (*Wyandot* I.) "Superior in intelligence."

Tap the forehead with the index, and make the sign for **Speak.** (*Apache* I.)

Wish; desire of possession.

Hook the forefinger over the nose. (*Arapaho* I.)

Right hand, with thumb and forefinger extended, is brought upward in front of the body, with the back of the hand outward, thumb and forefinger pointing toward the left, to the level of the breast, when the hand is quickly thrown upward, outward, and then slightly downward, *i. e.*, on a curve, so that the hand is horizontal with the palm upward. (*Dakota* I.) "I like it, wish it."

Wolf. (Compare **Dog.**)

Same sign as for **Dog.** (*Oto* I.)

Place the right hand, fingers joined and extended, above the mouth, pointing downward and forward. (*Ute* I.) "Long nose."

Woman. (Compare **Female.**)

The finger and thumb of the right hand, partly open, and placed as if laying hold of the breast. (*Dunbar.*)

The hands are passed from the top down each side of the head, indicating the parting of the hair on the top, and its flowing down each side. (*Long.*)

Pass the palm of the extended hand downward over the hair on the side of the head, or downward over the cheeks. (*Wied.*) Same as my description, but less precise. (*Matthews.*) The arms were flexed and the hands, fist-like, held at either side in the position of the female mammary glands, then sweeps semicircularly downward. There is no appreciable similarity in this sign and *Wied's*, the conception and execution of which are wholly different. (*Boteler.*) "One with prominent mammæ, who can bring forth young."

Pass the hand down both sides of the head, as if smoothing or stroking the long hair. (*Burton.*)

A finger directed toward the breast. (*Macgowan.*)

Turn the right hand about the right ear, as if putting the hair behind it. (*Dodge.*)

Draw the hand, the fingers separate and partially closed, palm toward the cheek, downward, as of combing the hair. (*Arapaho* I.)

Right-hand fingers close together, thumb lying along basal joint of forefinger, placed above the top and side of the head, bent and suddenly brought down and outward to the level and right of shoulder, finger ends still bent in toward the latter. (*Cheyenne* I.) "To express shortness as compared with man."

Pass the palm once down the face and the whole body. (*Ojibwa* I.) "The long, waving dresses [*sic*] or the graceful contour of the female body."

Hold the hands cup shaped over each breast. (*Sac, Fox, and Kickapoo* I.)

Pass the extended and flat right hand, back forward and outward, from the side of the crown downward toward the shoulder and forward. (*Absaroka* I; *Shoshoni and Banak* I.) "Represents the long hair."

The right hand brought to the top of the head and then carried out sidewise toward the right and downward as though drawing a comb through the long hair of a woman's head. (*Dakota* I.) "Long hair."

Right-hand fingers extended and joined (as in **T**), horizontal, held on the left side of the face, the fingers pointing to the rear, the thumb grasping and sliding downward to represent stroking the long braided hair of a squaw. (*Dakota* III.)

With the right hand, back forward, fingers slightly flexed and joined, thumb close to index, the little finger near the head, make a motion as if brushing the hair behind the ear by moving the hand backward and downward through an arc of about six inches. (*Dakota* IV.) "The women wear the hair behind the ears and plaited."

Pass the flat right hand, palm of extended fingers resting near the right side of the crown, and downward and to the front of the collarbone. (*Dakota* VI.) "Represents long hair."

The extended hands, palms backward, and pointing upward and inward, are held each near the temple of the same side. They are then swept simultaneously downward a foot or two. (*Mandan and Hidatsa* I.) "This is to indicate the mode of dressing the hair most common with women—a braid on each side."

Both hands are brought to a position corresponding with female mammæ. The hands are loosely clinched as in type (**F**) and laid loosely against chest on side corresponding with hand, although sometimes the arms are crossed and hands held in above positions on opposite sides. **My woman** is expressed by tapping the left breast by point of right index-finger in addition to above. (*Oto* I.) "A position indicative of female mammæ and connubial embrace."

Pass the extended and flat right hand, fingers joined, from the side of the crown downward and forward along the cheek to the front of the right side of the neck, the fingers pointing downward at termination of motion. Both hands are sometimes used. (*Kaiowa* I; *Comanche* III; *Apache* II; *Wichita* II.)

With the fingers and thumb of the right hand separated and partly bent or hooked, pass from the side of the head toward the front of the shoulder, gradually closing the hand in imitation of gathering and smoothing the lock of hair on that side. (*Pai-Ute* I.)

Touch the hair on the side of the head with the fingers of the right hand, then place the closed hand before the pubis, with the back forward, index and second fingers extended and separated, pointing downward; place the thumb against the palm so that the tip protrudes a little from the crotch thus formed by the fingers. (*Ute* I.) "Fourchette, glans clitoridis, and location of."

The left fore and second fingers are extended and separated, the remaining fingers closed; the thumb is then placed against the palm in such a manner that the tip is visible in the crotch formed by the extended fingers; the hand is then placed back forward in this position at the crotch. (*Apache* I.) "Resemblance to the *pudendum muliebre.*"

(1) Two fingers held downward (**N** reversed); (2) sweep hands up near legs and clasp them about the waist; (3) sweep hands from shoulder to waist loosely. (*Apache* III.) (1) "Human being (2) wearing shirts and (3) loose jackets."

Deaf-mute natural signs.—Take hold of the garment at the side below the hip and shake it to denote the skirt of a woman's dress. (*Ballard.*)

Point the hand to the rear side of the head, because of the combs the women wear as ornaments. (*Larson.*)

Instructed deaf-mutes generally mark the line of the bonnet-string down the check.

Italian sign.—Draw the hand down the cheek under the chin. (*Butler.*)

——— Old.

Make the sign for **Woman,** and then make the sign for **Progression with a staff.** (*Dakota* IV.) "Progression of a woman with a staff."

——— Young, girl.

Make the sign for woman, hands held in the same position, and brought from shoulder downward and outward in proportion to the height of the girl. (*Cheyenne* I.)

Deaf-mute natural sign.—Take hold of an imaginary garment below the thigh and shake it, and place the hand to the height of a girl. (*Ballard.*)

Touch the right ear with the finger, because of the ear-rings girls wear. (*Larson.*)

Wonder. (Compare **Admiration** and **Surprise.**)

Same as the sign for **Surprise.** Surprise and wonder seem to go hand in hand, but admiration and wonder do not seem to be necessarily connected. (*Dakota* I.)

Place right hand over mouth, the thumb being on the right and the fingers on the left of the nose; then shrink back. (*Omaha* I.) The gesture of placing the right hand before the mouth is seemingly involuntary with us, and appears also in the Egyptian hieroglyphs.

Deaf-mute natural sign.—Part the lips, raise the hand, and arch the eyebrows, each action in a slow manner. (*Ballard.*)

Raise apart the arms, with the hands open. (*Larson.*)

Wood.

Point to a piece of wood with right index extended. (*Dakota* I.)

Work, labor, etc.

The right hand, with fingers extended and joined, back of the hand outward, edge of fingers downward, is thrown from the level of the breast, forward, upward, and then downward, on a curve, so that the palm is brought upward, and then carried to the right side of the body, level of the face, where the extended fingers point upright, palm outward. (*Dakota* I.)

As work is a general term for manual exertion, the indefiniteness of this sign can be well understood. The arms and hands are extended before the body, the hands in type-position (**A**); the hands are then graspingly opened and shut as in seizing the plow-handles; the closed hands then approximate and forcibly strike as in working at mechanical pursuits. (*Oto and Missouri* I.) "The exertion required in different kinds of labor."

Hold both flat hands edgewise in front of the body, thumbs up, push forward with sudden interruptions, at each movement drawing back the fingers and throwing them forward at every rest. (*Kaiowa* I; *Comanche* III; *Apache* II; *Wichita* II.)

Make a sort of mild grasping motion with both hands in several directions downward. (*Apache* III.) "Suggestive of industrial activity, and supplemented by pantomime of sewing or chopping, if not promptly understood."

Wrap, To.

The left hand is held in front of the body, hand closed, horizontal, back upward, and the right hand, with fingers in position as though grasping something, is rotated around the stationary left. (*Dakota* I.) "From the act of wrapping."

Writing.

The act of writing is imitated by the finger in the palm of the opposite hand. (*Long.*)

(1) Left hand held up as if a piece of paper; (2) motion made with right hand as though writing. (*Cheyenne* I.)

The first part of the sign for **Book.** (*Dakota* I.)

Year.

Give the sign of **Rain** or **Snow.** (*Burton.*)

Sign for **Cold,** and then sign for **Counting**—one. (*Dakota* I.) "One winter."

Deaf-mute natural sign.—Point to shirt bosom and lower the extended fingers to signify *snow*, then raise the hand to denote the height or depth of the snow, and then depress the hands to signify *gone.* (*Ballard.*)

Yes. Affirmation. It is so. (Compare **Good** and **Truth.**)

The motion is somewhat like **Truth,** but the finger is held rather more upright, and is passed nearly straight forward from opposite the breast, and when at the end of its course it seems gently to strike something, though with a rather slow and not suddenly accelerated motion. (*Long.*)

Wave the hands straight forward from the face. (*Burton.*) This may be compared with the forward nod common over most of the world for assent, but that gesture is not universal, as the New Zealanders elevate the head and chin, and the Turks shake the head somewhat like our negative. *Rev. H. B. B. Barnum*, Harpoot, Turkey, in a contribution of signs received after the foregoing had been printed, denies the latter statement, but gives **Truth** as "gently bowing, with head inclined to the right."

Another: Wave the hand from the mouth, extending the thumb from the index and closing the other three fingers. (*Burton.*)

Gesticulate vertically downward and in front of the body with the extended forefinger (right hand usually), the remaining fingers and thumb closed, their nails down. (*Arapaho* I.)

Right hand elevated to the level and in front of the shoulder, two first fingers somewhat extended, thumb resting against the middle finger; sudden motion in a curve forward and downward. (*Cheyenne* I.) "The correspondence between this gesture and the one for **Sitting,** seemingly indicates that the origin of the motion for **Affirmation** is in imitation of resting, or settling a question."

Same as the sign for **Truth.** (*Dakota* I.) "But one tongue."

Extend the right index, the thumb against it, nearly close the other fingers, and from a position about a foot in front of the right breast, bend the hand from the wrist downward until the end of the index has passed about six inches through an arc. Some at the same time move the hand forward a little. (*Dakota* IV.) "A nod; the hand representing the head and the index the nose."

The right hand, with the forefinger (only) extended and pointing forward, is held before and near the chest. It is then moved forward one or two feet, usually with a slight curve downward. (*Mandan and Hidatsa* I.)

Bend the right arm, pointing toward the chest with the index-finger. Unbend, throwing the hand up and forward. (*Omaha* I.)

Another: Close the three fingers, close the thumb over them, extend forefinger, and then shake forward and down. This is more emphatic than the preceding, and signifies, *Yes, I know.* (*Omaha* I.)

The right arm is raised to head with the index-finger in type-position (**I** 1), modified by being more opened. From aside the head the hands sweep in a curve to the right ear as of something entering or hearing something; the finger is then more opened and is carried direct to the ground as something emphatic or direct. (*Oto and Missouri* I.) "'I hear,' emphatically symbolized."

The hand open, palm downward, at the level of the breast, is moved forward with a quick downward motion from the wrist, imitating a bow of the head. (*Iroquois* I.)

Throw the closed right hand, with the index extended and bent, as high as the face, and let it drop again naturally; but as the hand reaches its greatest elevation the index is fully extended and suddenly drawn in to the palm, the gesture resembling a beckoning from above toward the ground. (*Kaiowa* I; *Comanche* III; *Apache* II; *Wichita* II.)

Quick motion of the right hand forward from the mouth; first position about six inches from the mouth and final as far again away. In first position the index-finger is extended, the others closed; in final, the index loosely closed, thrown in that position as the hand is moved forward, as though hooking something with it; palm of hand out. (*Sahaptin* I.)

Another: Move right hand to a position in front of the body, letting arm hang loosely at the side, the thumb standing alone, all fingers hooked except forefinger, which is partially extended (**E** 1, with forefinger partially extended, palm upward). The sign consists in moving the forefinger from its partially extended position to one similar to the others, as though making a sly motion for some one to come to you. This is done once each time the assent is made. More emphatic than the preceding. (*Sahaptin* I.) "We are together, think alike."

Deaf-mute natural sign.—Indicate by nodding the head. (*Ballard.*)

Yesterday. See **Day.**

You.

The hand open, held upward obliquely, and pointing forward. (*Dunbar.*)

Is expressed by simply pointing at the persons. (*Long.*)

Point to or otherwise indicate the person designated. (*Arapaho* I.)

Point toward the person with the extended forefinger of the right hand, back upward, horizontal. (*Dakota* I.) "Designating the person."

——— To.

With the fingers and thumb extended, lying closely side by side, and pointing upward, palm toward individual addressed, slowly move the hand toward the hearer, the finger-tips slightly in advance of the wrist, as if laying something against the person. (*Kaiowa* I; *Comanche* III; *Apache* II; *Wichita* II.)

——— Yours.

The arm and hands are folded on the chest as in the sign for **Mine;** they are then thrown open from the breast toward another, palms outward. (*Oto and Missouri* I.) "Not mine, yours."

TRIBAL SIGNS.

Absaroka, Crow.

The hands held out each side, and striking the air in the manner of flying. (*Long.*)

Imitate the flapping of the bird's wings with the two hands, palms downward, brought close to the shoulder. (*Burton.*)

The sign for these Indians is the same as that for **Fly, to.** (*Dakota* I.) "Flight of the crow."

Another: The Crow Indians simply place the index upon the ridge of the nose, but this sign would be understood by the Sioux as meaning *Nose.* (*Dakota* I.) "From the Sioux idea that the heart is the seat of life; consequently my heart is I, is myself." (*Sic.*) The placing of the index upon the ridge of the nose is understood to signify personality "I, myself," and not to be a tribal sign.

Both hands extended, with fingers joined (**W**), held near the shoulders, and flapped to represent the wings of a crow. (*Dakota* III.)

At the height of the shoulders and a foot outward from them, move the opened hands forward and backward twice or three times from the wrist, palms forward, fingers and thumbs extended and separated a little; then place the back or the palm of the upright opened right hand against the upper part of the forehead; or half close the fingers, placing the end of the thumb against the ends of the fore and middle fingers, and then place the back of the hand against the forehead. (*Dakota* IV.) "To imitate the flying of a bird, and also indicate the manner in which the *Absaroka* wear their hair."

Place the flat hand as high as and in front or to the side of the right shoulder, move it up and down, the motion occurring at the wrist. For more thorough representation, both hands are sometimes employed. (*Dakota* V, VI, VII; *Kaiowa* I; *Comanche* III; *Apache* II; *Wichita* II.) "Bird's wing."

Make with the arms the motion of flapping wings. (*Kutine* I.)

The right hand, flattened, is held over and in front of the right shoulder, and quickly waved back and forth a few times as if fanning the side of the face. When made for the information of one ignorant of the common sign, both hands are used, and the hands are moved outward from the body, though still near the shoulder. (*Shoshoni and Banak* I.) "Wings, *i. e.*, of a crow."

Arikara. (Corruptly abbreviated **Ree.**)

With the right hand closed, curve the thumb and index, join their tips so as to form a circle, and place to the lobe of the ear. (*Absaroka* I; *Hidatsa* I.) "Big ear-rings."

Collect the fingers and thumb of the right hand nearly to a point, and make a tattooing or dotting motion toward the upper portion of the cheek. This is the old sign and was used by them previous to the adoption of the more modern one representing "corn-eaters. (*Arikara* I.)

Place the back of the closed right hand transversely before the mouth, and rotate it forward and backward several times. This gesture may be accompanied, as it sometimes is, by a motion of the jaws as if eating, to illustrate more fully the meaning of the rotation of the fist. (*Kaiowa* I; *Comanche* III; *Wichita* II; *Apache* II.) "Corn-eater; eating corn from the cob."

Signified by the same motions with the thumbs and forefingers that are used in shelling corn. The dwarf Ree (Arikara) corn is their peculiar possession, which their tradition says was given to them by God, who led them to the Missouri River and instructed them how to plant it. (Rev. C. L. Hall, in *The Missionary Herald*, April, 1880.) "They are the corn-shellers."

Apache. See also **Warm Spring.**

Make either of the signs for **Poor, in property.** (*Kaiowa* I; *Comanche* III; *Apache* II; *Wichita* II.) "It is said that when the first Apache came to the region they now occupy he was asked who or what he was, and not understanding the language he merely made the sign for **Poor,** which expressed his condition."

Rub the back of the extended forefinger from end to end with the extended index. (*Comanche* II; *Ute* I.) "Poor, poverty-stricken."

——— Mescalero.

Same sign as for **Lipan.** (*Kaiowa* I; *Comanche* III; *Apache* II; *Wichita* II.)

Arapaho.

The fingers of one hand touch the breast in different parts, to indicate the tattooing of that part in points. (*Long.*)

Seize the nose with the thumb and forefinger. (*The Prairie Traveler.* By Randolph B. Marcy, captain United States Army, p. 215. New York, 1859.)

Rub the right side of the nose with the forefinger: some call this tribe the "Smellers," and make their sign consist of seizing the nose with the thumb and forefinger. (*Burton.*)

Finger to side of nose. (*Macgowan.*)

Touch the left breast, thus implying what they call themselves, viz: the "Good Hearts." (*Arapaho* I.)

Hold the left hand, palm down, and fingers extended; then with the right hand, fingers extended, palm inward and thumb up, make a sudden stroke from left to right across the back of the fingers of the left hand, as if cutting them off. (*Sac, Fox, and Kickapoo.*)

Join the ends of the fingers (the thumb included) of the right hand, and, pointing toward the heart near the chest, throw the hand forward and to the right once, twice, or many times, through an arc of about six inches. (*Dakota* IV.) "Some say they use this sign because these Indians tattoo their breasts."

Rub the side of the extended index against the right side of the nose. (*Kaiowa* I; *Comanche* III; *Apache* II; *Wichita* II.)

Collect the fingers and thumb of the right hand to a point, and tap the tips upon the left breast briskly. (*Comanche* II; *Ute* I.) "Good-hearted." It was stated by members of the various tribes at Washington, in 1880, that this sign is used to designate the Northern Arapahos, while that in which the index rubs against, or passes upward alongside of the nose, refers to the Southern Arapahos, the reasons given for which will be referred to in a future paper.

Another: Close the right hand, leaving the index only extended; then rub it up and down, held vertically, against the side of the nose where it joins the cheek. (*Comanche* II; *Ute* I.)

The fingers and thumb of the right hand are brought to a point, and tapped upon the right side of the breast. (*Shoshoni and Banak* I.)

Assinaboin.

Make the sign of **Cutting the throat.** (*Kutine I.*)

With the right hand flattened, form a curve by passing it from the top of the chest to the pubis, the fingers pointing to the left, and the back forward. (*Shoshoni and Banak* I.) "Big bellies."

Atsina, Lower Gros Ventre.

Both hands closed, the tips of the fingers pointing toward the wrist and resting upon the base of the joint, the thumbs lying upon and extending over the middle joint of the forefingers; hold the left before the chest, pointing forward, palm up, placing the right, with palm down, just back of the left, and move as if picking small objects from the left with the tip of the right thumb. (*Absaroka* I; *Shoshoni and Banak* I.) "Corn-shellers."

Bring the extended and separated fingers and thumb loosely to a point, flexed at the metacarpal joints; point them toward the left clavicle, and imitate a dotting motion as if tattooing the skin. (*Kaiowa* I; *Comanche* III; *Apache* II; *Wichita* II.) "They used to tattoo themselves, and live in the country south of the Dakotas."

Banak.

Make a whistling sound "phew" (beginning at a high note and ending about an octave lower); then draw the extended index across the throat from the left to the right and out to nearly at arm's length. They used to cut the throats of their prisoners. (*Pai-Ute* I.)

Major Haworth states that the *Banaks* make the following sign for themselves: Brush the flat right hand backward over the forehead as if forcing back the hair. This represents the manner of wearing the tuft of hair backward from the forehead. According to this informant, the Shoshoni use the same sign for **Banak** as for themselves—**Snake.**

Blackfeet. (This title is understood to refer to the Algonkin Blackfeet, properly called **Satsika.**)

The finger and thumb encircle the ankle. (*Long.*)

Pass the right hand, bent spoon-fashion, from the heel to the little toe of the right foot. (*Burton.*)

The palmar surfaces of the extended fore and second fingers of the right hand (others closed) are rubbed along the leg just above the ankle. This would not seem to be clear, but these Indians do not make any sign indicating *black* in connection with the above. The sign does not, however, interfere with any other sign as made by the Sioux. (*Dakota* I.) "Blackfeet."

Touch the right foot with the right hand. (*Kutine* I.)

Close the right hand, thumb resting over the second joint of the forefinger, palm toward the face, and rotate over the cheek, though an inch or two from it. (*Shoshoni and Banak* I.) "From manner of painting the cheeks."

Caddo. (Compare **Nez Percés.**)

Pass the horizontally extended index from right to left under the nose. (*Kaiowa* I; *Comanche* III; *Apache* II; *Wichita* II.) "'Pierced noses,' from former custom of wearing rings in the septum."

Calispel. See **Pend d'Oreille.**

Cheyenne.

Draw the hand across the arm, to imitate cutting it with a knife. (*Marcy*, Prairie Traveller, *loc. cit.*, p. 215.)

Draw the lower edge of the right hand across the left arm as if gashing it with a knife. (*Burton.*)

With the index-finger of the right hand proceed as if cutting the left arm in different places with a sawing motion from the wrist upward, to represent the cuts or burns on the arms of that nation. (*Long.*)

Bridge palm of left hand with index-finger of right. (*Macgowan.*)

Draw the extended right hand, fingers joined, across the left wrist as if cutting it. (*Arapaho* I.)

Place the extended index at the right side of the nose, where it joins the face, the tip reaching as high as the forehead, and close to the inner corner of the eye. This position makes the thumb of the right hand rest upon the chin, while the index is perpendicular. (*Sac, Fox, and Kickapoo* I.) It is considered that this sign, though given to the collaborator as expressed, was an error. It applies to the Southern Arapahos.

As though sawing through the left forearm at its middle, with the edge of the right held back outward, thumb upward. Sign made at the left side of the body. (*Dakota* I.) "Same sign as for a **Saw.** The Cheyenne Indians are known to the Sioux by the name of 'The Saws.'"

Right-hand fingers and thumb extended and joined (as in **S**), outer edge downward, and drawn sharply across the other fingers and forearm as if cutting with a knife. (*Dakota* III.)

Draw the extended right index or the ulnar (inner) edge of the opened right hand several times across the base of the extended left index, or across the left forearm at different heights from left to right. (*Dakota* IV.) "Because their arms are marked with scars from cuts which they make as offerings to spirits."

Draw the extended index several times across the extended forefinger from the tip toward the palm, the latter pointing forward and slightly toward the right. From the custom of striping arms transversely with colors. (*Kaiowa* I; *Comanche* II, III; *Apache* II; *Ute* I; *Wichita* II.)

Another: Make the sign for **Dog** and that for **To Eat.** This sign is generally used, but the other and more common one is also employed, especially so with individuals not fully conversant with the sign-language as employed by the Comanches, &c. (*Kaiowa* I; *Comanche* III; *Apache* II; *Wichita* II.) "Dog-eaters."

Draw the extended index across the back of the left hand and arm as if cutting it. The index does not touch the arm as in signs given for the same tribe by other Indians, but is held at least four or five inches from it. (*Shoshoni and Banak* I.)

Chippeway. See **Ojibwa.**

Comanche.

Imitate, by the waving of the hand or forefinger, the forward crawling motion of a snake. (*Burton*, also *Blackmore* in introduction to *Dodge's* Plains of the Great West, p. xxv. New York, 1877.) The same sign is used for the Shoshoni, more commonly called "Snake" Indians, who as well as the Comanche belong to the Numa linguistic family. "The silent stealth of the tribe." (Thirty Years of Army Life on the Border. By Col. R. B. Marcy, p. 33. New York, 1866.) But see **Shoshoni** for distinction between the signs.

Motion of a snake. (*Macgowan.*)

Hold the elbow of the right arm near the right side, but not touching it; extend the forearm and hand, palm inward, fingers joined on a level with the elbow, then with a shoulder movement draw that forearm and hand back until the points of the fingers are behind the body; at the same time that the hand is thus being moved back, turn it right and left several times. (*Sac, Fox, and Kickapoo* I.) "Snake in the grass. A snake drawing itself back in the grass instead of crossing the road in front of you."

Another: The sign by, and for the Comanches themselves is made by holding both hands and arms upward from the elbow, both palms inward, and passing both hands with their backs upward along the lower end of the hair to indicate *long hair*, as they never cut it. (*Sac, Fox, and Kickapoo* I.)

Right hand horizontal, flat, palm downward (**W**), advanced to the front by a motion to represent the crawling of a snake. (*Dakota* III.)

Extend the closed right hand to the front and left; extend the index, palm down, and rotate from side to side while drawing it back to the right hip. (*Kaiowa* I; *Comanche* III; *Apache* II; *Wichita* II.)

Make the reverse gesture for **Shoshoni**, *i. e.*, begin away from the body, drawing the hand back to the side of the right hip while rotating it. (*Comanche* II.)

Cree. Knisteno. Kristeneaux.

Sign for **Wagon** and then the sign for **Man.** (*Dakota* I.) "This indicates the Red River half-breeds, with their carts, as these people are so known from their habit of traveling with carts."

Dakota. Sioux.

The edge of the hand passed across the throat, as in the act of cutting that part. (*Long; Marcy*, Army Life, *loc. cit.*, p. 33.)

Draw the lower edge of the hand across the throat. (*Burton.*)

Draw the extended right hand across the throat. (*Arapaho* I.) "The cut-throats."

Draw the forefinger of the left hand from right to left across the throat. (*Sac, Fox, and Kickapoo* I.) "A cut-throat."

Forefinger and thumb of right hand extended (others closed) is drawn from left to right across the throat as though cutting it. The Dakotas have been named the "cut-throats" by some of the surrounding tribes (*Dakota* I.) "Cut-throats."

Right hand horizontal, flat, palm downward (as in **W**), and drawn across the throat as if cutting with a knife. (*Dakota* III.)

Draw the opened right hand, or the right index, from left to right horizontally across the throat, back of hand upward, fingers pointing toward the left. (*Dakota* IV.) "It is said that after a battle the Utes took many Sioux prisoners and cut their throats; hence the sign "cut-throats."

Pass the flat hand, with the palm down, from left to right across the throat. (*Dakota* VI.)

Draw the extended right hand, palm downward, across the throat from left to right. (*Kaiowa* I; *Comanche* II, III; *Shoshoni and Banak* I; *Ute* I; *Apache* II; *Wichita* II.) "Cut-throats."

——— Blackfoot (Sihasapa).

Pass the right hand quickly over the right foot from the great toe outward, turn the heel as if brushing something therefrom. (*Dakota* V.)

Pass the widely separated thumb and index of the right hand over the lower leg, from just below the knee nearly down to the heel. (*Kaiowa* I; *Comanche* III; *Apache* II; *Wichita* II.)

——— Brulé.

Rub the upper and outer part of the right thigh in a small circle with the opened right hand, fingers pointing downward. (*Dakota* IV.) "These Indians, it is said, were once caught in a prairie fire, many burned to death, and others badly burned about the thighs. Hence the name Si-cau-gu (burnt thigh) and the sign."

Brush the palm of the right hand over the right thigh, from near the buttock toward the front of the middle third of the thigh. (*Kaiowa* I; *Comanche* III; *Apache* II; *Wichita* II.)

——— Ogalala.

Fingers and thumb separated, straight (as in **R**), and dotted about over the face to represent the marks made by the small-pox. (*Dakota* III.) "This band suffered from the disease many years ago."

With the thumb over the ends of the fingers, hold the right hand upright, its back forward, about six inches in front of the face, or on one side of the nose near the face, and suddenly extend and spread all the fingers (thumb included). (*Dakota* IV.) "The word *Ogalala* means scattering or throwing at, and the name was given them, it is said, after a row in which they threw ashes into each other's faces."

Flathead, or **Selish.**

One hand placed on the top of the head, and the other on the back of the head. (*Long.*)

Place the right hand to the top of the head. (*Kutine* I.)

Pat the right side of the head above and back of the ear with the flat right hand. (*Shoshoni and Banak* I.) From the elongation of the occiput.

Fox, or **Outagami.**

Same sign as for **Sac.** (*Sac, Fox, and Kickapoo* I.)

Gros Ventre. See **Hidatsa.**

Hidatsa, Gros Ventre or **Minitari.**

Both hands flat and extended, palms toward the body, with the tips of the fingers pointing toward one another; pass from the top of the chest downward, outward, and inward toward the groin. (*Absaroka* I; *Shoshoni and Banak* I.) "Big belly."

Left and right hands in front of breast, left placed in position first, separated about four or five inches, left hand outside of the right, horizontal, backs outward, fingers extended and pointing left and right; strike the back of the right against the palm of the left several times, and then make the sign for **Go, Going.** (*Dakota* I.) "The Gros

Ventre Indians, Minnetarees (the Hidatsa Indians of *Matthews*), are known to the Sioux as the Indians who went to the mountains to kill their enemies; hence the sign."

Express with the hand the sign of a big belly. (*Dakota* III.)

Pass the flat right hand, back forward, from the top of the breast, downward, outward, and inward to the pubis. (*Dakota* VI; *Hidatsa* I; *Arikara* I.) "Big belly."

Indian (generically).

Rub the back of the extended left hand with the palmar surfaces of the extended fingers of the right. (*Comanche* II.) "People of the same kind; dark-skinned."

Rub the back of the left hand with the index of the right. (*Pai-Ute* I.)

Rub the back of the left hand lightly with the index of the right. (*Wichita* I.)

Kaiowa.

Make the signs of the **Prairie** and of **Drinking Water.** (*Burton; Blackmore* in Dodge's Plains of the Great West, xxiv. New York, 1877.)

Right-hand fingers and thumb extended and joined (as in **W**), placed in front of right shoulder, and revolving loosely at the wrist. (*Dakota* III.)

Place the flat hand with extended and separated fingers before the face, pointing forward and upward, the wrist near the chin; pass it upward and forward several times. (*Kaiowa* I; *Comanche* III; *Apache* II; *Wichita* II.)

Place the right hand a short distance above the right side of the head, fingers and thumb separated and extended; shake it rapidly from side to side, giving it a slight rotary motion in doing so. (*Comanche* II.) "Rattle-brained."

Same sign as (*Comanche* II), with the exception that both hands are generally used instead of the right one only. (*Ute* I.)

Kickapoo.

With the thumb and finger go through the motion of clipping the hair over the ear; then with the hand make a sign that the borders of the leggins are wide. (*Sac, Fox, and Kickapoo* I.)

Knisteno, or **Cree.**

Place the first and second fingers of the right hand in front of the mouth. (*Kutine* I.)

Kutine.

Place the index or second finger of the right hand on each side of the left index-finger to imitate riding a horse. (*Kutine* I.)

Hold the left fist, palm upward, at arm's length before the body, the right as if grasping the bowstring and drawn back. (*Shoshoni and Banak* I.) "From their peculiar manner of holding the long bow horizontally in shooting."

Lipan.

With the index and second fingers only extended and separated, hold the hand at arm's length to the front of the left side; draw it back in distinct jerks; each time the hand rests draw the fingers back against the inside of the thumb, and when the hand is again started on the next movement backward snap the fingers to full length. This is repeated five or six times during the one movement of the hand. The country which the Lipans at one time occupied contained large ponds or lakes, and along the shores of these the reptile was found which gave them this characteristic appellation. (*Kaiowa* I; *Comanche* III; *Apache* II; *Wichita* II.) "Frogs."

Mandan.

The first and second fingers of the right hand extended, separated backs outward (other fingers and thumb closed), are drawn from the left shoulder obliquely downward in front of the body to the right hip, (*Dakota* I.) "The Mandan Indians are known to the Sioux as 'The people who wear a scarlet sash, with a train,' in the manner above described."

Mexican. See VOCABULARY.

Minitari. See **Hidatsa.**

Negro. See VOCABULARY.

Since the VOCABULARY was printed the following has been received from Arapaho and Cheyenne Indians in Washington, as the sign for **Negro** as well as for **Ute.** Rub the back of the extended flat left hand with the extended fingers of the right, then touch some black object. Represents black skin. Although the same sign is generally used to signify **Negro,** an addition is sometimes made as follows: place the index and second fingers to the hair on the right side of the head, and rub them against each other to signify *Curly hair*. This addition is only made when the connection would cause a confusion between the "black skin" Indian (Ute) and negro.

Nez Percés. See **Sahaptin.**

Place the thumb and forefinger to the nostrils. (*Kutine* I.)

Ojibwa, or **Chippewa.**

Right hand horizontal, back outward, fingers separated, arched, tips pointing inward, is moved from right to left breast and generally over the front of the body with a trembling motion and at the same time a slight outward or forward movement of the hand as though drawing something out of the body, and then make the sign for **Man.** (*Dakota* I.) "Perhaps the first Chippewa Indian seen by a Sioux had an eruption on his body, and from that his people were given the name of the 'People with a breaking-out,' by which name the Chippewas have ever been known by the Sioux."

Osage, or **Wasaji.**

Pull at the eyebrows over the left eye with the thumb and forefinger of the left hand. This sign is also used by the Osages themselves. (*Sac, Fox, and Kickapoo* I.)

Hold the flat right hand, back forward, with the edge pointing backward, against the side of the head, then make repeated cuts, and the hand is moved backward toward the occiput. (*Kaiowa* I; *Comanche* III; *Apache* II; *Wichita* II.) "Former custom of shaving the hair from the sides of the head, leaving but an occipito-frontal ridge."

Pass the flat and extended right hand backward over the right side of the head, moving the index against the second finger in imitation of cutting with a pair of scissors. (*Comanche* II.) "Represents the manner of removing the hair from the sides of the head, leaving a ridge only from the forehead to the occiput."

Outagami. See **Fox.**

Pai-Utes, Head Chief of the.

Grasp the forelock with the right hand, palm backward; pass the hand upward about six inches, and hold it in that position a moment. (*Pai-Ute* I.) "Big chief."

Pai-Ute band, Chief of a.

Make the gesture as for **Pai-Ute, Head Chief,** but instead of holding the hand above the head lay it down over the right temple, resting it there a moment. (*Pai-Ute* I.) "Little chief."

Pani (Pawnee).

Imitate a wolf's ears with the two forefingers of the right hand extended together, upright, on the left side of the head. (*Burton.*)

Place a hand on each side of the forehead, with two fingers pointing to the front to represent the narrow, sharp ears of the wolf. (*Marcy,* Prairie Traveler, *loc. cit.*, p. 215.)

First and second fingers of right hand, straight upward and separated, remaining fingers and thumb closed (as in **N**), like the ears of a small wolf. (*Dakota* III.)

Place the closed right to the side of the temple, palm forward, leaving the index and second fingers extended and slightly separated, pointing upward. This is ordinarily used, though, to be more explicit, both hands may be used. (*Kaiowa* I; *Comanche* III; *Ute* I; *Apache* II; *Wichita* II.)

Extend the index and second fingers of the right hand upward from the right side of the head. (*Comanche* II.)

Pend d'Oreille, or **Calispel.**

Make the motion of paddling a canoe. (*Kutine* I.)

Both fists are held as if grasping a paddle vertically downward and working a canoe. Two strokes are made on each side of the body from the side backward. (*Shoshoni and Banak* I.)

Pueblo.

Place the clinched hand back of the occiput as if grasping the *queue*, then place both fists in front of the right shoulder, rotating them slightly to represent a loose mass of an imaginary substance. Represents the large mass of hair tied back of the head. This sign has been obtained from Arapaho and Cheyenne Indians while this paper was passing through the press.

Sac, or **Sauki.**

Pass the extended palm of the right hand over the right side of the head from front to back, and the palm of the left hand in the same manner over the left side of the head. (*Sac, Fox, and Kickapoo* I.) "Shaved-headed Indians."

Sahaptin, or **Nez Percés.**

Close the right hand, leaving the index straight but flexed at right angles with the palm; pass it horizontally to the left by and under the nose. This sign is made by the Nez Percés for themselves, according to Major Haworth. While this paper has been passing through the press information has been received from Arapaho and Cheyenne Indians, now in Washington, that this sign is also used to designate the Caddos, who practiced the same custom of perforating the nasal septum. The same informants also state that the Shawnees are sometimes indicated by the same sign. (*Comanche* II.) "Pierced nose."

Pass the extended index, pointing toward the left, remaining fingers and thumb closed, in front of and across the upper lip, just below the nose. The second finger is also sometimes extended. (*Shoshoni and Banak* I.) "From the custom of piercing the noses for the reception of ornaments." The Sahaptin, however, have not had that custom since being known to themselves.

Satsika. See **Blackfeet.**

Selish. See **Flathead.**

Shawnee. See **Nez Percés.**

Shoshoni, or **Snake.** (Compare **Comanche.**)

The forefinger is extended horizontally and passed along forward in a serpentine line. (*Long.*)

Right hand closed, palm down, placed in front of the right hip; extend the index and push it diagonally toward the left front, rotating it quickly from side to side in doing so. (*Absaroka* I; *Shoshoni and Banak* I.) "Snake."

Right hand, horizontal, flat, palm downward (**W**), advanced to the front by a motion to represent the crawling of a snake. (*Dakota* III.

With the right index pointing forward, the hand is to be moved for) ward about a foot in a sinuous manner, to imitate the crawling of a snake. (*Dakota* IV.)

Make the motion of a serpent with the right finger. (*Kutine* I.)

Place the closed right hand, palm down, in front of the right hip; extend the index, and move forward and toward the left, rotating the hand and finger from side to side in doing so. (*Kaiowa* I; *Comanche* III; *Apache* II; *Wichita* II.)

Place the closed right hand, palm down, in front of the right hip; extend the index, move it forward and toward the left, rotating the hand and index in doing so. (*Comanche* II.)

Close the right hand, leaving the index only extended and pointing forward, palm to the left, then move it forward and to the left. The rotary motion of the hand does not occur in this, as in the same sign given by other tribes for **Shoshoni** or **Snake.** (*Pai-Ute* I.)

——— Sheepeater. (Tukuarikai.)

Both hands, half closed, pass from the top of the ears backward, downward, and forward, in a curve, to represent a ram's horns; then, with the index only extended and curved, place the hand above and in front of the mouth, back toward the face, and pass it downward and backward several times. (*Shoshoni and Banak* I.) "Sheep," and "to eat."

Ute.

"They who live on mountains" have a complicated sign which denotes, "living in mountains" and is composed of the signs **Sit** and **Mountain.** (*Burton.*)

Left hand horizontal, flat, palm downward, and with the fingers of the right hand brush the other toward the wrist. (*Dakota* III.)

Place the flat and extended left hand at the height of the elbow before the body, pointing to the front and right, palm toward the ground; then pass the palmar surface of the flat and extended fingers of the right hand over the back of the left from near the wrist toward the tips of the fingers. (*Kaiowa* I; *Comanche* III; *Apache* II; *Wichita* II.) "Those who use sinew for sewing, and for strengthening the bow."

Indicate the color **Black,** then separate the thumbs and forefingers of both hands as far as possible, leaving the remaining fingers closed, and pass upward over the lower part of the legs. (*Shoshoni and Banak* I.) "Black or dark leggings."

Warm Spring Apache.

Hand curved (**Y,** more flexed) and laid on its back on top of the foot (*moccasins much curved up at toe*); then draw hands up legs to near knee, and cut off with edges of hands (*boot tops*). (*Apache* III.) "Those who wear booted moccasins with turn-up toes."

Wasija. See **Osage.**

White man; American. See VOCABULARY.

Wichita.

Indicate a circle over the upper portion of the right cheek, with the index or several fingers of the right hand. The statement of the Indian authorities for the above is that years ago the Wichita women painted spiral lines on the breasts, starting at the nipple and extending several inches from it; but after an increase in modesty or a change in the upper garment, by which the breast ceased to be exposed, the cheek has been adopted as the locality for the sign. (*Kaiowa* I; *Comanche* III; *Apache* II; *Wichita* II.)

Wyandot.

Pass the flat right hand from the top of the forehead backward over the head and downward and backward as far as the length of the arm. (*Wyandot* I.) "From the manner of wearing the hair."

PROPER NAMES, PHRASES, DIALOGUES, ETC.

PROPER NAMES.

President of the United States; Secretary of the Interior.

Close the right hand, leaving the thumb and index fully extended and separated; place the index over the forehead so that the thumb points to the right, palm toward the face; then draw the index across the forehead toward the right; then elevate the extended index, pointing upward before the shoulder or neck; pass it upward as high as the top of the head; make a short turn toward the front and pass it pointing downward toward the ground, to a point farther to the front and a littlelower than at the beginning. (*Absaroka* I; *Dakota* VI, VII; *Shoshoni and Banak* I; *Ute* I; *Apache* I.) "White [man] chief."

Make the same signs for **White man** and **Chief**, and conclude by making that for **Parent** by collecting the fingers and thumb of the right hand nearly to a point and drawing them forward from the left breast. (*Kaiowa* I; *Comanche* III; *Apache* II; *Wichita* II.) "White man; chief; father."

Washington, City of.

The sign for **Go**; the sign for **House** or **Wigwam**; the sign for **Cars**, and the sign for **Council**. The sign for **Father** is briefly executed by passing the open hand down and from the loins, then bringing it erect before the body; then the sign for **Cars**, combined of **Go** and **Wagon**, making with the mouth the noise of an engine; the hands then raised before the eyes and approximated at points, as in the sign for **House**; then diverge to indicate **Extensive**; this being followed by the sign for **Council**. (*Oto and Missouri* I.) "The home of our fathers, where we go on the puffing wagon to council."

Missouri River.

Make the sign for **Water** and the sign for **Large**, and then rapidly rotate the right hand from right to left several times, its back upward, fingers spread and pointing forward to show that it is stirred up or muddy. (*Dakota* IV.)

Eagle Bull (a Dakota chief).

Place the clinched fists to either side of the head; then extend the left hand, flat, palm down, before the left side, fingers pointing forward; the outer edge of the flat and extended right hand is then laid transversely across the back of the left hand, and slid forward over the fingers. (*Dakota* VI; *Arikara* I.) "Bull and eagle—'*Haliaëtus leucocephalus*, (*Linn.*) *Sav.*'"

Rushing Bear (a Dakota chief).

Place the right fist in front of the right side of the breast, palm down; extend and curve the thumb and little finger so that their tips point toward one another before the knuckles of the remaining closed fingers, then reach forward a short distance and pull toward the body several times rather quickly; suddenly push the fist, in this form, forward to arm's length twice. (*Dakota* VI; *Arikara* I.) "Bear and rushing."

Spotted Tail (a Dakota chief).

With the index only of the right hand extended, indicate a line or curve from the sacrum (or from the right buttock) downward, backward, and outward toward the left; then extend the left forefinger, pointing forward from the left side, and with the extended index draw imaginary lines transversely across the left forefinger. (*Absaroka* I; *Shoshoni* I; *Dakota* VI, VII; *Arikara* I.) "Tail; spotted."

Stumbling Bear (a Kaiowa chief).

Place the right fist in front of the right side of the breast, palm down; extend and curve the thumb and little finger so that their tips point toward one another before the knuckles of the remaining closed fingers; then place the left flat hand edgewise before the breast, pointing to the right; hold the right hand flat pointing down nearer the body; move it forward toward the left, so that the right-hand fingers strike the left palm and fall downward beyond the left. (*Kaiowa* I.) "Bear; stumble or stumbling."

Swift Runner (a Dakota Indian).

Place the right hand in front of the right side, palm down; close all the fingers excepting the index, which is slightly curved, pointing forward; then push the hand forward to arm's length twice, very quickly. (*Dakota* VI; *Arikara* I.) "Man running rapidly or swiftly."

Wild Horse (a Comanche chief).

Place the extended and separated index and second fingers of the right hand astraddle the extended forefinger of the left hand. With the right hand loosely extended, held as nigh as and nearly at arm's length before the shoulder, make several cuts downward and toward the left. (*Comanche* III.) "Prairie or wild horse."

PHRASES.

Where is your mother?

After placing the index into the mouth (*mother*), point the index at the individual addressed (*your*); then separate and extend the index and second fingers of the right hand; hold them, pointing forward, about twelve or fifteen inches before the face, and move them from side to side, eyes following the same direction (*I see*); then throw the flat right hand in a short curve outward to the right until the back points toward the ground (*not*), and look inquiringly at the individual addressed. (*Ute* I.) "Mother your I see not; where is she?"

Are you brave?

Point to the person and make sign for **Brave**, at same time looking with an inquiring expression. (*Absaroka* I; *Shoshoni and Banak* I.)

Bison, I have shot a.

Move the open left hand (palm to the front) toward the left and away from the body slowly (motion of the buffalo when chased). Move right hand on wrist as axis, rapidly (man on pony chasing buffalo); then extend left hand to the left, draw right arm as if drawing a bow, snap the forefinger and middle finger of left hand, and thrust the right forefinger over the left hand. (*Omaha* I.)

You gave us many clothes, but we don't want them.

Lean forward, and, holding the hands concavo-convex, draw them up over the limbs severally, then cross on the chest as wrapping a blanket. The arms are then extended before the body, with the hands in type-position (**W**), to a height indicating a large pile. The right hand then sweeps outward, showing a negative state of mind. The index of right hand finally touches the chest of the second party and approaches the body, in position (**I**), horizontal. (*Oto and Missouri* I.) "Something to put on that I don't want from you."

Question.

Hold the extended and flattened right hand, palm forward, at the height of the shoulder or face, and about fifteen inches from it, shaking the hand from side to side (at the wrist) as the arm is slightly raised, resembling the outline of an interrogation mark (?) made from below upward. (*Absaroka* I; *Dakota* V, VI, VII; *Hidatsa* I; *Kaiowa* I; *Arikara* I; *Comanche* II, III; *Pai-Ute* I; *Shoshoni and Banak* I; *Ute* I; *Apache* I, II; *Wichita* II.)

——— What? What is it?

First attract the person's notice by the sign for **Attention**, and then the right-hand, fingers extended, pointing forward or outward, fingers joined, &c., horizontal, is carried outward, obliquely in front of the right breast, and there turned partially over and under several times. (*Dakota* I.)

——— What are you doing? What do you want?

Throw the right hand about a foot from right to left several times, describing an arc upward, palm inward, fingers slightly bent and separated, and pointing forward. (*Dakota* IV.)

——— What are you? *i. e.*, What tribe do you belong to?

Shake the upright opened right hand four to eight inches from side to side a few times, from twelve to eighteen inches in front of the chin, the palm forward, fingers relaxed and a little separated. (*Dakota* IV.)

Place the flat right hand at some distance in front of and as high as the shoulder, palm forward and downward, then shake the hand from side to side, passing it slightly forward and upward at the same time. (*Dakota* VII.)

Pass the right hand from left to right across the face. (*Kutine* I.)

——— What do you want?

The arm is drawn to front of chest and the hand in position (**N** 1), modified by palms being downward and hand horizontal. From the chest center the hand is then passed spirally forward toward the one addressed; the hand's palm begins the spiral motion with a downward and ends in an upward aspect. (*Oto* I.) "To unwind or open."

——— Who are you? or, what is your name?

The right or left hand approximates close to center of the body; the arm is flexed and hand in position (**D**), or a little more closed. From inception of sign near center of body the hand slowly describes the arc of a quadrant, and fingers unfold as the hand recedes. We think the proper intention is, for the inception of sign to be located at the heart, but it is seldom truly, anatomically thus located. (*Oto* I.) "To unfold one's self or make known."

——— Are you through?

With arms hanging at the side and forearms horizontal, place the fists near each other in front of body; then with a quick motion separate them as though breaking something asunder. (*Sahaptin* I.)

——— Do you know?

Shake the right hand in front of the face, a little to the right, the whole arm elevated so as to throw the hand even with the face, and the forearm standing almost perpendicular. Principal motion with hand, slight motion of forearm, palm out. (*Sahaptin* I.)

——— How far is it?

Sign for **Do you know?** followed with a precise movement throwing right hand (palm toward face) to a position as far from body as convenient, signifying "far?"; then with the same quick, precise motion, bring the hand to a position near the face—near? (*Sahaptin* I.)

——— How will you go—horseback or in wagon?

First make the sign for **Do you know?** then throw right hand forward—"go or going,"; then throw fore and middle fingers of right astride the forefinger of the left hand, signifying, "will you ride?"; then swing the forefingers of each hand around each other, sign of wheel running, signifying, "or will you go in wagon?" (*Sahaptin* I.)

DIALOGUES, ETC.

The following conversation took place at Washington, in April, 1880, between TENDOY, chief of the Shoshoni and Banak Indians of Idaho, and HUERITO, one of the Apache chiefs from New Mexico, in the presence of Dr. W. J. Hoffman. Neither of these Indians spoke any language known to the other, or had ever met or heard of one another before that occasion:

HUERITO.—**Who are you?**

Place the flat and extended right hand, palm forward, about twelve inches in front of and as high as the shoulder, then shake the hand from side to side as it is moved forward and upward—*question, who are you?*

TENDOY.—**Shoshoni chief.**

Place the closed right hand near the right hip, leaving the index only extended, palm down; then pass the hand toward the front and left, rotating the hand from side to side—*Shoshoni;* then place the closed hand, with the index extended and pointing upward, near the right cheek, pass it upward as high as the head, then turn it forward and downward toward the ground, terminating with the movement a little below the initial point—*chief.*

HUERITO.—**How old are you?**

Clinch both hands and cross the forearms before the breast with a trembling motion—*cold—winters, years;* then elevate the left hand as high as the neck and about twelve or fifteen inches before it, palm toward the face, with fingers extended and pointing upward; then, with the index, turn down one finger after another slowly, beginning at the little finger, until three or four are folded against the palm, and look inquiringly at the person addressed—*how many.*

TENDOY.—**Fifty-six.**

Close and extend the fingers and thumbs of both hands, with the palms forward, five times—*fifty;* then extend the fingers and thumb of the left hand, close the right, and place the extended thumb alongside of and near the left thumb—*six.*

HUERITO.—**Very well. Are there any buffalo in your country?**

Place the flat right hand, pointing to the left, with the palm down, against the breast-bone; then move it forward and slightly to the right and in a curve upward; make the gesture rather slowly and nearly to arm's length (otherwise, *i. e.*, if made hastily and but a short distance, it would only mean *good*)—*very good;* place both closed hands to their respective sides of the head, palms toward the hair, leaving the forefingers curved—*buffalo;* then reach out the fist to arm's length toward the west, and throw it forcibly toward the ground for a distance of about six inches, edge downward—*country, away to the west;* then point the curved index rather quickly and carelessly toward the person addressed—*your.*

TENDOY.—**Yes; many black buffalo.**

Pass the closed right hand, with the index partly flexed, to a position about eight inches before the right collar-bone, and, as the hand reaches that elevation, quickly close the index—*yes;* then make the same sign as in the preceding question for *buffalo;* touch the hair on the right side of the head with the palms of the extended fingers of the right hand—*black;* spread the curved fingers and thumb of both hands, place them before either thigh, pointing downward; then draw them toward one another and upward as high as the stomach, so that the fingers will point toward one another, or may be interlaced—*many.*

TENDOY.—**Did you hear anything from the Secretary? If so, tell me.**

Close the right hand, leaving the index and thumb widely separated; pass it by the ear from the back of the ear downward and toward the chin, palm toward the head—*hear;* point to the individual addressed—*you;* close the hand again, leaving the index and thumb separated as in the sign for **Hear** and placing the palmar surface of the finger horizontally across the forehead, pointing to the left, allow the thumb to rest against the right temple; then draw the index across the forehead from left to right, leaving the thumb touching the head—*white man;* then place the closed hand, with elevated index, before the right side of the neck or in front of the top of the shoulder; pass the index, pointing upward, as high as the top of the head; turn it forward and downward as far as the breast—*chief;* pass the extended index, pointing upward and forward, forward from the mouth twice—*talk;*

then open and flatten the hand, palm up, outer edge toward the face, place it about fifteen inches in front of the chin, and draw it horizontally inward until the hand nearly touches the neck—*tell me.*

HUERITO.—**He told me that in four days I would go to my country.**

Close the right hand, leaving the index curved; place it about six inches from the ear and move it in toward the external meatus—*told me;* with the right hand still closed, form a circle with the index and thumb by allowing their tips to touch; pass the hand from east to west at arm's length—*day;* place the left hand before the breast, the fingers extended, and the thumb resting against the palm, back forward, and, with the index, turn down one finger after another, beginning at the little finger—*four;* touch the breast with the tips of the finger and thumb of the left hand collected to a point; drop the hand a short distance and move it forward to arm's length and slightly upward until it points above the horizon—*I, go to**; then, as the arm is extended, throw the fist edgewise toward the ground—*my country.*

TENDOY.—**In two days I go to my country just as you go to yours. I go to mine where there is a great deal of snow, and we shall see each other no more.**

Place the flat hands, horizontally, about two feet apart, move them quickly in an upward curve toward one another until the right lies across the left—*night;* repeat this sign—*two nights* (literally, *two sleeps hence*); point toward the individual addressed with the right hand—*you;* and in a continuous movement pass the hand to the right, *i. e.*, toward the south, nearly to arm's length—*go;* then throw the fist edgewise toward the ground at that distance—*your country;* then touch the breast with the tips of the fingers of the left hand—*I;* move the hand off slowly toward the left, *i. e.*, toward the north, to arm's length—*go to**; and throw the clinched hand toward the ground—*my country;* then hold both hands toward the left as high as the head, palms down, with fingers and thumbs pendent and separated; move them toward the ground two or three times—*rain;* then place the flat hands horizontally to the left of the body about two feet from the ground—*deep;* (literally, *deep rain*) *snow*—and raise them until about three feet from the ground—*very deep—much;* place the hands before the body about twelve inches apart, palms down, with forefingers only extended and pointing toward one another; push them toward and from one another several times—*see each other;* then hold the flat right hand in front of the breast, pointing forward, palm to the left, and throw it over on its back toward the right—*not, no more.*

EXPLANATORY NOTE.—Where the asterisks appear in the above dialogue the preposition *to* is included in the gesture. After touching the breast for *I*, the slow movement forward signifies *going to*, and *country* is signified by locating it at arm's length toward the west, to the left of the gesturer, as the stopping-place, also *possession* by the clinched fist being directed toward the ground. It is the same as for *my* or *mine*, though made before the body in the latter signs. The direction of Tendoy's hands, first to the south and afterward to the north, was understood not as pointing to the exact locality of the two parts of the country, but to the difference in their respective climates.

PATRICIO'S NARRATIVE.

This narrative was obtained in July, 1880, by Dr. FRANCIS H. ATKINS, acting assistant surgeon, United States Army, at South Fork, New Mexico, from TI-PE-BES-TLEL (Sheepskin-leggings), habitually called Patricio, an intelligent young Mescalero Apache. It gives an account of what is locally termed the "April Round-up," which was the disarming and imprisoning by a cavalry command of the United States Army, of the small Apache subtribe to which the narrator belonged. The references to signs not described are to the contributions of Dr. ATKINS, marked in the VOCABULARY (*Apache* III).

(1) Left hand on edge, curved, palm forward, extended backward length of arm toward the West (*far westward*).

(2) Arms same, turned hand, tips down, and moved it from north to south (*river*).

(3) Dipped same hand several times above and beyond last line (*beyond*).

(4) Hand curved (**Y**, more flexed) and laid on its back on top of his foot (*moccasins much curved up at toe*); then drew hands up legs to near knee, and cuts off with edges of hands (*boot tops*). (*Warm Spring Apaches*, who wear booted moccasins with turn-up toes.)

(5) Hands held before him, tips near together, fingers gathered (**U**); then alternately opened and gathered fingers of both hands (**P** to **U**, **U** to **P**), and thrusting them toward each other a few times (*shot or killed many*).

(6) Held hands six inches from side of head, thumbs and forefingers widely separated (*Mexican, i. e., wears a broad hat*).

(7) Held right hand on edge, palm toward him, threw it on its back, forward and downward sharply toward earth (**T** on edge to **X**), (*dead, so many dead*).

(8) Put thumbs to temples and indices forward, meeting in front, other fingers closed (*soldiers, i. e., cap-visor.*)

(9) Repeated No. 5 and No. 7 (*were also shot dead*).

(10) Placed first and second fingers of right hand (others closed) astride of left index, held horizontally (*horses*).

(11) Held hands on edge and forward (**T** on edge forward), pushed them forward, waving vertically (*marching*, which see; also, *travel* or

fight, i. e., ran off with soldiers' horses or others). N. B.—Using both hands indicates double ranks of troops marching also.

(12) Struck right fist across in front of chin from right to left sharply (*bad*).

(13) Repeated No. 4 (*Warm Spring Apache*).

(14) Moved fist, thumb to head, from center of forehead to right temple and a little backward (*fool*).

(15) Repeated No. 8 and No. 11 (*soldiers riding in double column*).

(16) Thrust right hand down over and beyond left, both palms down (**W**) (*came here*).

(17) Repeated No. 8 (*soldier*).

(18) Touched hair (*hair*).

(19) Touched tent (*quite white*).

(20) Touched top of shoulder (*commissioned officer, i. e., shoulder-straps*).

(21) Thrust both hands up high (*high rank*).

(22) Right forefinger to forehead; waved it about in front of face and rolled head about (primarily *fool*, but qualified in this case by the interpreter as *no sabe much*).

(23) Drew hands up his thighs and body and pointed to himself (*Mescalero Indian, q. v.*).

(24) Approximated hands before him, palms down, with thumbs and indices widely separated, as if inclosing a circle (*captured, i. e., corralled, surrounded*).

(25) Placed tips of hands together, wrists apart, held them erect (**T**, both hands inclined), (*house;* in this case *the agency*).

(26) Threw both hands, palms back, forward and downward, moving from knuckles (metacarpo-phalangeal joint) only, several times (*issuing rations*).

(27) Thrust two fingers (**N**) toward mouth and downward (*food*).

(28) Repeated No. 25 (*house*); outlined a hemispherical object (*wick-i-up*); repeated these several times, bringing the hands with emphasis several times down toward the earth (*village permanently here*).

(29) Repeated No. 25 several times and pointed to a neighboring hillside (*village over there*).

(30) Repeated Nos. 17 to 21, inclusive (*General X*).

(31) Thrust two fingers forward from his eyes (primarily *I see;* also I *saw*, or *there were*).

(32) Repeated No. 11 (*toward said hillside*), (*troops went over there with General X*).

(33) Repeated No. 4, adding swept indices around head and touched red paper on a tobacco wrapper (*San Carlos Apaches*, scouts especially distinguished by wearing a red fillet about the head); also added, drew indices across each cheek from nose outward (*were much painted*).

(34) Repeated No. 24 and No. 23 (*to capture the Mescalero Indians*).

(35) Repeated No. 31 (*there were*).

(36) Repeated No. 33 (*San Carlos scouts*).

(37) Repeated No. 8 (*and soldiers*).

(38) Clasped his hands effusively before his breast (*so many! i. e., a great many*).

(39) Repeated No. 31 (*I saw*).

(40) Repeated No. 23 (*my people*).

(41) Brought fists together under chin, and hugged his arms close to his breast, with a shrinking motion of body (*afraid*).

(42) Struck off half of left index with right index (*half*, or *a portion*).

(43) Waved off laterally and upward with both hands briskly (*fled*).

(44) Projected circled right thumb and index to eastern horizon, thence to zenith (*next morning*, *i. e.*, sunrise to noon).

(45) Repeated No. 23 (*the Mescaleros*).

(46) Held hands in position of aiming a gun—left oblique—(*shoot*).

(47) Waved right index briskly before right shoulder (*no, did not; negation*).

(48) Swept his hand from behind forward, palm up (**Y**) (*the others came*).

(49) Repeated No. 5 (*and shot*).

(50) Repeated No. 23 (*the Mescaleros*).

(51) Repeated No. 7 (*many dead*).

(52) Repeated No. 8 (*soldiers*).

(53) Repeated No. 10 (*horse, mounted*).

(54) Hand forward, palm down (**W**) moved forward and up and down (*walking, i. e., infantry*).

(55) Beckoned with right hand, two fingers curved (**N** horizontal and curved) (*came*).

(56) Repeated No. 11 (*marching*).

(57) Repeated No. 28 (*to this camp*, or *village*).

(58) Repeated No. 23 (*with Mescaleros*).

(59) Repeated No. 24 (*as prisoners, surrounded*).

(60) Repeated No. 33 (*San Carlos scouts*).

(61) Placed hands, spread out (**R** inverted), tips down, about waist (*many cartridges*).

(62) Repeated No. 46 (*and guns*).

(63) Repeated No. 5 (*shot many*).

(64) Repeated No. 4 (*Warm Spring Apaches*).

(65) Repeated No. 23 (*and Mescaleros*).

(66) Moved fist—thumbs to head— across his forehead from right to left, and cast it toward earth over left shoulder (*brave, i. e., the San Carlos scouts are brave*).

CONTINUOUS TRANSLATION OF THE ABOVE.

Far westward beyond the Rio Grande are the Warm Spring Apaches, who killed many Mexicans and soldiers and stole their horses. They (the United States soldiers) are bad and fools.

Some cavalry came here under an aged officer of high rank, but of inferior intelligence, to capture the Mescalero Indians.

The Mescaleros wished to have their village permanently here by the agency, and to receive their rations, *i. e.*, were peacefully inclined.

Our village was over there. I saw the general come with troops and San Carlos scouts to surround (or capture) the Mescalero Indians. There were a great many San Carlos scouts and soldiers.

I saw that my people were afraid, and half of them fled.

Next morning the Mescaleros did not shoot (were not hostile). The others came and killed many Mescaleros. The cavalry and infantry brought us (the Mescaleros) to this camp as prisoners.

The San Carlos scouts were well supplied with ammunition and guns, and shot many Warm Spring Indians and Mescaleros.

The San Carlos scouts are brave men.

TSODIÁKO'S REPORT.

The following statement was made to Dr. W. J. Hoffman by TSODIÁKO (*Shaved head Boy*), chief of the Wichitas in Indian Territory, while on a visit to Washington, D. C., in June, 1880.

The Indian being asked whether there was any timber in his part of the Territory, replied in signs as follows:

(1) Move the right hand, fingers loosely extended, separated and pointing upward, back to the front, upward from the height of the waist to the front of the face–*tree;* repeat this two or three times—*trees;* (2) then hold the hand, fingers extended and joined, pointing upward, with the back to the front, and push it forward toward different points on a level with the face—*standing at various places;* (3) both hands, with spread and slightly curved fingers, are held about two feet apart, before the thighs, palms facing, then draw them toward one another horizontally and gradually upward until the wrists cross, as if grasping a bunch of grass and pulling it up—*many;* (4) point to the southwest with the index, elevating it a little above the horizon—*country;* (5) then throw the fist edgewise toward the surface, in that direction—*my, mine;* (6) place both hands, extended, flat, edgewise before the body, the left below the right, and both edges pointing toward the ground a short distance to the left of the body, then make repeated cuts toward that direction from different points, the termination of each cut ending at nearly the same point—*cut down;* (7) hold the left hand with the fingers and thumb collected to a point, directed horizontally forward, and make several cutting motions with the edge of the flat right hand transversely by the tips of the left, and upon the wrist—*cut off the ends;* (8) then cut upon the left hand, still held in the same position, with the right, the cuts being parallel to the longitudinal axis of the palm—*split;* (9) both hands closed in front of the body, about four inches apart, with forefingers and thumbs approximating half circles,

palms toward the ground, move them forward so that the back of the hand comes forward and the half circles imitate the movement of wheels—*wagon ;* (10) hold the left flat hand before the body, pointing horizontally forward, with the palm down, then bring the right flat hand from the right side and slap the palm upon the back of the left several times—*load upon ;* (11) partly close the right hand as if grasping a thick rod, palm toward the ground, and push it straight forward nearly to arm's length—*take ;* (12) hold both hands with fingers naturally extended and slightly separated nearly at arm's length before the body, palms down, the right lying upon the left, then pass the upper forward and downward from the left quickly, so that the wrist of the right is raised and the fingers point earthward—*throw off ;* (13) cut the left palm repeatedly with the outer edge of the extended right hand—*build ;* (14) hold both hands edgewise before the body, palms facing, spread the fingers and place those of one hand into the spaces between those of the left, so that the tips of one protrude beyond the backs of the fingers of the other—*log house ;* (15) then place the flat right hand, palm down and fingers pointing to the left, against the breast and move it forward, and slightly upward and to the right—*good.*

ANALYSIS OF THE FOREGOING.

[There is] much (3) | timber (1, 2) | [in] my (5) | country (4) | [of which I] cut down (6) [some], | trimmed, (7) | split, (8) | loaded it upon (10) | [a] wagon (9) [and] | took it (11) away, | [where I] threw (12) [it] off | [and] built (13) | [a] good (15) | house (14) | .

NOTES.—As will be seen, the word **timber** is composed of signs No. 1 and 2, signifying **trees standing.** Sign No. 3, for **many,** in this instance, as in similar other examples, becomes **much.** The word **in,** in connection with **country** and **my,** is expressed by the gesture of pointing (passing the hand less quickly than in ordinary sign language), before making sign No. 5. That sign, commonly given for **possession,** would, without the prefix of indication, imply **my country,** and with that prefix signifies **in my country.** Sign No. 7, **trimmed,** is indicated by chopping off the ends, and facial expression denoting **satisfaction.** In sign Nos. 11 and 12, the gestures were continuous, but at the termination of the latter the narrator straightened himself somewhat, denoting that he had overcome the greater part of the labor. Sign No. 14, denotes **log-house** from the manner of interlacing the finger-ends, thus representing the corner of a log-house, and the arrangement of the ends of the same. **Indian lodge** would be indicated by another sign, although the latter is often used as an abbreviation for the former, when the subject of conversation is known to all present.

SIGNALS.

The collaborators in the present work have not generally responded to the request to communicate material under this head. It is, however, hoped that by now printing some extracts from published works and the few unpublished statements recently procured, the attention of observers will be directed to the further prosecution of research in this direction.

The term "signal" is here used in distinction from the signs noted in the VOCABULARY, as being some action or manifestation intended to be seen at a distance, and not allowing of the minuteness or detail possible in close converse. Signals may be executed, first, exclusively by bodily action; second, by action of the person in connection with objects, such as a blanket, or a lance, or in the direction imparted to a horse; third, by various devices, such as smoke or fire-arrows, when the person of the signalist is not visible. They are almost entirely conventional, and while their study has not the same kind of importance as that of gesture-signs, it possesses some peculiar interest.

SIGNALS EXECUTED BY BODILY ACTION.

Some of these will probably be found to be identical, or nearly so, with the gesture-signs used by the same people.

Alarm. See notes on Cheyenne and Arapaho signals.

Anger.

Close the hand, place it against the forehead, and turn it back and forth while in that position. (*Thirty Years of Army Life on the Border*, by Col. R. B. Marcy, U. S. A., p. 34, New York, 1866.)

Come here.

The right hand is to be advanced about eighteen inches at the height of the navel, horizontal, relaxed, palm downward, thumb in the palm; then draw it near the side and at the same time drop the hand to bring the palm backward. The farther away the person called is, the higher the hand is raised. If very far off, the hand is raised high up over the head and then swung forward and downward, then backward and downward to the side. (*Dakota* IV.)

Danger. (There is something dangerous in that place.)

Right-hand index-finger and thumb forming a curve, the other fingers closed; move the right hand forward, pointing in the direction of the dangerous place or animal. (*Omaha* I.)

Defiance.

Right-hand index and middle fingers open; motion toward the enemy. "I do not fear you." Reverse the motion, bringing the hand toward the subject. "Do your worst to me." (*Omaha* I.)

Direction. Pass around that object or place near you—she-í-he ti-dhá-ga.

When a man is at a distance, I say to him "Go around that way." Describe a curve by raising the hand above the head, forefinger open, move to right or left according to direction intended and hand that is used, *i. e.*, move to the left, use right hand; move to the right, use left hand. (*Omaha* I; *Ponka* I.)

Halt! (To inquire disposition.)

Raise the right hand with the palm in front and gradually push it forward and back several times; if they are not hostile it will at once be obeyed. (*The Prairie Traveler*, by Randolph B. Marcy, p. 214, New York, 1859.)

——— Stand there! He is coming to you.

Right hand extended, flat, edgewise, moved downward several times. (*Omaha* I.)

——— He is going toward you.

Hold the open right hand, palm to the left, with the tips of the fingers toward the person signaled to; thrust the hand forward in either an upward or downward curve. (*Omaha* I; *Ponka* I.)

——— Lie down flat where you are (she-dhu bis-pé zha$^{n\prime}$-ga).

Extend the right arm in the direction of the person signaled to, having the palm down; move downward by degrees to about the knees. *Omaha* I; *Ponka* I.)

Peace; Friendship.

Hold up palm of hand.—Observed as made by an Indian of the Kansas tribe in 1833. (*Indian Sketches*, by John T. Irving, vol. ii, p. 253, Philadelphia, 1835.)

Elevate the outstretched hands wide open and fingers parted above and on either side of the head at arm's length.—Observed by Dr. W. J. Hoffman, as made in Northern Arizona in 1871 by the Mojave and Seviches. "No arms"—corresponding with "hands up" of road-agents.

The right hand held aloft, empty. (*My Life on the Plains*, by General G. A. Custer, p. 238, New York, 1874.)

Question. (I do not know you. Who are you?)

After halting a party coming: Right hand raised, palm in front and slowly moved to the right and left. [Answered by tribal sign.] (Marcy's *Prairie Traveler*, *loc. cit.*, 214.)

——— To inquire if coming party is peaceful.

Raise both hands, grasped in the manner of shaking hands, or by locking the two forefingers firmly while the hands are held up. If friendly they will respond with the same signal. (Marcy's *Prairie Traveler*, *loc. cit.*, 214.)

——— Whence come you?

First the sign for **You,** then the hand extended open and drawn to the breast, and lastly the sign for **Bringing.** (*Dunbar.*)

Submission.

The United States steamer Saranac in 1874, cruising in Alaska waters, dropped anchor in July, 1874, in Freshwater Harbor, back of Sitka, in latitude 59° north. An armed party landed at a T'linkit village deserted by all the inhabitants except one old man and two women, the latter seated at the feet of the former. The man was in great fear, turned his back and held up his hands as a sign of utter helplessness. (Extract from notes kindly furnished by Lieutenant-Commander Wm. Bainbridge Hoff, U. S. N., who was senior aid to Rear-Admiral Pennock, on the cruise mentioned.)

Surrender.

The palm of the hand is held toward the person [to whom the surrender is made]. (*Long.*)

SIGNALS IN WHICH OBJECTS ARE USED IN CONNECTION WITH PERSONAL ACTION.

Buffalo discovered. See also notes on Cheyenne and Arapaho signs.

When the Ponkas or Omahas discover buffalo the watcher stands erect on the hill, with his face toward the camp, holding his blanket with an end in each hand, his arms being stretched out (right and left) on a line with shoulders. (*Omaha* I; *Ponka* I.)

Come! To beckon to a person.

Hold out the lower edge of the robe or blanket, then wave it in to the legs. This is made when there is a desire to avoid general observation. (*Matthews.*)

Come back!

Gather or grasp the left side of the unbuttoned coat (or blanket) with the right hand, and, either standing or sitting in position so that the signal can be seen, wave it to the left and right as often as may be necessary for the sign to be recognized. When made standing the person should not move his body. (*Dakota* I.)

Danger. See also notes on Cheyenne and Arapaho signals.

Horseman at a distance, galloping, passing and repassing, and crossing each other—*enemy comes.* But for notice of herd of buffalo, they gallop back and forward abreast—do not cross each other. (*Views of Louisiana*, by H. M. Brackenridge, p. 250, Pittsburgh, 1814.)

Riding rapidly round in a circle. "Danger! Get together as quickly as possible." (*The Plains of the Great West*, &c., by Richard Irving Dodge, lieutenant-colonel United States Army, p. 368, New York, 1877.)

Discovery of enemies, or of other game than buffalo. See also notes on Cheyenne and Arapaho signals.

When enemies are discovered, or other game than buffalo, the sentinel waves his blanket over his head up and down, holding an end in each hand. (*Omaha* I; *Ponka* I.)

Drill, Military.

It is done by signals, devised after a system of the Indian's own invention, and communicated in various ways.

Wonderful as the statement may appear, the signaling on a bright day, when the sun is in the proper direction, is done with a piece of looking-glass held in the hollow of the hand. The reflection of the sun's rays thrown on the ranks communicates in some mysterious way the wishes of the chief. Once standing on a little knoll, overlooking the valley of the South Platte, I witnessed almost at my feet a drill of about one hundred warriors by a Sioux chief, who sat on his horse on a knoll opposite me, and about two hundred yards from his command in the plain below. For more than half an hour he commanded a drill, which for variety and promptness of action could not be equaled by any civilized cavalry of the world. All I could see was an occasional movement of the right arm. He himself afterwards told me that he used a looking-glass. (*The Plains of the Great West*, &c., by Richard Irving Dodge, lieutenant-colonel United States Army, pp. 307, 308. New York, 1877.)

Halt! Stand there! He is coming that way.

Grasp the end of the blanket or robe; wave it downward several times. (*Omaha* I.)

Peace, coupled with invitation.

Motion of spreading a real or imaginary robe or skin on the ground Noticed by Lewis and Clark on their first meeting with the Shoshoni in 1805. (*Lewis and Clark's Travels*, &c., London, 1817, vol. ii, p. 74.)

Question.

The ordinary manner of opening communication with parties known or supposed to be hostile is to ride toward them in zigzag manner, or to ride in a circle. (*My Life on the Plains*, &c., by Gen. G. A. Custer, U. S. A., p. 58. New York, 1874.)

This author mentions (p. 202) a systematic manner of waving a blanket, by which the son of Satana, the Kaiowa chief, conveyed information to him, and a similar performance by Yellow Bear, a chief of the Arapahos (p. 219), neither of which he explains in detail.

Safety. All quiet. See notes on Cheyenne and Arapaho signals.

SIGNALS MADE WHEN THE PERSON OF THE SIGNALIST IS NOT VISIBLE.

Those noted consist of **Smoke, Fire,** or **Dust** signals.

SMOKE SIGNALS GENERALLY.

"Their systems of telegraphs are very peculiar, and though they might seem impracticable at first, yet so thoroughly are they understood by the savages that it is availed of frequently to immense advantage. The most remarkable is by raising smokes, by which many important facts are communicated to a considerable distance and made intelligible by the manner, size, number, or repetition of the smokes, which are commonly raised by firing spots of dry grass. When traveling, they will also pile heaps of stones upon mounds or conspicuous points, so arranged as to be understood by their passing comrades; and sometimes they set up the bleached buffalo heads, which are everywhere scattered over those plains, to indicate the direction of their march, and many other facts which may be communicated by those simple signs." (*Commerce of the Prairies*, by Josiah Gregg, vol. ii, p. 286. New York, 1844.)

The highest elevations of land are selected as stations from which signals with smoke are made. These can be seen at a distance of from twenty to fifty miles. By varying the number of columns of smoke different meanings are conveyed. The most simple as well as the most varied mode, and resembling the telegraphic alphabet, is arranged by building a small fire, which is not allowed to blaze; then by placing an armful of partially green grass or weeds over the fire, as if to smother it, a dense white smoke is created, which ordinarily will ascend in a continuous vertical column for hundreds of feet. Having established a current of smoke, the Indian simply takes his blanket and by spreading it over the small pile of weeds or grass from which the smoke takes its source, and properly controlling the edges and corners of the blanket, he confines the smoke and is in this way able to retain it for several moments. By rapidly displacing the blanket, the operator is enabled to cause a dense volume of smoke to rise, the length or shortness of which, as well as the number ánd frequency of the columns, he can regulate perfectly, simply by a proper use of the blanket. (*Custer's Life on the Plains*, *loc. cit.*, p. 187.)

They gathered an armful of dried grass and weeds, which were placed and carried upon the highest point of the peak, where, everything being in readiness, the match was applied close to the ground; but the blaze was no sooner well lighted and about to envelop the entire amount of grass collected than it was smothered with the unlighted portion. A slender column of gray smoke then began to ascend in a perpendicular column. This was not enough, as it might be taken for the smoke rising from a simple camp fire. The smoldering grass was then covered with a blanket, the corners of which were held so closely to the ground as to almost completely confine and cut off the column of smoke. Waiting a few moments, until the smoke was beginning to escape from beneath, the blanket was suddenly thrown aside, when a beautiful balloon-shaped column puffed upward like the white cloud of smoke which attends the discharge of a field-piece. Again casting the blanket on the pile of grass, the column was interrupted as before, and again in due time released, so that a succession of elongated, egg-shaped puffs of smoke kept ascending toward the sky in the most regular manner. This bead-like column of smoke, considering the height from which it began to ascend, was visible from points on the level plain fifty miles distant. (*Ib.*, p. 217.)

SMOKE SIGNALS OF THE APACHES.

The following information was obtained by Dr. W. J. HOFFMAN, from the Apache chiefs named on page 15, under the title of TINNEAN, *Apache* I:

The materials used in making smoke of sufficient density and color consist of pine or cedar boughs, leaves and grass, which can nearly always be obtained in the regions occupied by the Apaches of Northern New Mexico. These Indians state that they employ but three kinds of signals, each of which consists of columns of smoke, numbering from one to three or more.

Alarm.

This signal is made by causing three or more columns of smoke to ascend, and signifies danger or the approach of an enemy, and also requires the concentration of those who see them. These signals are communicated from one camp to another, and the most distant bands are guided by their location. The greater the haste desired the greater the number of columns of smoke. These are often so hastily made that they may resemble puffs of smoke, and are caused by throwing heaps of grass and leaves upon the embers again and again.

Attention.

This signal is generally made by producing one continuous column, and signifies attention for several purposes, viz, when a band had become tired of one locality, or the grass may have been consumed by the ponies, or some other cause necessitating removal; or should an enemy be reported, which would require further watching before a decision as to future action would be made, the intention or knowledge of anything unusual would be communicated to neighboring bands by causing one column of smoke to ascend.

Establishment of a camp; Quiet; Safety.

When a removal of camp has been made, after the signal for **Attention** has been given, and the party have selected a place where they propose to remain until there may be a necessity or desire for their removal. two columns of smoke are made, to inform their friends that they propose to remain at that place. Two columns are also made at other times during a long-continued residence, to inform the neighboring bands that a camp still exists, and that all is favorable and quiet.

FOREIGN SMOKE SIGNALS.

The following examples of smoke signals in foreign lands are added for comparison.

Miss Haigh, speaking of the *Guanches* of the Canary Islands at the time of the Spanish conquest, says: "When an enemy approached, they alarmed the country by raising a thick smoke or by whistling, which was repeated from one to another. This latter method is still in use among the people of Teneriffe, and may be heard at an almost incredible distance." (*Trans. Eth. Soc. Lond.* vii, 1869, sec. ser., pp. 109, 110.)

"The natives have an easy method of telegraphing news to their distant friends. When Sir Thomas Mitchell was traveling through Eastern Australia he often saw columns of smoke ascending through the trees in the forests, and he soon learned that the natives used the smoke of fires for the purpose of making known his movements to their friends. Near Mount Frazer he observed a dense column of smoke, and subsequently other smokes arose, extending in a telegraphic line far to the south, along the base of the mountains, and thus communicating to the natives who might be upon his route homeward the tidings of his return.

"When Sir Thomas reached Portland Bay he noticed that when a whale appeared in the bay the natives were accustomed to send up a column of smoke, thus giving timely intimation to all the whalers. If the whale should be pursued by one boat's crew only, it might be taken; but if pursued by several, it would probably be run ashore and become food for the blacks." (*Eastern Australia*, by Maj. T. L. Mitchell, F. G. S., vol. ii, p. 241.)

Jardine, writing of the natives of Cape York, says that a communication between the islanders and the natives of the mainland is frequent; and the rapid manner in which news is carried from tribe to tribe, to great distances, is astonishing. I was informed of the approach of Her Majesty's Steamer Salamander, on her last visit, two days before her arrival here. Intelligence is conveyed by means of fires made to throw up smoke in different forms, and by messengers who perform long and rapid journeys." (Quoted by *Smyth*, *loc. cit.*, vol. 1, p. 153, from *Overland Expedition*, p. 85.)

Messengers in all parts of Australia appear to have used this mode of signaling. In Victoria, when traveling through the forests, they were accustomed to raise smoke by filling the hollow of a tree with green boughs and setting fire to the trunk at its base; and in this way, as they always selected an elevated position for the fire when they could, their movements were made known.

When engaged in hunting, when traveling on secret expeditions, when approaching an encampment, when threatened with danger, or when foes menaced their friends, the natives made signals by raising a smoke, and their fires were lighted in such a way as to give forth signals that would be understood by people of their own tribe and by friendly tribes. They exhibited great ability in managing their system of telegraphy; and in former times it was not seldom used to the injury of the white settlers, who at first had no idea that the thin column of smoke rising through the foliage of the adjacent bush, and perhaps raised by some feeble old woman, was an intimation to the warriors to advance and attack the Europeans. (*The Aborigines of Victoria*, vol. i, by R. Brough Smyth, F. L. S., F. G. S., Assoc. Inst. C. E., etc., pp. 152, 153.)

FIRE ARROWS.

"Travelers on the prairie have often seen the Indians throwing up signal lights at night, and have wondered how it was done. * * * They take off the head of the arrow and dip the shaft in gunpowder, mixed with glue. * * * The gunpowder adheres to the wood, and coats it three or four inches from its end to the depth of one-fourth of an inch. Chewed bark mixed with dry gunpowder is then fastened to the stick, and the arrow is ready for use. When it is to be fired, a warrior places it on his bowstring and draws his bow ready to let it fly; the point of the arrow is then lowered, another warrior lights the dry bark, and it is shot high in the air. When it has gone up a little distance, it bursts out into a flame, and burns brightly until it falls to the ground. Various meanings are attached to these fire-arrow signals. Thus, one arrow meant, among the Santees, 'The enemy are about'; two arrows from the same point, 'Danger'; three, 'Great danger'; many, 'They are too strong, or we are falling back'; two arrows sent up at the same moment, 'We will attack'; three, 'Soon'; four, 'Now'; if shot diagonally, 'In that direction.' These signals are constantly changed, and are always agreed upon when the party goes out or before it separates. The Indians send their signals very intelligently, and seldom make mistakes in telegraphing each other by these silent monitors. The amount of information they can communicate by fires and burning arrows is perfectly wonderful. Every war party carries with it bundles of signal arrows." (*Belden, The White Chief; or Twelve Years among the Wild Indians of the Plains*, pp. 106, 107. Cincinnati and New York, 1871.)

With regard to the above, it is possible that white influence has been felt in the mode of signaling as well as in the use of gunpowder, but it would be interesting to learn if any Indians adopted a similar expedient before gunpowder was known to them.

DUST SIGNALS.

When any game or an enemy is discovered, and should the sentinel be without a blanket, he throws a handful of dust up into the air. When the Brulés attacked the Ponkas, in 1872, they stood on the bluff and threw up dust. (*Omaha* I; *Ponka* I.)

There appears to be among the Bushmen a custom of throwing up sand or earth into the air when at a distance from home and in need of help of some kind from those who were there. (Miss L. C. Lloyd, *MS. Letter*, dated July 10, 1880, from Charlton House, Mowbray, near Cape Town, Africa.)

NOTES ON CHEYENNE AND ARAPAHO SIGNALS.

The following information was obtained from WÁ-U^{n} (*Bobtail*), MO-HÍ-NUK-MA-HÁ-IT (*Big Horse*), Cheyennes, and Ó-CHÓ-HIS-A (*The Mare*, better known as "Little Raven"), and NÁ-UATSH (*Left Hand*), Arapahos, chiefs and members of a delegation who visited Washington, D. C., in September, 1880, in the interest of their tribes located in Indian Territory:

A party of Indians going on the war-path leave camp, announcing their project to the remaining individuals and informing neighboring friends by sending runners. A party is only systematically organized when several days away from their headquarters, unless circumstances should require immediate action. The pipe-bearers are appointed, who precede the party while on the march, carrying the pipes, and no one is allowed to cross ahead of these individuals, or to join the party by riding up before the head of the column, as it would endanger the success of the expedition. All new arrivals fall in from either side or the rear. Upon coming in sight of any elevations of land likely to afford a good view of the surrounding country the party come to a halt and secrete themselves as much as possible. The scouts, who have already been selected, advance just before daybreak to within a moderate distance of the elevation to ascertain if any of the enemy have preceded them. This is only discovered by carefully watching the summit to see if any objects are in motion; if not, the flight of birds is observed, and if any should alight upon the hill or butte it would indicate the absence of anything that might ordinarily scare them away. Should a large bird, as a raven, crow, or eagle, fly toward the hill-top and make a sudden swerve to either side and disappear, it would indicate the presence of something sufficient to require further examination. When it is learned that there is reason to suspect an enemy, the scout, who has all the time been closely watched by the party in the rear, makes a signal for them to lie still, signifying **Danger** or **Caution.** It is made by grasping the blanket with the right hand and waving it earthward from a position in front of and as high as the shoulder. This is nearly the same as we use the hand for a similar purpose in battle or hunting to direct "lie quiet!"

Should the hill, however, be clear of any one the Indian will ascend slowly, and under cover as much as possible, and gain a view of the country. If there is no one to be seen, the blanket is grasped and waved horizontally from right to left and back again repeatedly, showing a clear surface. If the enemy is discovered, the scout will give the **Alarm** by running down the hill upon a side visible to the watchers, in a zigzag manner which communicates the state of affairs.

Should any expedition or advance be attempted at night, the same signals as are made with the blanket are made with a firebrand, which is constructed of a bunch of grass tied to a short pole.

When a war party encamp for a night or a day or more a piece of wood is stuck into the ground, pointing in the direction pursued, with a number of cuts, notches, or marks corresponding to the number of days which the party spent after leaving the last camp until leaving the present camp, serving to show to the recruits to the main party, the course to be followed, and the distance.

A hunting party take the same precautions in advancing as a war party, so as not to be surprised by an enemy. If a scout ascends a prominent elevation and discovers no game, the blanket is grasped and waved horizontally from side to side at the height of the shoulders or head; and if game is discovered the Indian rides back and forth (from left to right) a short distance so that the distant observers can view the maneuver. If a large herd of buffalo is found, the extent traveled over in going to and fro increases in proportion to the size of the herd. A quicker gait is traveled when the herd is very large or haste on the part of the hunters is desired.

It is stated that these Indians also use mirrors to signal from one elevation to another, but the system could not be learned, as they say they have no longer use for it, having ceased warfare (?).

ADDRESS

BY

COL. GARRICK MALLERY, U. S. A.

THE GESTURE SPEECH OF MAN.

Anthropology tells the march of mankind out of savagery. In that march some peoples have led with the fleet course of videttes or the sturdy stride of pioneers, some have only plodded on the roads opened by the vanguard, while others still lag in the unordered rear, mere dragweights to the column. All commenced their progress toward civilization from a point of departure lower than the stage reached by the lowest of the tribes now found on earth, and all, even the most advanced, have retained marks of their rude origin. These marks are of the same kind, though differing in distinctness, and careful search discovers the fact that none are missing, showing that there is a common source to all the forms of intellectual and social development, notwithstanding their present diversities. Perhaps the most notable criterion of difference is in the copiousness and precision of oral language, and in the unequal survival of the communication by gesture signs which, it is believed, once universally prevailed. The phenomena of that mode of human utterance, wherever it still appears, require examination as an instructive vestige of the prehistoric epoch. In this respect the preëminent gesture system of the North American Indians calls for study in comparison with other less developed or more degenerate systems. It may solve problems in psychologic comparative philology not limited to the single form of speech, but embracing all modes of expressing ideas. Perhaps, therefore, a condensed report of such study pursued with advantages possessed by few persons even in this country will, on this occasion, be an acceptable contribution as illustrating the gesture speech of man.

So far as the use of gesture signs continued, however originating, in the necessity for communication between peoples of different oral speech, North America shows more favorable conditions for its development than any other thoroughly explored part of the world. In that great continent the precolumbian population was, as is now believed, scanty, and so subdivided dialectically, that the members of but few bands could readily converse with others. The number of now defined stocks or families of Indian languages within the territory of the United States amounts to sixty-five, and these differ among themselves as radically as each differs from the Hebrew, Chinese, or English. In each of these linguistic families there are several, sometimes as many as twenty, separate languages, which also differ from each other as much as do the English, French, German and Persian divisions of the Aryan linguistic stock.

The conditions upon which the survival of sign language among the Indians has depended are well shown by those attending its discontinuance among certain tribes. The growth of the mongrel tongue, called the Chinook jargon, arising from the same causes that produced the pigeon-English, or *lingua franca* of the Orient, explains the known recent disuse of systematic signs among the Kalapuyas and other tribes of the North Pacific coast. The Alaskan tribes also generally used signs not more than a generation ago. Before the advent of the Russians the coast Alaskans traded their dried fish and oil for the skins and paints of the eastern tribes by visiting the latter, whom they did not allow to come to the coast, and this trade was conducted mainly in sign language. The Russians brought a better market, so the travel to the interior ceased, and with it the necessity for the signs, which therefore gradually died out, and are little known to the present generation on the coast, though still continuing in the interior where the inhabitants are divided by dialects.

No explanation is needed for the gradual disuse of signs for the special purpose of intertribal communication when the speech of surrounding civilization becomes known as the best common medium. When that has become general, and there is a compelled end both to hunting and warfare, signs, as systematically employed before, fade away, or survive only in formal oratory and impassioned conversation.

THEORIES ENTERTAINED RESPECTING INDIAN SIGNS.

It is not now proposed to pronounce upon theories. The mere collection of facts cannot, however, be prosecuted to advantage without predetermined rules of direction, nor can they be classified at all without the adoption of some principle which involves a tentative theory. Now, also, since the great principle of evolution has been brought to general notice, no one will be satisfied with knowing a fact without also trying to establish its relation to other facts. Therefore a working hypothesis, which shall not be held to with tenacity, is not only allowable but necessary. It is likewise proper to examine with respect the theories advanced by others.

NOT CORRELATED WITH MEAGERNESS OF LANGUAGE.

The ever unconfirmed report of travellers that certain languages cannot be clearly understood in the dark by their possessors, using their mother tongue between themselves, when asserted, as it often has been, in reference to any of the tribes of North America, is absolutely false. It must be attributed to the error of visitors, who seldom see the natives except when trying to make themselves intelligible to them by a practice which they have found by experience to have been successful with strangers to their tongue. Captain Burton specially states that the Arapahos possess a very scanty vocabulary, pronounced in a quasi-unintelligible way, and can hardly converse with one another in the dark. The truth is that their vocabulary is by no means scanty, and they do converse with each other with perfect freedom without any gestures when they so please. The same distinguished explorer also gives a story "of a man who, being sent among the Cheyennes to qualify himself for interpreting, returned in a week and proved his competency; all he did, however, was to go through the usual pantomine with a running accompaniment of grunts." And he might as well have omitted the grunts, for obviously he only used sign language.

The similar accusation made against the Shoshonian stock, that their tongue, without signs, was too meager for understanding, is refuted by my own experience. When Ouray, the late head chief of the Utes, was last at Washington, after an interview with the

Secretary of the Interior, he made report of it to the others of the delegation who had not been present. He spoke without pause in his own language for nearly an hour, in a monotone and without a single gesture. The reason for this depressed manner was undoubtedly because he was very sad at the result, involving loss of land and change of home; but the fact remains that full information was communicated on a complicated subject without the aid of a manual sign, and also without even such change of inflection of voice as is common among Europeans. All theories based upon the supposed poverty of American languages must be abandoned.

The grievous accusation against foreign people that they have no intelligible language is venerable and general. With the Greeks the term ἄγλωσσος, "tongueless," was used synonymous with βάρβαρος, "barbarian," of all who were not Greek. The name "Slav," assumed by a grand division of the Aryan family, means "the speaker," and is contradistinguished from the other peoples of the world, such as the Germans, who are called in Russian "Njemez," that is, "speechless." In Isaiah (xxxiii, 19) the Assyrians are called a people "of a stammering tongue, that one cannot understand." The common use of the expressions "tongueless" and "speechless," so applied, has probably given rise to the mythical stories of actually speechless tribes of savages, and the instances now presented tend to discredit the many other accounts of languages which are incomplete without the help of gesture. The theory that sign language was in whole or in chief the original utterance of mankind would be strongly supported by conclusive evidence to the truth of such travellers' tales, but does not depend upon them. Nor, considering the immeasurable period during which, in accordance with modern geologic views, man has been on the earth, is it probable that any existing peoples can be found among whom speech has not obviated the absolute necessity for gesture in communication between themselves. The signs survive for convenience, used together with oral language, and for special employment when language is unavailable.

ITS ORIGIN FROM ONE TRIBE OR REGION.

My correspondents in the Indian country have often contended that sign language was invented by a certain tribe in a particular region from which its knowledge spread among other tribes inversely as their distance from, and directly as their intercourse with, the alleged inventors. Unfortunately there is no agreement as to the latter, and probably the accident that the several correspondents met, in certain tribes, specially skilful sign-talkers, determined their opinions. The theory also supposes a comparatively recent origin of sign language, whereas so far as can be traced, the conditions favorable to it existed very long ago and were coëxtensive with the territory of North America occupied by any of the tribes. Some writers confine its use to the Great Plains. It is, however, ascertained to have prevailed among the Iroquois, Wyandots, Ojibwas, and at least three generations back among the Crees and the Mandans and other far northern Dakotas. Some of these and many other tribes of the United States never habiting the Plains, as also the Kutchins of eastern Alaska and the Kutine and Selish of British Columbia, use signs now. Instead of referring to some past period when they did not use signs, many Indians examined speak of a time when they or their fathers employed them more freely and copiously than at present.

Perhaps the most salutary criticism to be offered regarding the theory would be in the form of a query whether sign language has ever been invented by any one body of people at any one time, and whether it is not simply a phase in evolution, surviving and reviving when needed. Not only does the burden of proof rest unfavorably upon the attempt to establish one parent stock for sign language in North America, but it also comes under the stigma now fastened upon the immemorial effort to name and locate the original oral speech of man. It is only next in difficulty to the old persistent determination to decide upon the origin of the whole Indian "race," in which most peoples of antiquity in the eastern hemisphere, including the lost tribes of Israel, the Gipsies, and the Welsh, have figured conspicuously as putative parents.

SIGN LANGUAGE NOT UNIFORM.

The general report that there is but one sign language in North America, any deviation from which is either blunder, corruption, or a dialect in the nature of provincialism, originated with sign talkers in several regions. Now a mere sign talker is often a bad authority upon principles and theories. He may not be liable to the satirical compliment of Dickens' "brave courier," who "understood all languages indifferently ill;" but many men speak some one language fluently, and yet are wholly unable to explain or analyze its words and forms so as to teach it to another, or even to give an intelligent summary or classification of their own knowledge. What such a sign talker has learned is by memorizing, as a child learns English, and though both the sign talker and the child may be able to give some separate items useful to a philologist or foreigner, such items are spoiled when colored by the attempt of ignorance to theorize. A German who has studied English to thorough mastery, except in the mere facility of speech, may in a discussion upon some of its principles be contradicted by any mere English speaker, who insists upon his superior knowledge because he actually speaks the language and his antagonist does not, but the student will probably be correct and the talker wrong. It is an old adage about oral speech that a man who understands but one language understands none. The science of a sign talker possessed by a restrictive theory is like that of Mirabeau, who was greater as an orator than as a philologist, and who on a visit to England gravely argued that there was something seriously wrong in the British mind because the people would persist in saying "give me some bread" instead of "*donnez-moi du pain*," which was so much easier and more natural. When a sign is presented which such a sign talker has not before seen, he will at once condemn it as bad, just as a United States Minister to Vienna, who had been nursed in the mongrel Dutch of Berks County, Pennsylvania, declared that the people of Germany spoke very bad German.

An argument for the uniformity of the signs of Indians is derived from the fact that those used by any of them are generally understood by others. But signs may be understood without being identical with any before seen. It is a common experience that when Indians find a sign which has become conventional among their tribe not to be understood by an interlocutor, a self-expressive sign is substituted for it, from which a visitor may form the impression that there are no conventional signs. It may likewise occur that the self-expressive sign substituted will be met with by a visitor in several localities, different Indians, in their ingenuity, taking the best and the same means of reaching the exotic intelligence.

There is some evidence that where sign language is now found among Indian tribes it has become more uniform than ever before, simply because many tribes have for some time past been forced to dwell near together at peace. The resulting uniformity in these cases might either be considered as a jargon or as the natural tendency to a compromise for mutual understanding—the unification so often observed in oral speech coming under many circumstances out of former heterogeneity. The rule is that dialects precede languages and that out of many dialects comes one language.

The process of the formation and introduction of signs is the same among Indians as often observed among deaf-mutes. When a number of those unfortunate persons, possessed only of such crude signs as were used by each among his speaking relatives come together for a considerable time, they are at first only able to communicate on a few subjects, but the number of those and the general scope of expression will be continually enlarged. Each one commences with his own conception and his own presentment of it, but the universality of the medium used makes it sooner or later understood. This independent development often renders the first interchange of thought between strangers slow, for the signs must be self-interpreting. There can be no natural universal language which is absolute and arbitrary. When used without convention, as sign language alone of all modes of utterance can be, it must be tentative, experimental, and flexible. The mutes will also resort to the invention of new signs for new ideas as they arise, which will be made intelligible, if necessary, through the illustration and definition given by signs formerly adopted. The fittest signs will in due course be evolved, after rivalry and trial, and will survive. But there may not always be such a preponderance of fitness that all but one of the rival signs shall die out, and some being equal in value to express the same idea or object, will continue to be used indifferently, or as a matter of individual taste, without confusion. A multiplication of the numbers confined together, either of deaf-mutes or of Indians whose speech is diverse, will not decrease the resulting uniformity, though it will increase both the copiousness and the precision of the vocabulary. The Indian use of signs, though maintained by linguistic diversities, is not coincident with any linguistic boundaries. The tendency is to their uniformity among groups of people who from any cause are brought into contact with each other while still speaking different languages. The longer and closer such contact, while no common tongue is adopted, the greater will be the uniformity of signs.

Some writers take a middle ground with regard to the identity of the sign language of the North American Indians, comparing it with the dialects and provincialisms of the English language, as spoken in England, Ireland, Scotland and Wales.

But those dialects are the remains of actually diverse languages, which to some speakers have not become integrated. In England alone the provincial dialects are traceable as the legacies of Saxons, Angles, Jutes, and Danes, with a varying amount of Norman influence. A thorough scholar in the composite tongue, now called English, will be able to understand all the dialects and provincialisms of English in the British Isles, but the uneducated man of Yorkshire is not able to communicate readily with the equally uneducated man of Somersetshire. This is the true distinction. A thorough sign talker would be able to to talk with several Indians who have no signs in common, and who, if their knowledge of signs were only memorized, could not communicate with each other. So, also, as an educated Englishman will understand the attempts of a foreigner to speak in very imperfect and broken English, a good Indian sign-expert will apprehend the feeblest efforts of a tyro in gestures. But the inference that there is but one true Indian sign language, just as there is but one true English language, is not correct unless it can be shown that a much larger proportion of the Indians who use signs at all, than present researches show to be the case, use identically the same signs to express the same ideas. It would also seem necessary to the parallel that the signs so used should be absolute, if not arbitrary, as are the words of an oral language, and not independent of preconcert and self-interpreting at the instant of their invention or first exhibition, as all true signs must originally have been and still measurably remain.

ARE SIGNS CONVENTIONAL OR INSTINCTIVE?

There has been much discussion on the question whether gesture signs were originally invented, in the strict sense of that term, or whether they result from a natural connection between

them and the ideas represented by them, that is, whether they are conventional or instinctive. Cardinal Wiseman (*Essays*, III, 537) thinks they are of both characters; but referring particularly to the Italian signs and the proper mode of discovering their meaning, he observes that they are used primarily with words and form the usual accompaniment of certain phrases. "For these the gestures become substitutes, and then by association express all their meaning, even when used alone." This would be the process only where systematic gestures had never prevailed or had been so disused as to be forgotten, aud were adopted after elaborate oral phrases and traditional oral expressions had become common. Sign language as a product of evolution has been developed rather than invented, and yet it seems probable that each of the separate signs, like the several steps that lead to any true invention, had a definite origin arising out of some appropriate occasion, and the same sign may in this manner have had many independent origins due to identity in the circumstances, or, if lost, may have been reproduced.

Another form of the query is whether signs are arbitrary or natural. An unphilosophic answer will often be made in accordance with what the observer considers to be natural to himself. A common sign among both deaf-mutes and Indians for *woman* consists in designating the arrangement of the hair, but such a represented arrangement of hair familiar to the gesturer as had never been seen by the person addressed would not seem "natural" to the latter. It would be classed as arbitrary, and could not be understood without context or explanation, indeed without translation such as is required from foreign oral speech. Signs most naturally, that is appropriately, expressing a conception of the thing signified, are first adopted from circumstances of environment, and afterwards modified so as to appear, without full understanding, conventional and arbitrary, yet they are as truly "natural" as the signs for hearing, seeing, eating, and drinking, which continue all over the world as they were first formed because there is no change in those operations.

Perhaps no signs in common use are in their origin conventional. What appears to be conventionality largely consists in the form of abbreviation which is agreed upon. When the signs of the Indians have from ideographic form become demotic they may be roughly called conventional, but still not arbitrary.

SOME NATURAL SIGNS CONVENTIONALIZED.

But while all Indians, as all gesturing men, have many natural signs in common, they use many others which have become conventional in the sense that their origin and conception are not now known or regarded by the persons using them. The conventions by which the latter were established occurred during long periods, when the tribes forming them were so separated as to have established altogether diverse customs and mythologies, and when the several tribes were exposed to such different environments as to have formed varying conceptions needing appropriate sign expression. The old error that the North American Indians constitute one homogeneous race is now abandoned. Nearly all the characteristics once alleged as segregating them from the rest of mankind have proved not to belong to the whole of the precolumbian population, but only to those portions of it first explored. The practice of scalping is not now universal, if it ever was, even among the tribes least influenced by civilization, and therefore the cultivation of the scalp-lock separated from the rest of the hair of the head, or with the removal of all other hair, is not a general feature of their appearance. The arrangement of the hair is so different among tribes as to be one of the most convenient modes for their pictorial distinction. The war paint, red in some tribes, was black in others; the mystic rites of the calumet were in many regions unknown, and the use of wampum was by no means extensive. The wigwam is not the type of native dwellings, which show as many differing forms as those of Europe. In color there is a great variety, and even admitting that the term "race" is properly applied, no competent observer would characterize it as red, still less copper-colored. Some tribes differ from each other in all respects nearly as much as either of them do from the lazzaroni of Naples, and more than either do from certain tribes of Australia. It would therefore be expected, as is the case, that the conventional signs of different stocks and regions differ as do the words of English, French and German, which, nevertheless, have sprung from the same linguistic roots. No one of those languages is a dialect of any of the others; and although the sign systems of the several tribes have greater generic unity with less specific variety than oral languages, no one of them is necessarily the dialect of any other. To insist that sign language is uniform were to assert that it is perfect—" That faultless monster that the world ne'er saw."

GENERAL ANCIENT USE OF THE SYSTEM IN N. A.

The supposition that the systematic use of signs once existed among all Indian tribes receives support from the fact that in nearly all instances where such existence has been at first denied, further research has discovered the remains, even if not the practice, of sign language. This has been even among tribes long exposed to European influence and officially segregated from all others. Collections have been obtained from the Iroquois, Ojibwas, Alaskans, Apaches, Zuñi, Pimas, Papagos and Maricopas, after army officers, missionaries, Indian agents and travellers had denied them to be possessed of any knowledge on the subject.

One of the most interesting proofs of the general knowledge of sign language, even when seldom used, was given in the visit of five Jicarilla Apaches to Washington in April, 1880, under the charge of their agent. The latter said he had never heard of any use of signs among them. But it happened that there was a delegation of Absaroka (Crows) at the same hotel, and the two parties, from regions one thousand miles apart, not knowing a word of each other's language, immediately began to converse in signs, resulting in a decided sensation. One of the Crows asked the Apaches whether they ate horses, and it happening that the sign for *eating* was misapprehended for that known by the Apaches for *many*, the question was supposed to be whether the latter had many horses, which was answered in the affirmative. Thence ensued a misunderstanding on the subject of hippophagy, which was curious both as showing the general use of signs as a practice and the diversity in special signs for particular meanings. The surprise of the agent at the unsuspected accomplishment of his charges was not unlike that of a hen which, having hatched a number of duck eggs, is perplexed at the instinct with which the brood takes to the water.

The denial of the use of signs is sometimes faithfully though erroneously reported from the distinct statements of Indians to that effect. In that, as in other matters, they are often provokingly reticent about their old habits and traditions. Chief Ouray asserted to me, that his people, the Utes, had not the practice of sign talk, and had no use for it. This was much in the proud spirit in which an Englishman would have made the same statement, as the idea involved an accusation against the civilization of his people, whom he wished to appear highly advanced. Within the same week I took seven Utes, members of the delegation then with Ouray, to the National Deaf-Mute College, and they showed not only perfect familiarity with, but expertness in, signs.

The studies thus far pursued lead to the conclusion that at the time of the discovery of North America all its inhabitants practised sign language, though with different degrees of expertness; and that, while under changed circumstances, it was disused by some, others, especially those who after the acquisition of horses became nomads of the Great Plains, retained and cultivated it to the high development now attained.

PERMANENCE OF SIGNS.

It is important to inquire into the permanence of particular gesture signs to express a special idea or object when the system has been long continued. The gestures of classic times are still in use by the modern Italians with the same signification; indeed the former, on Greek vases or reliefs, or in Herculanean bronzes, can only be interpreted by the latter. In regard to the signs of instructed deaf-mutes in this country there appears to be a permanence beyond expectation. A pupil of the Hartford Institute half a century ago lately stated that the signs used by teachers and pupils at Hartford, Philadelphia, Washington, Council Bluffs and Omaha, were nearly the same as he had learned. "We still adhere to the old sign for President from Monroe's three-cornered hat, and for governor we designate the cockade worn by that dignitary on grand occasions three generations ago."

Specific comparisons made of the signs reported by the Prince of Wied, in 1832, with those now used by the same tribes from whom he obtained them, show a remarkable degree of permanence. If they have persisted for half a century their age is probably much greater. In general it is believed that signs, constituting as they do a natural mode of expression, though enlarging in scope as new ideas and new objects require to be included and though abbreviated variously, do not readily change in their essentials.

I do not present any Indian signs as precisely those of primitive man, not being so carried away by enthusiasm as to suppose them possessed of immutability and immortality not found in any other mode of human utterance. Signs as well as words, animals, and plants have had their growth, development and change, their births and deaths, and their struggle for existence with survival of the fittest. Yet when signs, which are general among Indian tribes, are also prevalent in other parts of the world, they probably are of great antiquity. The use of derivative meanings to a sign only enhances this presumption. At first there might not appear to be any connection between the ideas of *same* and *wife*, expressed by the sign of horizontally extending the two forefingers side by side. The original idea was doubtless that given by the Welsh captain in Shakspere's Henry V: "'Tis so like as my fingers is to my fingers;" and from this similarity comes "equal," "companion," and subsequently the close life-companion "wife." The sign is used in each of these senses by different Indian tribes, and sometimes the same tribe applies it in all of the senses as the context determines. It appears also in many lands with all the significations except that of "wife."

Many signs but little differentiated were unstable, while others that have proved the best modes of expression have survived as definite and established. A note may be made in this connection of the large number of diverse signs for *horse*, all of which must have been invented within a comparatively recent period, and the small variation in the signs for *dog*, which are probably ancient.

IS THE INDIAN SYSTEM SPECIAL AND PECULIAR?

While denying the uniformity of Indian signs, it is proper to inquire whether their system, as a whole, is special and peculiar to themselves. This may be determined by comparing that system with those of other peoples and of deaf-mutes.

COMPARISONS WITH FOREIGN SIGNS.

My researches during several years show a surprising number of signs for the same idea which are substantially identical, not only among savage tribes, but among all peoples that use gesture signs with any freedom. Men, in groping for a mode of communication with each other, and using the same general methods, have been under many varying conditions and circumstances which have determined differently many conceptions and their semiotic execution, but there have also been many of both which were similar. North American Indians have no special superstition concerning the evil-eye like the Italians, nor have they been long familiar with the jackass so as to make him, with more or less propriety, emblematic of stupidity; therefore signs for those concepts are not cisatlantic, but many are substantially in common between our Indians and Italians. Many other Indian signs are identical, not only with those of the Italians and the classic Greeks and Romans, but of other peoples of the Old World, both savage and civilized. The generic uniformity is obvious, while the occasion of specific varieties can be readily understood.

The same remark applies to the collections of signs already obtained by correspondence from among the Turks, Armenians and Koords, the Bushmen of Africa, the Fijians, the Redjangs and Lelongs of Sumatra, the Chinese and the Australians. The results of researches in Ceylon, India, South America and several other parts of the world, are not yet sufficient to allow of their classification. Much interesting material is expected from inquiries recently instituted through the medium of Mr. Hyde Clarke, Vice President of the Anthropological Institute of Great Britain and Ireland, into the sign language of the mutes of the Seraglio at Constantinople. That they had a system of communication was noticed by Sibscota, in 1670, without his giving any details. It appears not only to be known to the inmates themselves, but to high officials, eunuchs and other persons connected with the Sublime Porte. As it is supposed that the Osmanli Sultans followed the Byzantine emperors in the employment of mutes, and that they adopted them from Persian kings, it is possible that the signs, now in systematic, though limited, use, have been regularly transmitted from high oriental antiquity.

COMPARISON WITH DEAF-MUTE SIGNS.

The Indians who have been brought to the eastern states have often held happy intercourse by signs with white deaf-mutes, who surely have no semiotic code preconcerted with any of the plain-roamers. While many of their signs were identical, and all sooner or later were mutually understood, it has been noticed that the signs of the deaf-mutes were more readily understood by the Indians than were theirs by the deaf-mutes, and that the latter greatly excelled in pantomimic effect. What is to the Indian a mere adjunct or accomplishment is to the deaf-mute the natural mode of utterance. The "action, action, action," of Demosthenes is their only oratory, not mere heightening of it, however valuable.

The result of the comparisons is that the so-called sign language of Indians is not properly speaking one language, but that it and the gesture systems of deaf-mutes and of all peoples constitute together one language—the gesture speech of mankind—of which each system is a dialect.

GESTURES AIDING ARCHÆOLOGICAL RESEARCH.

The most interesting light in which the Indians of North America can be regarded is in their present representation of a stage of evolution once passed through by our own ancestors. Their signs, as well as their myths and customs, form a part of the palæontology of humanity to be studied in the history of the latter, as the geologist, with similar object, studies all the strata of the physical world. At this time it is only possible to state that gesture signs have been applied to elucidate pictographs, and also to discover religious, sociologic, and historic ideas preserved in themselves, as has been done with great success in the radicals of oral speech.

SIGNS CONNECTED WITH PICTOGRAPHS.

The picture writing of Indians is the sole form in which they recorded events and ideas that can ever be interpreted without the aid of a traditional key, such as is required for the signification of the wampum belts of the northeastern tribes and the *quippus* of Peru. Strips of bark, tablets of wood, dressed skins of animals, and the smooth surfaces of rocks have been and still are used for such records, those most ancient, and therefore most interesting, being the rock etchings; but they can only be deciphered by the ascertained principles on which the more modern and obvious are made. Many of the widespread rock carvings are mere idle sketches of natural objects, mainly animals, and others are as strictly mnemonic as is the wampum. But where there has existed a rude form of graphic representation, and at the same time a system of ideographic gesture signs prevailed, it would be expected that the form of the latter would appear in the former. That this is the fact among North American Indians will be shown in a paper to be read before the section by my collaborator Dr. W. J. Hoffman, and at greater length in a report by myself to form part of the first Annual Report of the Bureau of Ethnology, now in press. This fact is of great archæologic importance, as the reproduction of gesture lines in the pictographs made by Indians has, for obvious reasons, been most frequent in the attempt to convey those subjective ideas which were beyond the range of an artistic skill limited to the direct representation of objects, so that the part of the pictographs which is the most difficult of interpretation is the one which the study of sign language can elucidate. Traces of the same signs used by Indians found in the ideographic pictures of the Egyptians, and in Chinese and Aztec characters, are also exhibited by illustrations in the Report above mentioned.

HISTORY OF GESTURE LANGUAGE.

There is ample evidence of record, besides that derived from other sources, that the systematic use of gesture speech was of great antiquity. Livy so declared, and Quintilian specifies that the "*lex gestus* * * * *ab illis temporibus heroicis orta est.*" Athenæus tells that gestures were reduced to distinct classification with appropriate terminology. One of these classes was adapted by Bathyllus to pantomime.

While the general effect of the classic pantomimes is often mentioned, there remain but few detailed descriptions of them. Apuleius, however, in his *Metamorphosis* gives sufficient details of the performance of the Judgment of Paris to show that it resembled the best form of modern ballet opera. The popularity of these exhibitions continued until the sixth century, and it is evident from a decree of Charlemagne that they were not lost, or, at least, had been revived in his time. Those of us who have enjoyed the performance of the original Ravel troupe will admit that the art still survives, though not with the magnificence or perfection, especially with reference to serious subjects, which it exhibited in the age of imperial Rome.

Quintilian gave most elaborate rules for gestures in oratory, which are specially noticeable from the importance attached to the manner of disposing the fingers. He attributed to each particular disposition a significance or suitableness which is not now obvious. The value of these digital arrangements is, however, exhibited by their use among the modern Italians, to whom they have directly descended. Their curious elaboration appears in the volume by the canon Andrea de Jorio, *La Mimica degli Antichi investigata nel Gestire Napoletano*, *Napoli*, 1832. The canon's chief object was to interpret the gestures of the ancients as exhibited in their works of art and described in their writings, by the modern gesticulations of the Neapolitans, and he has shown that the general system of gesture once prevailing in ancient Italy is substantially the same as now observed. With an understanding of the existing language of gesture the scenes on the most ancient Greek vases and reliefs obtain a new and interesting significance and form a connecting link between the present and prehistoric times.

USE BY MODERN ACTORS.

Less of practical value can be learned of sign language, considered as a system, from the study of the gestures used by actors, than would appear without reflection. The pantomimist, indeed, who uses no words whatever, is obliged to avail himself of every natural or imagined connection between thought and gesture, and depending wholly upon the latter, makes himself intelligible. With speaking actors, however, words are the main reliance, and gestures generally serve for rhythmic movement and to display personal grace.

When many admirers of Ristori, who were wholly unacquainted with the language in which her words were delivered, declared that her gesture and expression were so perfect that they understood every sentence, it is to be doubted if they would have been so delighted if they had not been thoroughly familiar with the plots of Queen Elizabeth and Mary Stuart. This view is confirmed by the case of a deaf-mute, known to me, who had prepared to enjoy Ristori's acting by reading in advance the advertised play, but on his reaching the theatre another play was substituted and he could derive no idea from its presentation. A crucial test on this subject was made at the representation at Washington last April, of *Frou-Frou* by Sara Bernhardt and the excellent French company supporting her. Several persons of special intelligence and familiar with theatrical performances, but who did not understand spoken French, and had not heard or read the play or even seen an abstract of it, paid close attention to ascertain what they could learn of the plot and incidents from the gestures alone. This could be determined in the special play the more certainly as it is not founded on historic events or any known facts. The result was that from the entrance of the heroine during the first scene in a peacock-blue riding habit to her death in a black walking-suit, three hours or five acts later, none of the students formed any distinct conception of the plot. This want of apprehension extended even to uncertainty whether *Gilberte* was married or not; that is, whether her adventures were those of a disobedient daughter or a faithless wife, and, if married, which of the half dozen male personages was her husband. There were gestures enough, indeed rather a profusion of them, and they were thoroughly appropriate to the words (when those were understood) in which fun, distress, rage, and other emotions were expressed, but in no cases did they interpret the motive for those emotions. They were the dressing for the words of the actors as the superb millinery was that of their persons, and perhaps acted as varnish to bring out dialogues and soliloquies in heightened effect. But though varnish can bring into plainer view dull or faded characters, it cannot introduce into them significance where none before existed. The simple fact was that the gestures of the most famed histrionic school, the Comédie Française, were not significant, far less self-interpreting, and though praised as the perfection of art, have diverged widely from nature.

However numerous and correct may be the actually significant gestures made by a great actor in the representation of his part, they must be in small proportion to the number of gestures not at all significant, and which are no less necessary to give to his declamation precision, grace and force. Histrionic perfection is, indeed, more shown in the slight shades of movement of the head, glances of the eye, and poises of the body than in violent attitudes; but these slight movements are wholly unintelligible apart from the words uttered with them. Even in the expression of strong emotion the same gesture will apply to many and utterly diverse conditions of fact. Its fitness consists in being the same which the hearer of the expository words would spontaneously assume if yielding to the same emotions, and which therefore by association tends to induce a sympathetic yielding. The greatest actor in telling that his father was dead can convey his grief with a shade of difference from that which he would use if saying that his wife had run away, his son been arrested for murder, or his house burned down; but that shade would not without words inform any person, ignorant of the supposed event, which of the four misfortunes had occurred. A true sign language, however, would fully express the exact circumstances, either with or without any exhibition of the general emotion appropriate to them.

Even among the best sign-talkers, whether Indian or deaf-mute, it is necessary to establish some *rapport* relating to theme or subject-matter, since many gestures, as indeed is the case in a less degree with spoken words, have widely different significations, according to the object of their exhibition, as well as the context. Rabelais (*Pantagruel*, Book III, ch. xiv) hits the truth upon this point, however ungallant in his application of it to the fair sex. Panurge is desirous to consult a dumb man, but says it would be useless to apply to a woman, for "whatever it be that they see they do always represent unto their fancies, and imagine that it hath some relation to love. Whatever signs, shows or gestures, we shall make, or whatever our behavior, carriage or demeanor, shall happen to be in their view and presence, they will interpret the whole in reference to androgynation." A story is told to the same point by Guevara, in his fabulous life of the Emperor Marcus Aurelius. A young Roman gentleman encountering at the foot of Mount Celion a beautiful Latin lady, who from her very cradle had been deaf and dumb, asked her in gesture what senators in her descent from the top of the hill she had met with, going up thither. She straightway imagined that he had fallen in love with her and was eloquently proposing marriage, whereupon she at once threw herself into his arms in acceptance. The experience of travellers of the Plains is to the same general effect, that signs commonly used to men are understood by women in a sense so different as to occasion embarrassment.

RESULTS SOUGHT IN THE STUDY OF SIGN LANGUAGE.

These may be divided into (1) its practical application, (2) its aid to philologic researches, and (3) its archæologic relations.

PRACTICAL APPLICATION.

The most obvious application of sign language will for its practical utility depend upon the correctness of the view submitted that it is not a mere semaphoric repetition of motions to be memorized from a limited traditional list, but is a cultivable art, founded upon principles which can be readily applied by travellers. This advantage is not merely theoretical, but has been demonstrated to be practical by a professor in a deaf mute college who, lately visiting several of the wild tribes of the plains, made himself understood among all of them without knowing a word of any of their languages, and by another who had a similar experience in Italy and southern France. It must, however, be observed that the use of signs is only of great assistance in communicating with foreigners, whose speech is not understood, when both parties agree to cease all attempt at oral language, relying wholly upon gestures. So long as words are used at all, signs will be made only as their accompaniment, and they will not always be ideographic.

POWERS OF SIGNS COMPARED WITH SPEECH.

Sign language is superior to all others in that it permits every one to find in nature an image to express his thoughts on the most needful matters intelligibly to any other person. The direct or substantial natural analogy peculiar to it prevents a confusion of ideas. It is possible to use words without understanding them which yet may be understood by those addressed, but it is hardly possible to use signs without full comprehension of them. Separate words may be comprehended by persons hearing them without the whole connected sense of the words taken together being caught, but signs are more intimately connected. Even those most appropriate will not be understood if the subject is beyond the comprehension of their beholders. They would be as unintelligible as the wild clicks of his instrument, in an electric storm, would be to the telegrapher, or as the semaphore, driven by wind, to the signalist. In oral speech even onomatopes are arbitrary, the most strictly natural sounds striking the ear of different individuals and nations in a manner wholly diverse. The instances given by Sayce are in point. Exactly the same sound was intended to be reproduced in the "*bilbit* amphora" of Nævius, the "*glut glut* murmurat unda sonans" of the Latin Anthology, and the "*puls*" of Varro. The Persian "*bulbul*," the "*jugjug*" of Gascoigne, and the "*whitwhit*" of others are all attempts at imitating the note of the nightingale. But successful signs must have a much closer analogy and establish a concord between the talkers far beyond that produced by the mere sound of words. The merely emotional sounds or interjections may be advantageously employed in connection with merely emotional gestures, but whether with or without them, they would be useless for the explicit communication of facts and opinions of which signs by themselves are capable. The combinations which can be made by signs are infinite and their enthusiastic teachers may be right in claiming that if they had been elaborated by the secular labor devoted to spoken language, man could, by his hands, arms and fingers, with facial and bodily accentuation, express any idea that could be conveyed by words. As, however, sign language has been chiefly used during historic time either as a scaffolding around a more valuable structure, to be thrown aside when the latter was completed, or as an occasional substitute, such development was not to be expected.

A comparison sometimes drawn between sign language and that of the North American Indians, founded on the statement of their common poverty in abstract expressions, is not just to either. Deeper study into Indian tongues has ascertained that they are by no means so confined to the concrete as was once believed, and the process of forming signs to express abstract ideas is only a variant from that of oral speech, in which the words for the most abstract ideas, such as law, virtue, infinitude, and immortality, are shown by Max Müller to have been derived and deduced, that is, abstracted, from sensuous impressions. This is done by selecting what is and rejecting what is not in common to the concrete ideas. Concepts of the intangible and invisible are only learned through precepts of tangible and visible objects, whether finally expressed to the eye or to the ear, in terms of sight or of sound. In the use of signs the countenance and manner as well as the tenor decide whether objects themselves are intended, or the forms, positions, qualities, and motions of other objects which are suggested; and signs for moral and intellectual ideas, founded on analogies, are common all over the world as well as among deaf-mutes. The very concepts of *plurality*, *momentum* and *righteousness*, selected by Tylor as the result of combined and compared thought which requires words, can be clearly expressed by signs, and it is not understood why those signs could not have attained their present abstract significance through the thoughts arising from the combination and comparison of other signs, without the actual intervention of words.

The elements of sign language are natural and universal, by recurring to which the less natural signs adopted dialectically or for expedition can always, with some circumlocution, be explained. This power of interpreting itself is a peculiar advantage over spoken languages, which, unless explained by gestures or indications, can only be interpreted by means of some other spoken language. When highly cultivated, its rapidity on familiar subjects exceeds that of speech and approaches to that of thought itself. This statement may be startling to those who do not consider that oral speech is now wholly conventional, and that with the similar development of sign language conventional expressions with hands and body could be made more quickly than with the vocal organs, because more organs could be worked at once. At the same time it must be admitted that great increase in rapidity is chiefly obtained by a system of preconcerted abbreviations, and by the adoption of absolute forms, in which naturalness is sacrificed and conventionality established, as has been the case with all spoken languages in the degree in which they have become copious and convenient.

There is another characteristic of the gesture speech that, though it cannot be resorted to in the dark, nor where the attention of the person addressed has not been otherwise attracted, it has the countervailing benefit of use when the voice cannot be employed. This may be an advantage at a distance which the eye can reach, but not the ear, and still more frequently when silence or secrecy is desired. Dalgarno recommends it for use in the presence of great people, who ought not to be disturbed, and curiously enough "Disappearing Mist," the Iroquois chief, speaks of the former extensive employment of signs in his tribe by women and boys as a mark of respect to warriors and elders, their voices, in the good old days, not being uplifted in the presence of the latter. The decay of that wholesome state of discipline, he thinks, accounts partly for the disappearance of the use of signs among the modern impudent youth and the dusky claimants of woman's rights.

RELATIONS TO PHILOLOGY.

The aid to be derived from the study of sign language in prosecuting researches into the science of philology was pointed out by Leibnitz, in his *Collectanea Etymologica*, without hitherto exciting any thorough or scientific work in that direction, the obstacle to it probably being that scholars competent in other respects had no adequate data of the gesture speech of man to be used in comparison. The latter will, it is hoped, be supplied by the work now undertaken by me, under the direction of the Bureau of Ethnology, which extends to the collection and collation of signs from all parts of the world as well as those of North American Indians.

It is generally admitted that signs played an important part in giving meaning to spoken words, and that many primordial roots of language have been founded in the involuntary sounds accompanying certain actions. As, however, the action was the essential, and the concomitant or consequent sound the accident, it would be expected that a representation or feigned reproduction of the action would have been used to express the idea before the sound associated with that action would have been separated from it. Philology, therefore, comparing the languages of earth in their radicals, must henceforth include the graphic or manual presentation of thought, and compare the elements of ideography with those of phonics. Etymology now examines the ultimate roots, not the fanciful resemblances between oral forms, in the different tongues; the internal, not the mere external parts of language. A marked peculiarity of sign language consists in its limited number of radicals and the infinite combinations into which those radicals enter while still remaining distinctive. It is therefore a proper field for etymologic study.

It is possible to ascertain the included gesture even in many English words. The class represented by the word *supercilious* will occur to all, but one or two examples may be given not so obvious and more immediately connected with the gestures of Indians. *Imbecile*, generally applied to the weakness of old age, is derived from the Latin *in*, in the sense of on, and *bacillum*, a staff, which at once recalls the Cheyenne sign for *old man*, viz.: holding the right hand forward, bent at elbow with the fist closed sidewise, as if holding a staff. So *time* appears more nearly connected with the Greek τείνω, to stretch, when information is given of the sign for *long time*, viz.,: placing the thumbs and forefingers as if a small thread were held between the thumb and forefinger of each hand, the hands first touching each other, and then slowly moving apart, as if *stretching* a piece of gum-elastic.

Some special resemblances exist between the language of signs and the character of the oral languages found on this continent. Dr. J. Hammond Trumbull remarks of the composition of the words that they were "so constructed as to be thoroughly self-defining and immediately intelligible to the hearer." In another connection the remark is further enforced. "Indeed, it is a requirement of the Indian languages that every word shall be so framed as to admit of immediate resolution to its significant elements by the hearer It must be thoroughly *self-defining*, for (as Max Müller has expressed it) 'it requires tradition, society, and literature to maintain words which can no longer be analyzed at once.' * * * In the ever-shifting state of a nomadic society no debased coin can be tolerated in language, no obscure legend accepted on trust. The metal must be pure and the legend distinct."

Indian languages, like those of higher development, sometimes exhibit changes of form by the permutation of vowels, but often an incorporated particle, whether suffix, affix, or infix, shows the etymology which often, also, exhibits the same objective conception that would be executed in gesture. There are, for instance, different forms for standing, sitting, lying, falling, and for standing, sitting, lying on or falling from the same level or a higher or lower level. This resembles the pictorial conception and execution of signs.

Indian languages exhibit the same fondness for demonstration which is necessary in sign language. The two forms of utterance are alike in their want of power to express certain words, such as the verb "to be," and in the criterion of organization, so far as concerns a high degree of synthesis and imperfect differentiation, they bear substantially the same relation to the English language.

It may be added that as not only proper names but nouns generally in Indian languages are connotive, predicating some attribute of the object, they can readily be expressed by gesture signs, and therefore among them, relations may be established between the words and the signs. Such have also been noticed, especially by my valued correspondent, Mr. Hyde Clarke, to exist between signs and the words of old Asiatic and African languages, showing the same operation of conditions in the same psychologic horizon.

DIVISIONS OF GESTURE SPEECH.

Gesture speech is composed of corporeal motion and facial expression. An attempt has been made by some writers to discuss these general divisions separately, and its success would be practically convenient if it were always understood that their connection is so intimate that they can never be altogether severed. A play of feature, whether instinctive or voluntary, accentuates and qualifies all motions intended to serve as signs, and strong instinctive facial expression is generally accompanied by action of the body or some of its members. But, so far as a distinction can be made, expressions of the features are the result of emotional, and corporeal gestures, of intellectual action. The former in general and the small number of the latter that are distinctively emotional are nearly identical among men from physiological causes which do not affect with the same similarity the processes of thought. The large number of corporeal gestures expressing intellectual operations require and admit of more variety and conventionality. Thus the features and the body among all mankind act almost uniformly in exhibiting fear, grief, surprise and shame, but all objective conceptions are varied and variously portrayed. Even such simple indications as those for "no" and "yes" appear in several different motions. While, therefore, the terms sign language and gesture speech necessarily include and suppose facial expression when emotions are in question, they refer more particularly to corporeal motions and attitudes. For this reason much of the valuable contribution of Darwin in his *Expression of the Emotions in Man and Animals* is not directly applicable to sign language. His analysis of emotional gestures into those explained on the principles of serviceable associated habits, of antithesis, and of constitution of the nervous system, should, nevertheless, always be remembered. The earliest gestures were doubtless emotional, preceding those of a pictorial, metaphoric, and, still subsequent, conventional character.

THE ORIGIN OF SIGN LANGUAGE.

When examining into the origin of sign language through its connection with that of oral speech, it is necessary to be free from the vague popular impression that some oral language, of the general character of that now used among mankind, is "natural" to mankind. It will be admitted that all the higher oral languages were at some past time less opulent and comprehensive than they are now, and as each particular language has been thoroughly studied it has become evident that it grew out of some other and less advanced form.

Oral language consists of variations and mutations of vocal sounds produced as signs of thought and emotion. But it is not enough that those signs should be available as the vehicle of the producer's own thoughts. They must be also efficient for the communication of such thoughts to others. It has been, until of late years, generally held that thought was not possible without oral language, and that, as man was supposed to have possessed from the first the power of thought, he also from the first possessed and used oral language substantially as at present. That the latter, as a special faculty, formed the main distinction between man and the brutes, has been and still is the prevailing doctrine. It may, however, be doubted if there is any more necessary connection between ideas and sounds, the mere signs of thought that strike the ear, than there is between the same ideas and signs addressed only to the eye.

The point most debated for centuries has been, not whether there was any primitive oral language, but what that language was. Some literalists have indeed argued from the Mosaic narrative that the primitive language had been taken away as a disciplinary punishment, as the Paradisiac Eden had been earlier lost, and that, therefore, the search for it was as fruitless as to attempt the passage of the flaming sword. More liberal Christians have been disposed to regard the Babel story as allegorical, if not mythical, and have considered it to represent the disintegration of tongues out of one which was primitive. Though its quest has led into error, it has, like those of the philosopher's stone, of perpetual motion and of other phantasms in other directions of thought, been of great indirect utility. It has stimulated philologic science, the advance of which has successively shifted back the postulated primitive language from Hebrew to Sanscrit, thence to Aryan, and now it is attempted to evoke from the vasty deeps of antiquity the ghosts of other rival claimants for precedence in dissolution.

The discussion is now, however, varied by the suggested possibility that man at some time may have existed without any oral language. It is of late conceded that mental images or representations can be formed without any connection with sound, and may at least serve for thought, if not for expression. It is certain that concepts, however formed, can be expressed by other means than sound. One mode of this expression is by gesture, and there is less reason to believe that gestures commenced as the interpretation of or substitute for words, than that the latter originated in and served to translate gestures. Many arguments have been advanced to prove that gesture language preceded articulate speech and formed the earliest attempt at communication, resulting from the interacting subjective and objective conditions to which primitive man was exposed. Some of the facts on which deductions have been based, made in accordance with well-established modes of scientific research from study of the lower animals, children, individuals in mental disorder or isolated from their fellows, and the lower types of mankind, are of great interest, but it is only possible now to examine those relating to deaf-mutes.

UNINSTRUCTED DEAF-MUTES.

The signs made by congenital and uninstructed deaf-mutes are either those originating in or invented by individuals, or those of a colloquial character used by such mutes where associated. The accidental or merely suggestive signs peculiar to families, one member of which happens to be a mute, are too much affected by the other members of the family to be of certain value. Those, again, which are taught in institutions have become conventional and were designedly adapted to translation into oral speech, although founded by the abbé de l'Épée, followed by the abbé Sicard, in the natural signs first above mentioned.

A great change has doubtless occurred in the estimation of congenital deaf-mutes since the Justinian Code which consigned them forever to legal infancy, as incapable of intelligence, and classed them with the insane. Yet most modern writers, for instance, Archbishop Whately and Max Müller, have declared that deaf-mutes could not think until after having been instructed. It cannot be denied that the deaf-mute thinks after his instruction either in the ordinary gesture signs or in the finger alphabet, or more lately in artificial speech. By this instruction he has become master of a highly-developed language, such as English or French, which he can read, write, and actually talk, but that foreign language he has obtained through the medium of signs. This is a conclusive proof that signs constitute a real language and one which admits of thought, for no one can learn a foreign language unless he had some language of his own, whether by descent or acquisition, by which it could be translated, and such translation into the new language could not even be commenced unless the mind had been already in action and intelligently using the original language for that purpose. In fact the use by deaf-mutes of signs originating in themselves exhibits a creative action of mind and innate faculty of expression beyond that of ordinary speakers who acquired language without conscious effort.

GESTURES OF FLUENT TALKERS.

The command of a copious vocabulary common to both speaker and hearer undoubtedly tends to a phlegmatic delivery and disdain of subsidiary aid. An excited speaker will, however, generally make a free use of his hands without regard to any effect of that use upon auditors. Even among the gesture-hating English, when they are aroused from torpidity of manner, the hands are involuntarily clapped in approbation, rubbed with delight, wrung in distress, raised in astonishment, and waved in triumph. The fingers are snapped for contempt, the forefinger is vibrated to reprove or threaten, and the fist shaken in defiance. The brow is contracted with displeasure, and the eyes winked to show connivance. The shoulders are shrugged to express disbelief or repugnance, the eyebrows elevated with surprise, the lips bitten in vexation and thrust out in sullenness or displeasure. Quintilian becomes eloquent on the variety of motions of which the hands alone are capable.

"The action of the other parts of the body assists the speaker, but the hands speak themselves. By them do we not demand, promise, call, dismiss, threaten, supplicate, express abhorrence and terror, question and deny? Do we not by them express joy and sorrow, doubt, confession, repentance, measure, quantity, number, and time? Do they not also encourage, supplicate, restrain, convict, admire, respect?"

NATURAL PANTOMIME.

In the earliest part of man's history the subjects of his discourse must have been almost wholly sensuous, and therefore readily expressed in pantomime. Not only was pantomime sufficient for all the actual needs of his existence, but it is not easy to imagine how he could have used language such as is now known to us. If the best English dictionary and grammar had been miraculously furnished to him, together with the art of reading with proper pronunciation, the gift would have been valueless, because the ideas expressed by the words had not yet been formed.

That the early concepts were of a direct and material character is shown by what has been ascertained of the roots of language and there does not appear to be much difficulty in expressing by other than vocal instrumentality all that could have been expressed by those roots. Even now, with our vastly increased belongings of external life, avocations, and habits, nearly all that is absolutely necessary for our physical needs can be expressed in pantomime. Far beyond the mere signs for eating, drinking, sleeping, and the like, any one will understand a skilful representation in signs of a tailor, shoemaker, blacksmith, weaver, sailor, farmer, or doctor. So of washing, dressing, shaving, walking, driving, writing, reading, churning, milking, shooting, fishing, rowing, sailing, sawing, planing, boring, and, in short, an endless list.

Whether or not sight preceded hearing in order of development, it is difficult, in conjecturing the first attempts of man or his hypothetical ancestor at the expression either of percepts or concepts, to connect vocal sounds with any large number of objects, but it is readily conceivable that the characteristics of their forms and movements should have been suggested to the eye—highly exercised before the tongue—after the arms and fingers had become free for the requisite simulation or portrayal. It may readily be supposed that a troglodyte man would desire to communicate the finding of a cave in the vicinity of a pure pool, circled with soft grass, and shaded by trees bearing edible fruit. No sound of nature is connected with any of those objects, but the position and size of the cave, its distance and direction, the water, its quality, and amount, the verdant circling carpet, and the kind and height of the trees could have been made known by pantomime in the days of the mammoth, if articulate speech had not then been established, as Indians or deaf-mutes now communicate similar information by the same agency.

CONCLUSIONS.

It may be conceded that after man had attained to all his present faculties, he did not choose between the adoption of voice and gesture, and never, with those faculties, was in a state where the one was used to the absolute exclusion of the other. The epoch, however, to which our speculations relate, is that in which he had not reached the present symmetric development of his intellect and of his bodily organs, and the inquiry is, which mode of communication was earliest in adaptation to his simple wants and unformed intelligence. With the voice he could imitate distinctively but the few sounds of nature, while with gesture he could exhibit actions, motions, positions, forms, dimensions, directions and distances, with their derivatives and analogues. It would seem from this unequal division of capacity that oral speech remained rudimentary long after gesture had become an efficient instrument of thought and expression. With due allowance for all purely imitative sounds and for the spontaneous action of the vocal organs under excitement, it appears that the connection between ideas and words is only to be explained by a compact between the speaker and hearer which supposes the existence of a prior mode of communication. This was probably by gesture, which, in the happy phrase of Sayce, "like the rope-bridges of the Himalayas or the Andes, formed the first rude means of communication between man and man." At least we may gladly accept it as a clew leading us out of the labyrinth of philologic confusion, and as regulating the immemorial search for man's pristine speech.

INDEX OF NAMES